International Standard Book Number
1-56396-414-7

American Institute of Physics

500 Sunnyside Boulevard

Woodbury, New York 11797-2999

Printed in the United States of America

Journal of
Physical and Chemical Reference Data

Monograph No. 5

Atlas of Surface Structures: Volume 1B (1994) Based on the NIST Surface Structure Database (SSD)

P. R. Watson
Department of Chemistry
Oregon State University
Corvallis, Oregon 97331

M. A. Van Hove
Materials Sciences Division
Lawrence Berkeley Laboratory
University of California
Berkeley, California 94720

K. Hermann
Abteilung Theorie
Fritz-Haber-Institut
Faradayweg 4–6
D-14195 Berlin, Germany

Major Contributors:
S. Crampin, G. Darling, P. Pinkava, A. Barbieri, J. Batteas,
A. Kaduwela, N. Materer, U. Starke, W. Weiss

Consultants:
G. A. Somorjai and J. B. Pendry

Published by the **American Chemical Society**
and the **American Institute of Physics** for
the **National Institute of Standards and Technology**

Foreword

The *Journal of Physical and Chemical Reference Data* is published jointly by the American Institute of Physics and the American Chemical Society for the National Institute of Standards and Technology (NIST). Its objective is to provide critically evaluated physical and chemical property data, fully documented as to the original sources and the criteria used for evaluation. One of the principal sources of material for the journal is the NIST Standard Reference Data Program, a program promoting the compilation and critical evaluation of property data.

The regular issues of the *Journal of Physical and Chemical Reference Data* are published bimonthly and contain compilations and critical data reviews of moderate length. Longer works, volumes of collected tables, and other material unsuited to a periodical format have previously been published as *Supplements to the Journal*. Beginning in 1989 the generic title of these works has been changed to *Monograph*, which reflects their character as independent publications. This volume, ''Atlas of Surface Structures: Volume 1B (1994)'' by P.R. Watson, M.A. Van Hove, and K. Hermann is the second part of *Monograph No. 5* of the *Journal of Physical and Chemical Reference Data.*

Jean W. Gallagher, Editor
Journal of Physical and Chemical Reference Data

Atlas of Surface Structures: Volume 1B (1994)
Based on the NIST Surface Structure Database (SSD)

P.R. Watson
Department of Chemistry, Oregon State University, Corvallis, OR 97331

M.A. Van Hove
Materials Sciences Division, Lawrence Berkeley Laboratory, University of California, Berkeley, CA 94720

and

K. Hermann
Abteilung Theorie, Fritz-Haber-Institut, Faradayweg 4–6, D–14195 Berlin, Germany

Major Contributors:
S. Crampin, G. Darling, P. Pinkava, A. Barbieri, J. Batteas,
A. Kaduwela, N. Materer, U. Starke, W. Weiss

Consultants:
G. A. Somorjai and J. B. Pendry

These volumes contain listings of crystallographic information on almost 600 surface structures contained in the NIST Surface Structure Database (SSD). These are prefaced by explanatory text and are combined with high quality computer-generated views of the surface. The listings include full literature references, statements on experimental and theoretical methods used, bulk and surface unit cells, a complete list of atomic coordinates for overlayer, epilayer, interfacial and bulk atoms and selected bond lengths and angles.

Key words: electron diffraction; ion scattering; LEED; NEXAFS; photoelectron diffraction; SEXAFS; surface crystallography; surface structure; surface structure database; X-ray diffraction.

Contents

Volume 1A

Volume 1B

1. Introduction

This atlas of surface structures is aimed at scientists and students in physics, chemistry and materials sciences who wish to know and compare the detailed atomic-scale structures of surfaces obtained from experiment. It is a printed version of the Surface Structure Database (SSD) created, with partial NIST support, and published in 1993 in an electronic form for PCs and compatible computers. This printed version contains the same information (save for the correction of a few typographical errors).

The first edition attempts to cover all known surface structures since the inception of surface crystallography in the early 1970s through the end of 1991. The number of included structures is 597. Volume 1A provides extensive structural information about surface structures determined from experiment. A unified format is used to allow convenient direct comparisons of related but different structures, or of results obtained with different techniques for the same structure. Details of the format of these tables is provided in Sec. 1.4 of that volume.

The second volume of this atlas shows carefully selected views of the surface structures that complement the extensive numerical data tabulated in Volume 1A. These illustrations use a logical categorization scheme described in Sec. 2.1. In this way the reader can directly compare the illustration with the accompanying numerical data.

2. Structural Figures

The figures are ordered in a systematic manner starting with elemental metals, clean and with adsorbates or epitaxial layers, and then progressing to elemental and compound semiconductors, alloys and complex substrates. The figure numbers and most indented labels (e.g. 1a.1a.1) refer to figures shown in 2.2, which are ordered as in this index.

2.1. Categorized Index of Figures

2.1. Catagorized Index of Figures — Continued

2.1. Catagorized Index of Figures — Continued

2.1. Catagorized Index of Figures — Continued

2.1. Catagorized Index of Figures — Continued

2.2. Figures

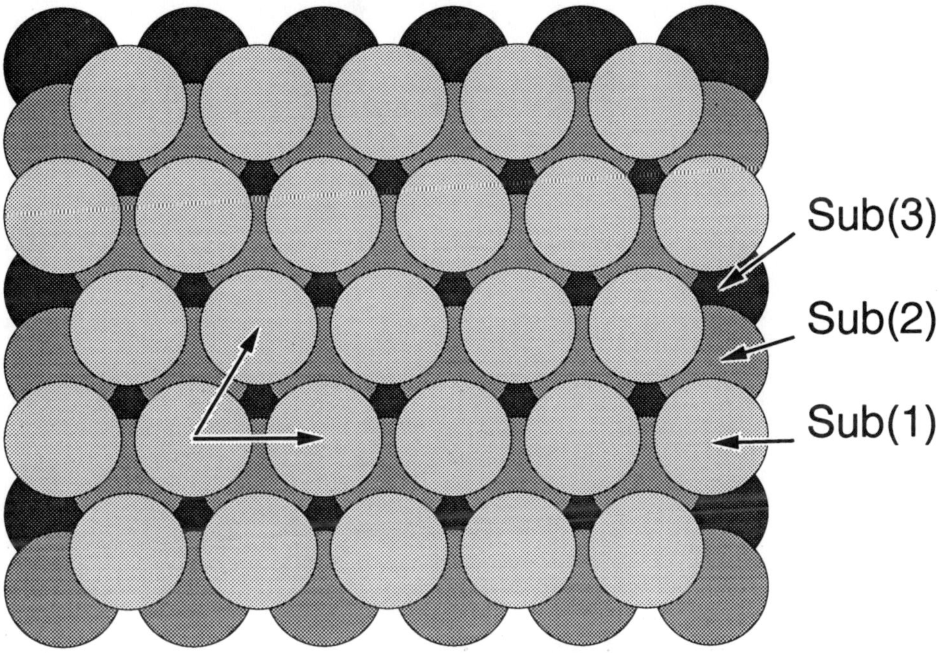

Fig. 1a : fcc(111)-(1x1) (top view)

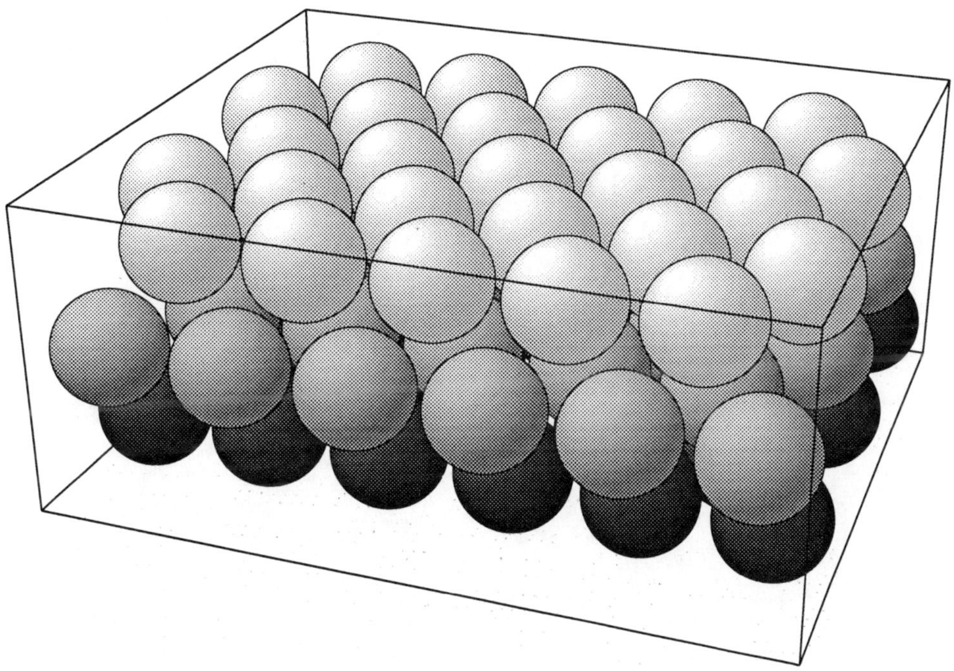

Fig. 1b : fcc(111)-(1x1) (perspective)

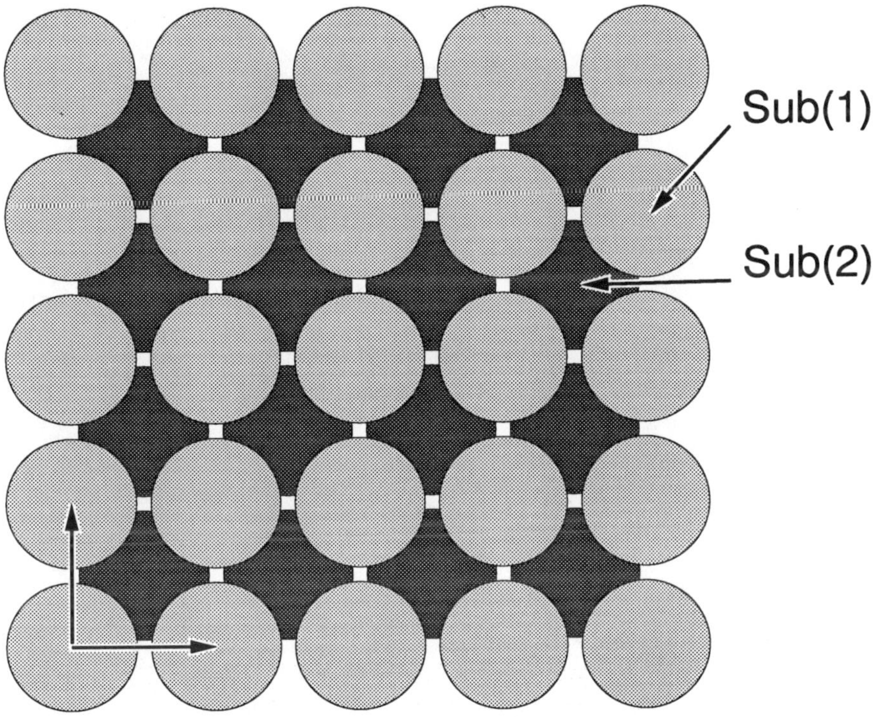

Fig. 2a : fcc(100)-(1x1) (top view)

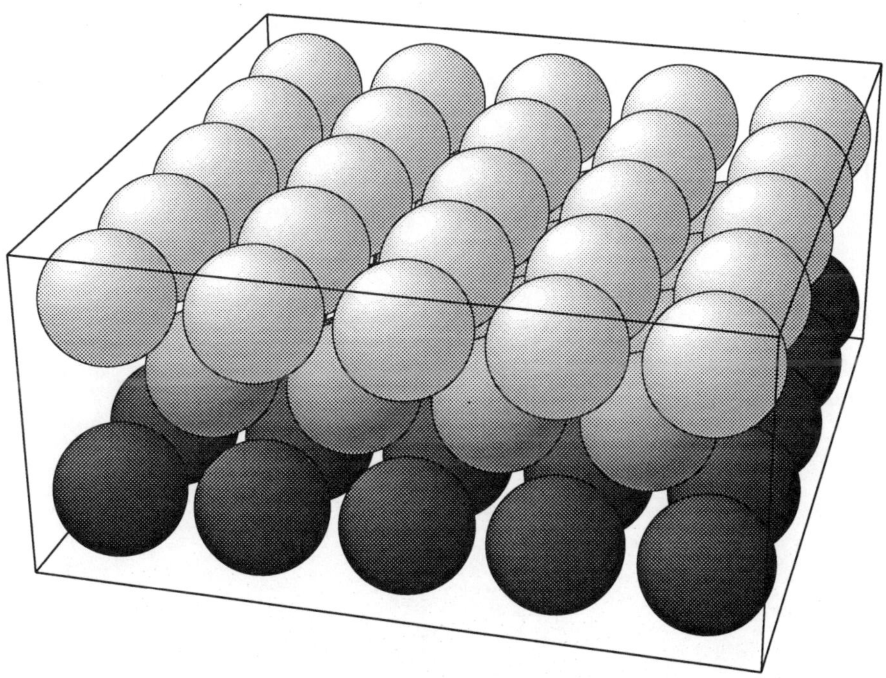

Fig. 2b : fcc(100)-(1x1) (perspective)

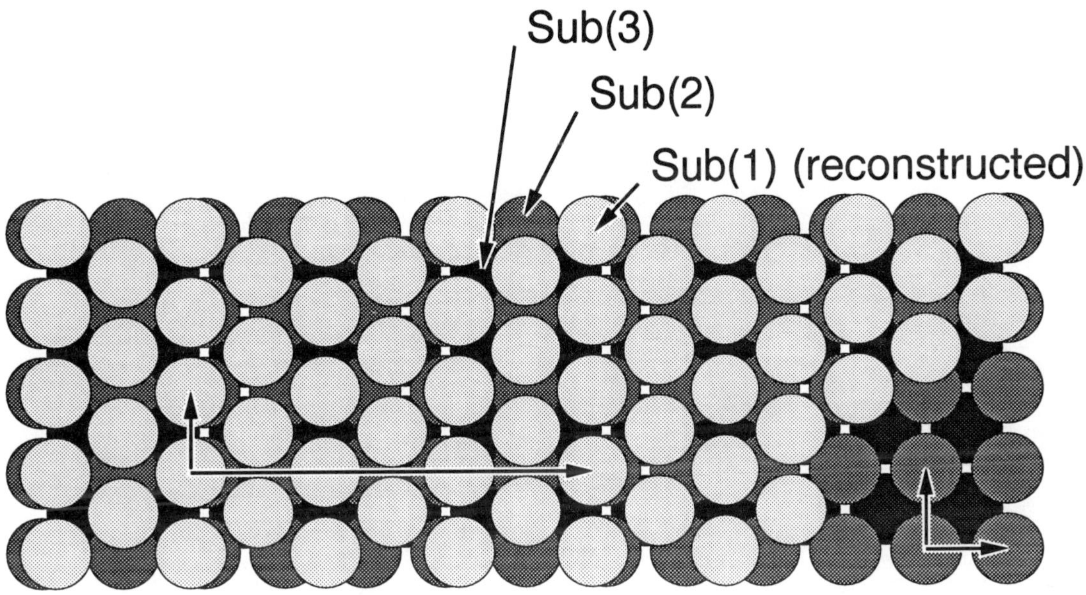

Fig. 3a : Ir(100)-(1x5) (top view)

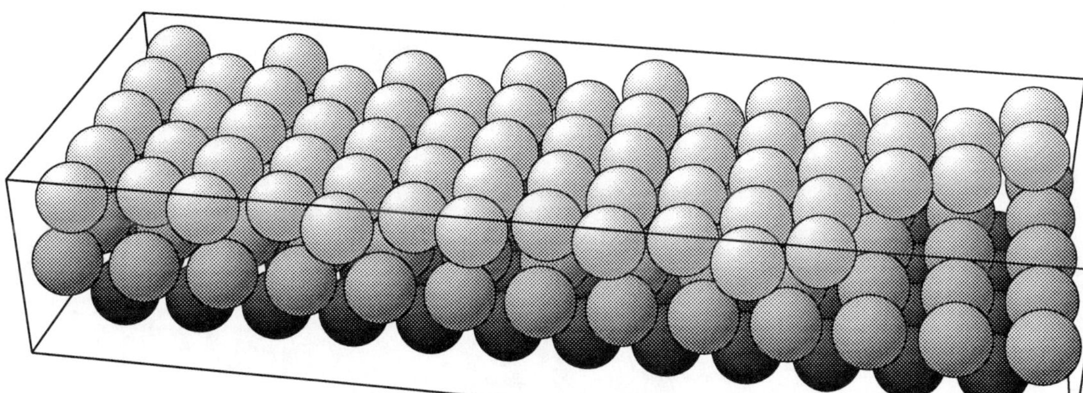

Fig. 3b : Ir(100)-(1x5) (perspective)

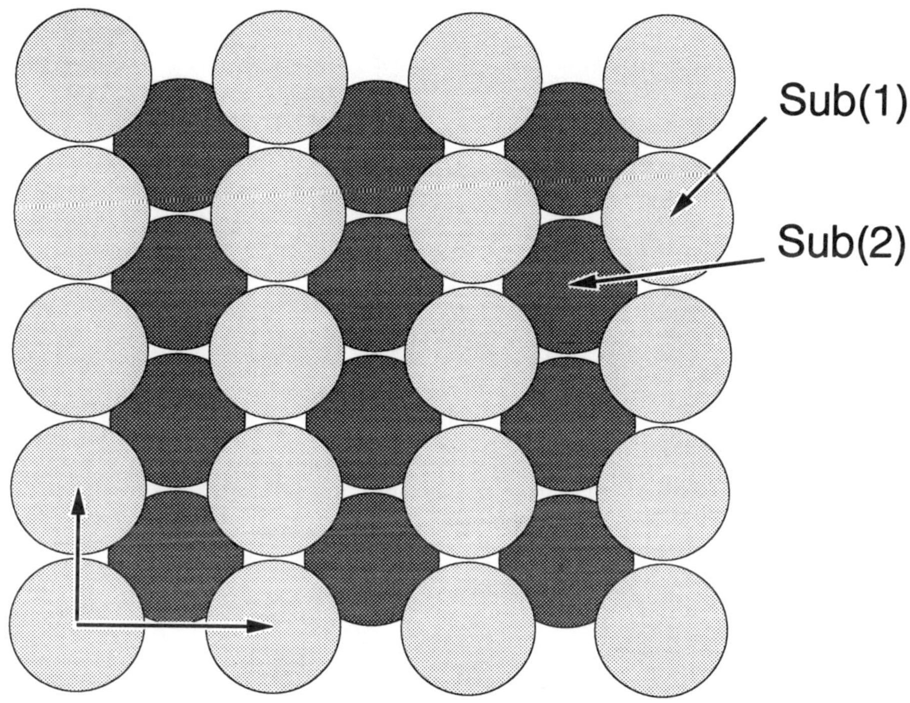

Fig. 4a : fcc(110)-(1x1) (top view)

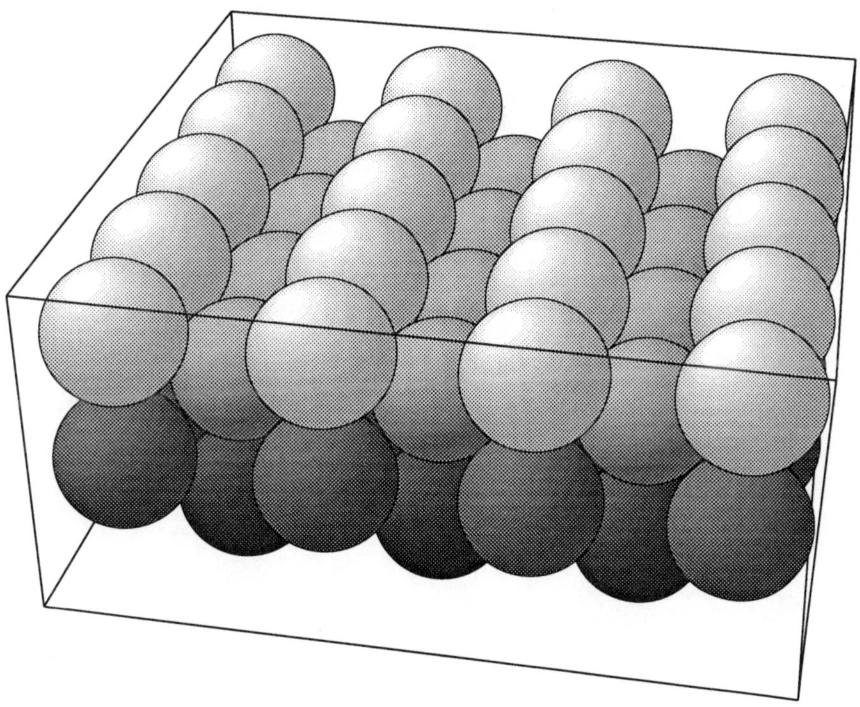

Fig. 4b : fcc(110)-(1x1) (perspective)

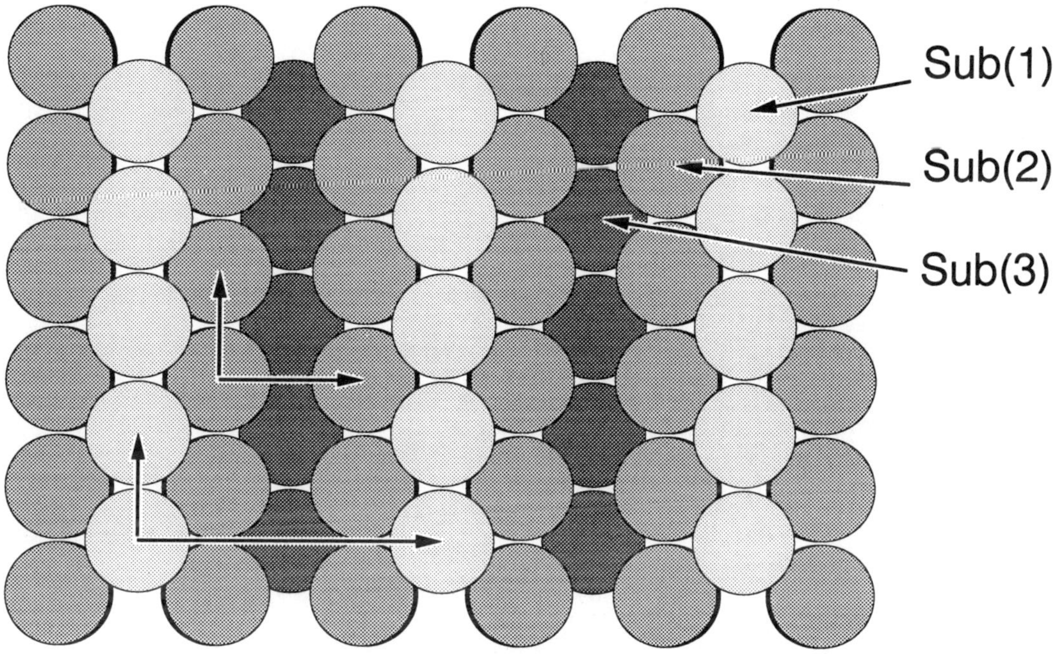

Fig. 5a : fcc(110)-(1x2) missing row (top view)

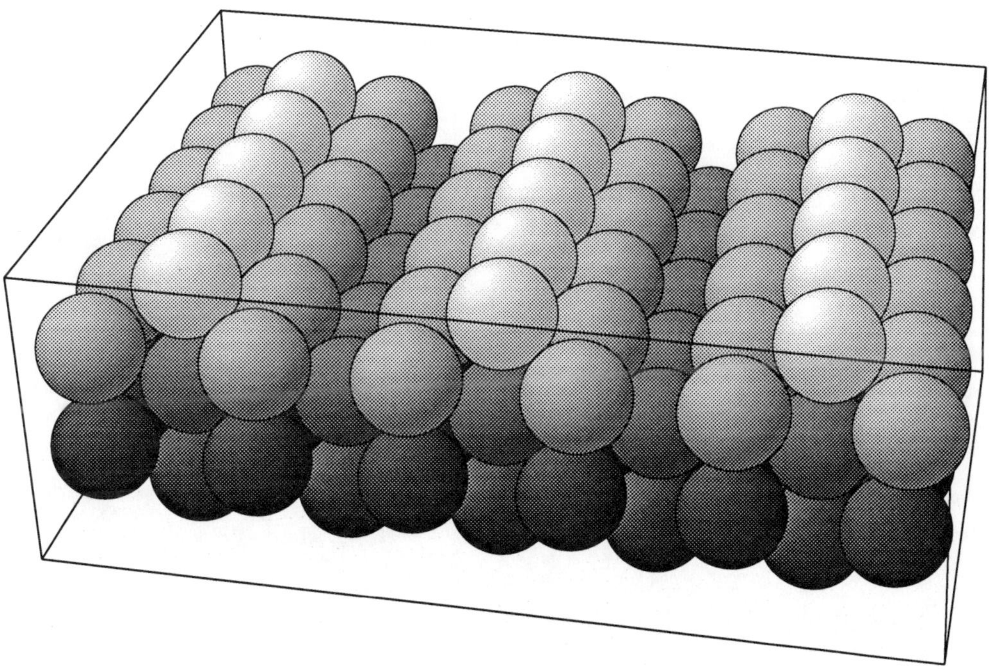

Fig. 5b : fcc(110)-(1x2) missing row (perspective)

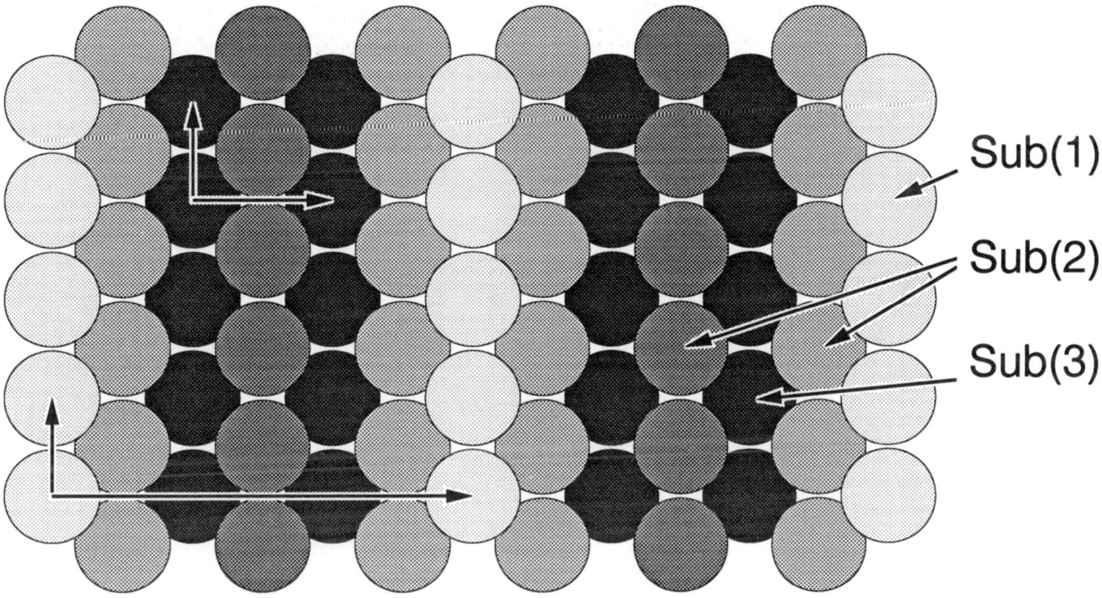

Fig. 6a : fcc(110)-(1x3) facets I (top view)

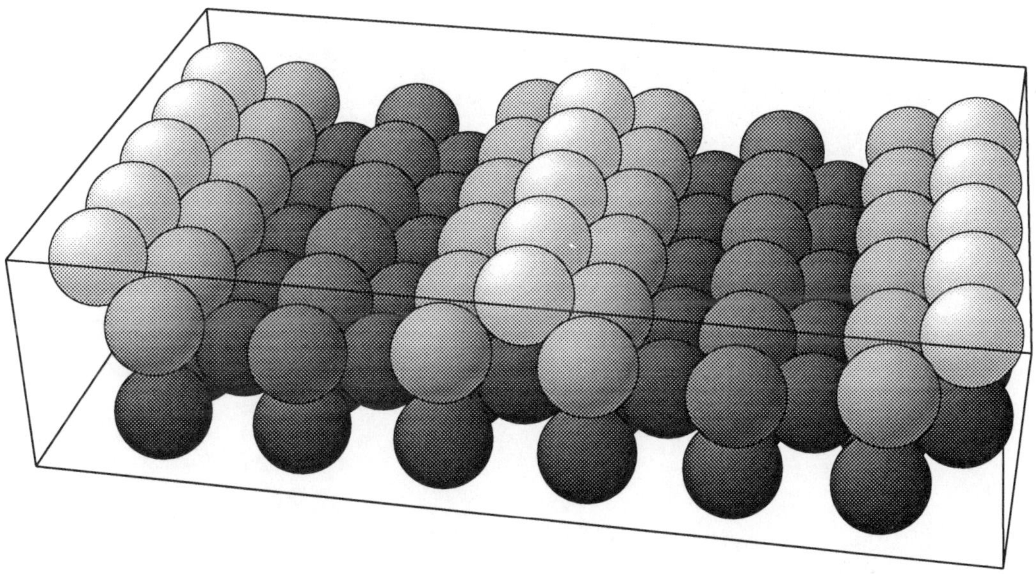

Fig. 6b : fcc(110)-(1x3) facets I (perspective)

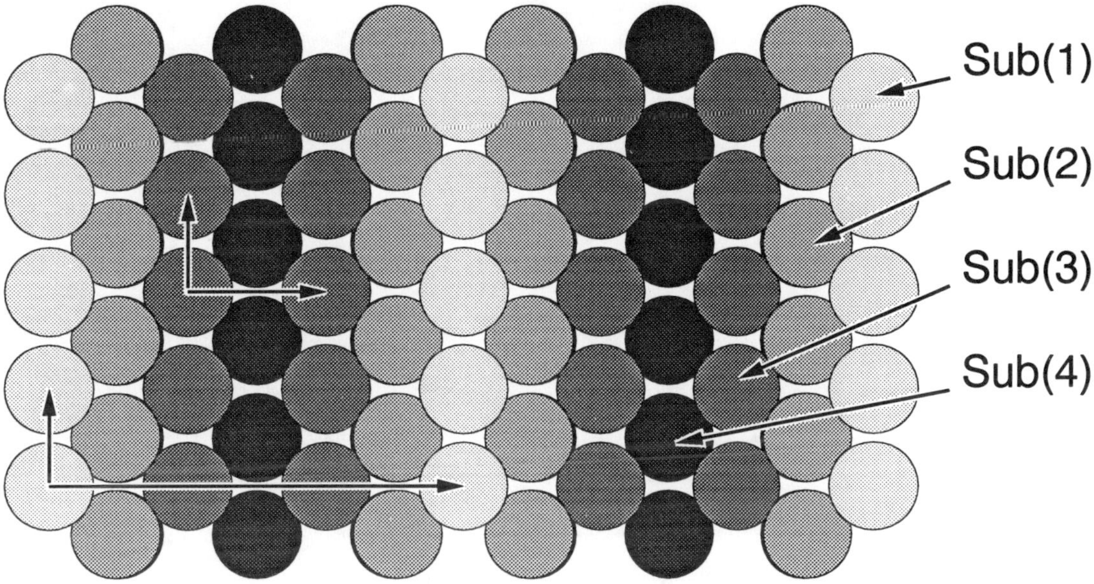

Fig. 7a : fcc(110)-(1x3) facets II (top view)

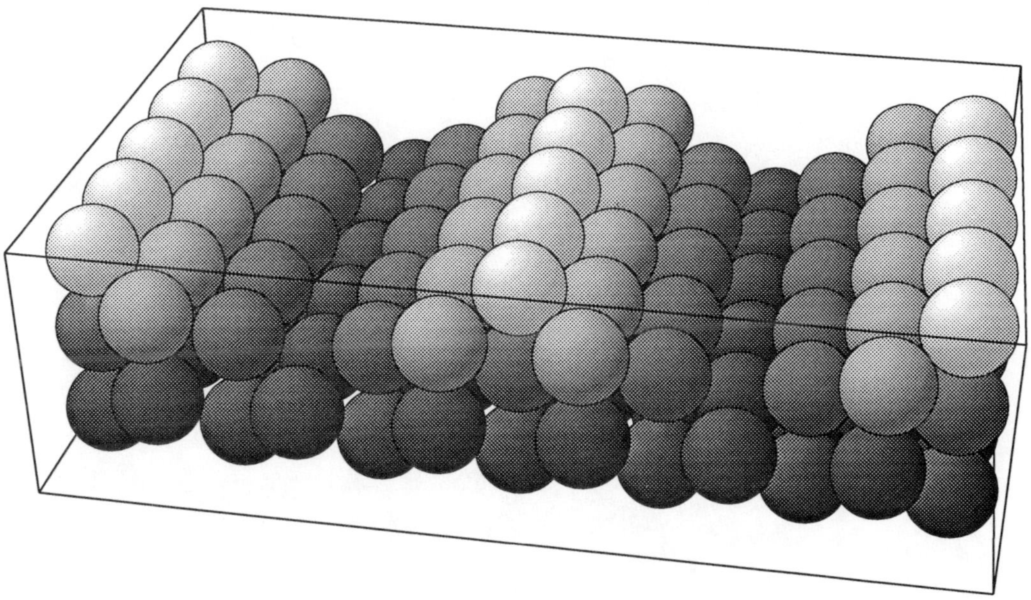

Fig. 7b : fcc(110)-(1x3) facets II (perspective)

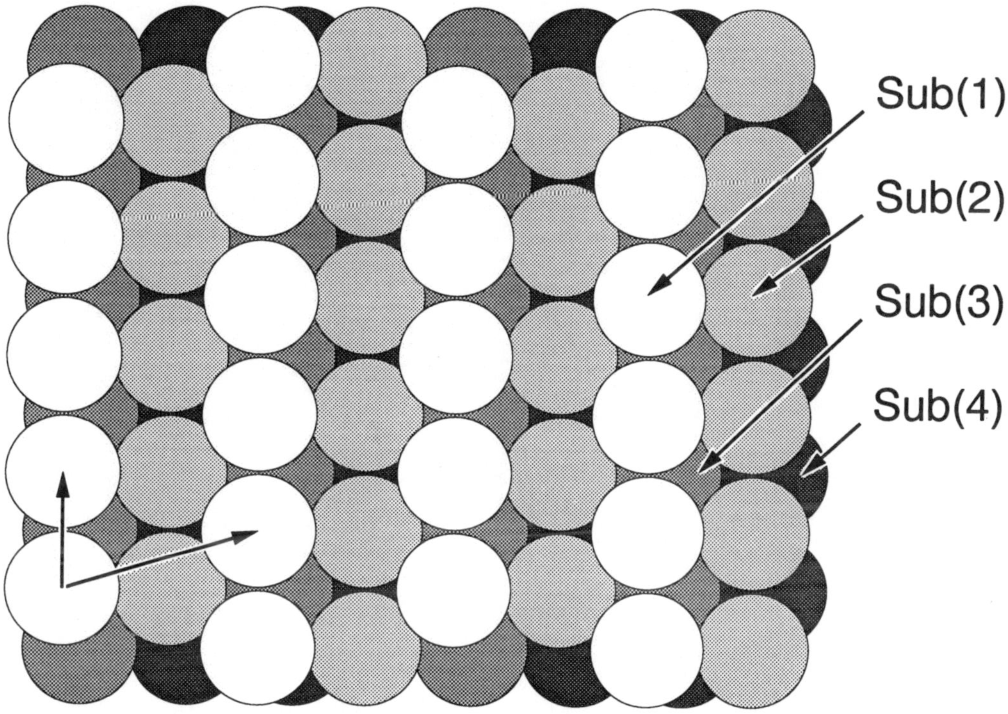

Fig. 8a : fcc(311)-(1x1) (top view)

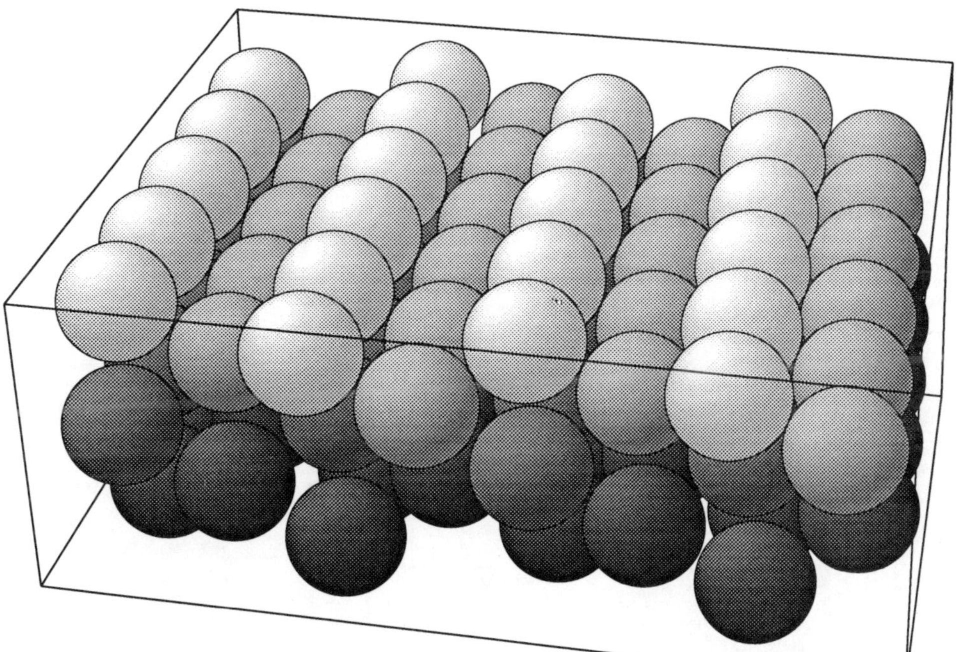

Fig. 8b : fcc(311)-(1x1) (perspective)

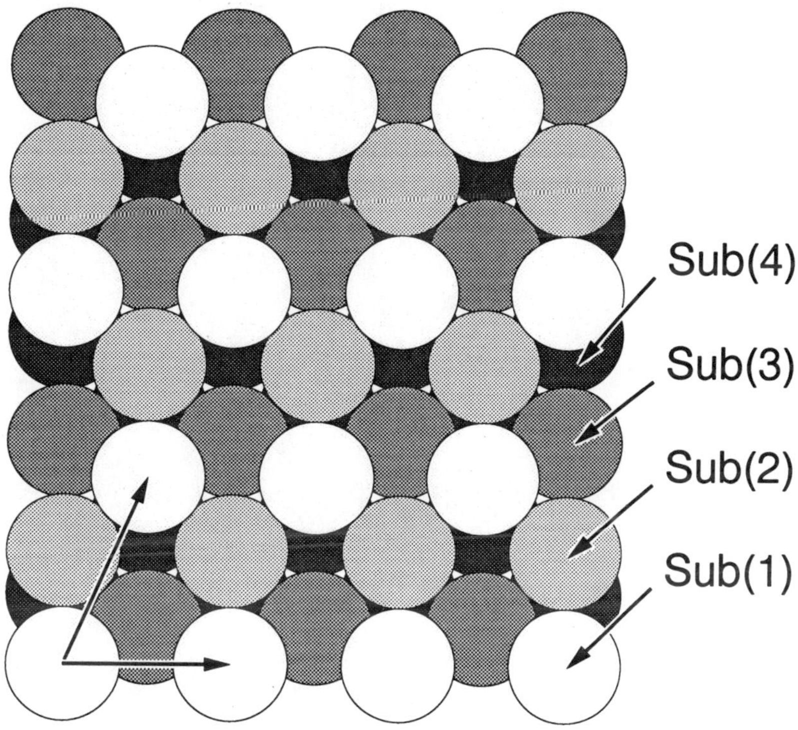

Fig. 9a : fcc(210)-(1x1) (top view)

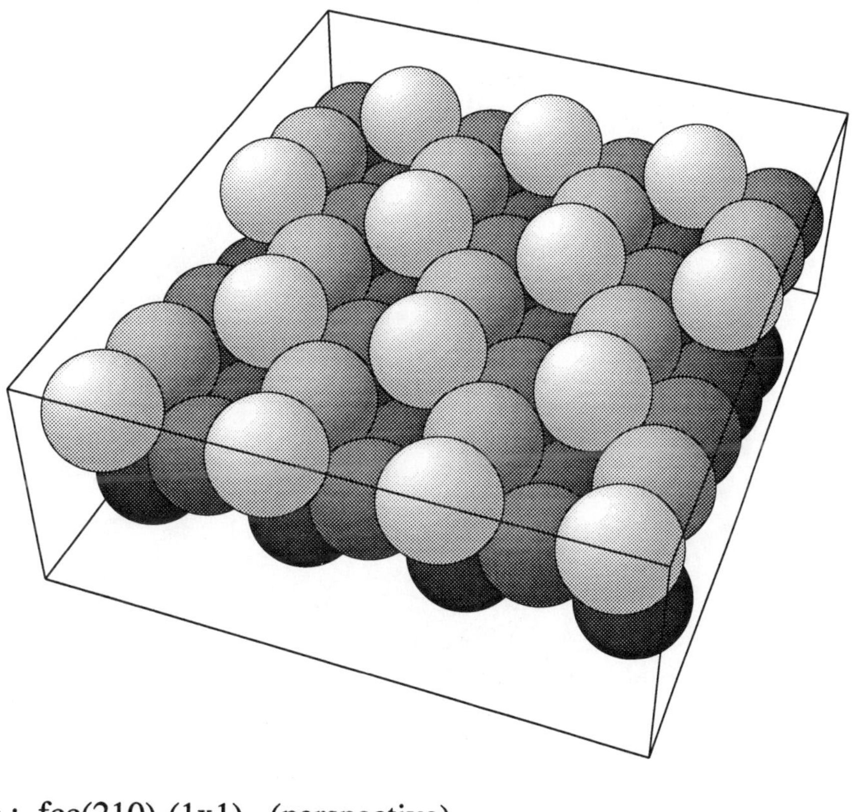

Fig. 9b : fcc(210)-(1x1) (perspective)

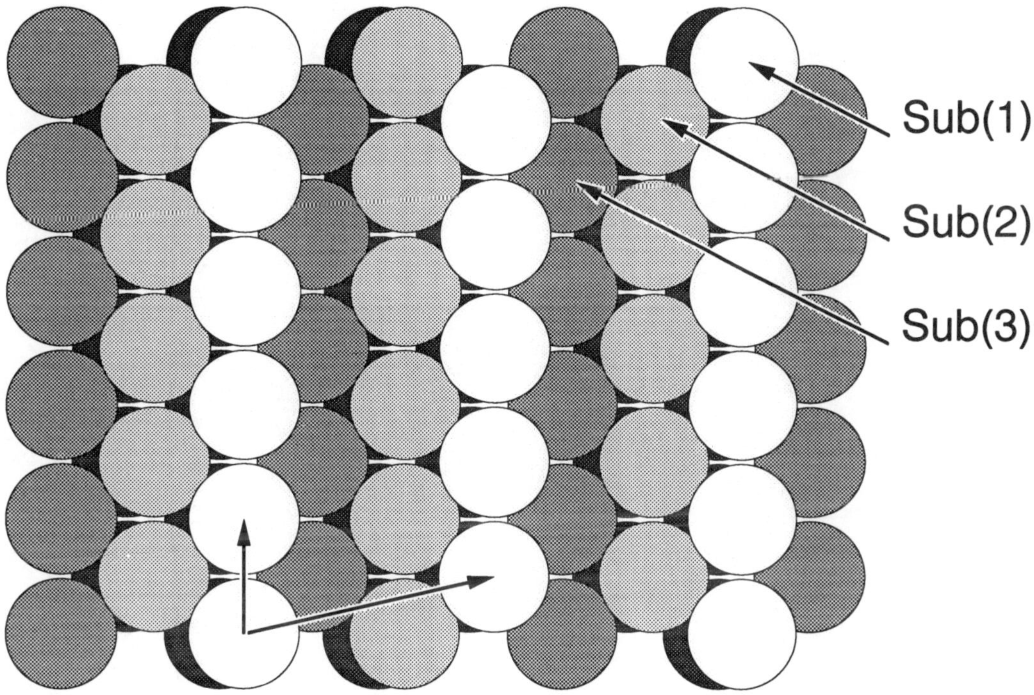

Fig. 10a : fcc(331)-(1x1) (top view)

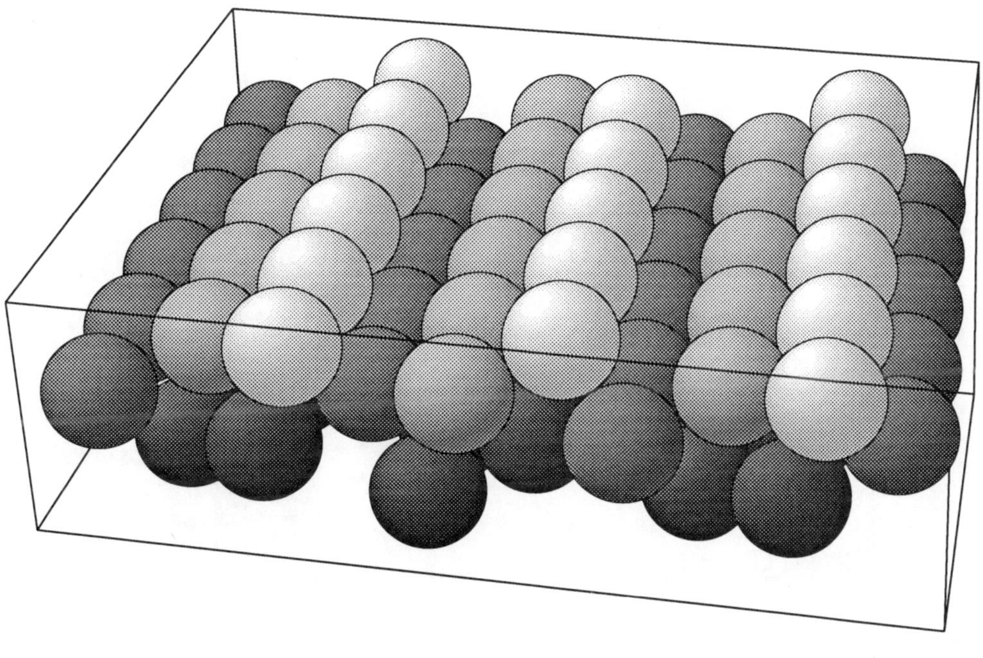

Fig. 10b : fcc(331)-(1x1) (perspective)

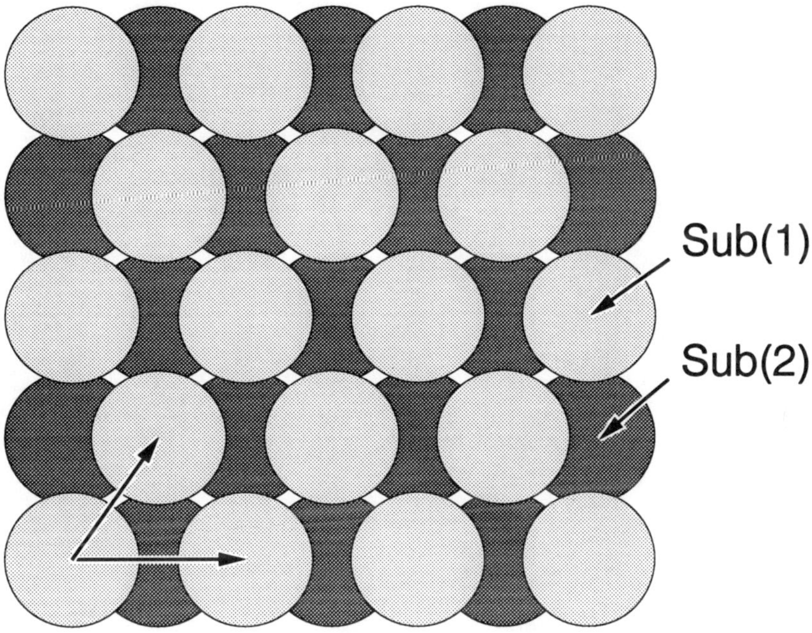

Fig. 11a : bcc(110)-(1x1) (top view)

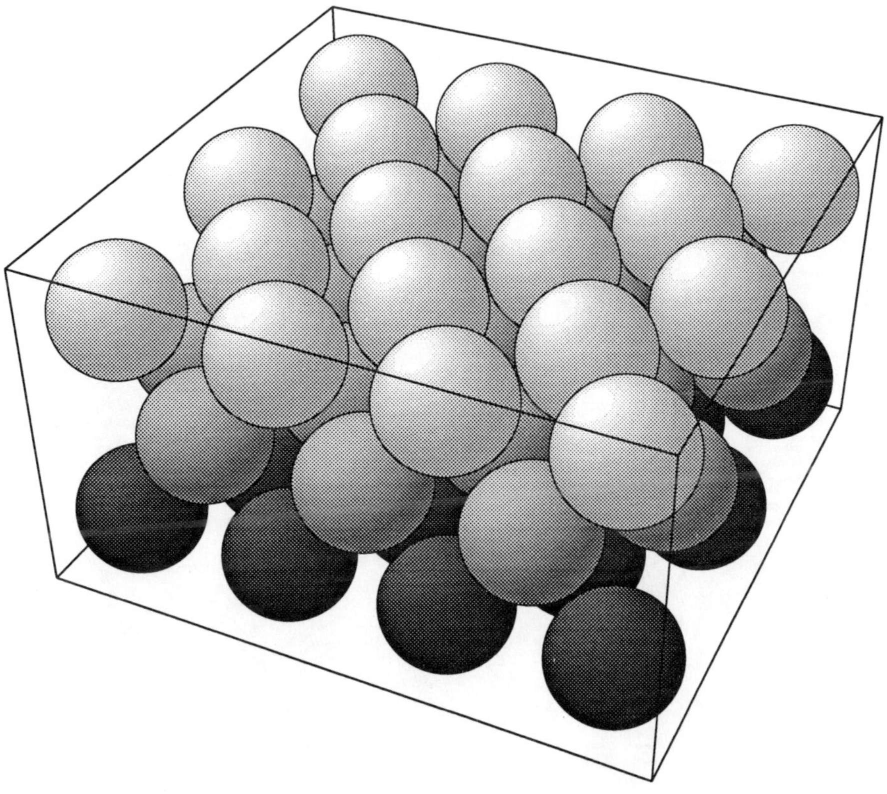

Fig. 11b : bcc(110)-(1x1) (perspective)

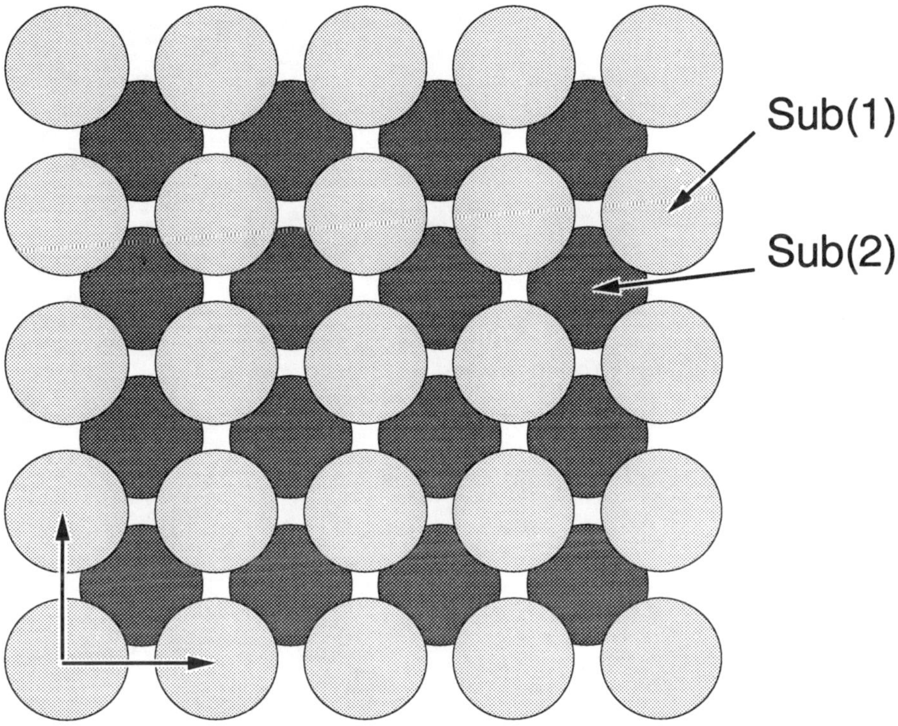

Fig. 12a : bcc(100)-(1x1) (top view)

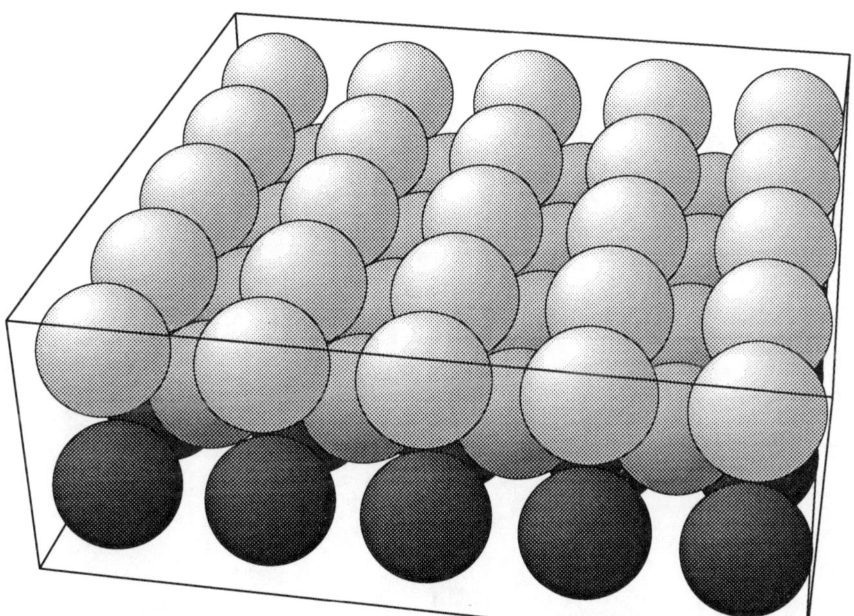

Fig. 12b : bcc(100)-(1x1) (perspective)

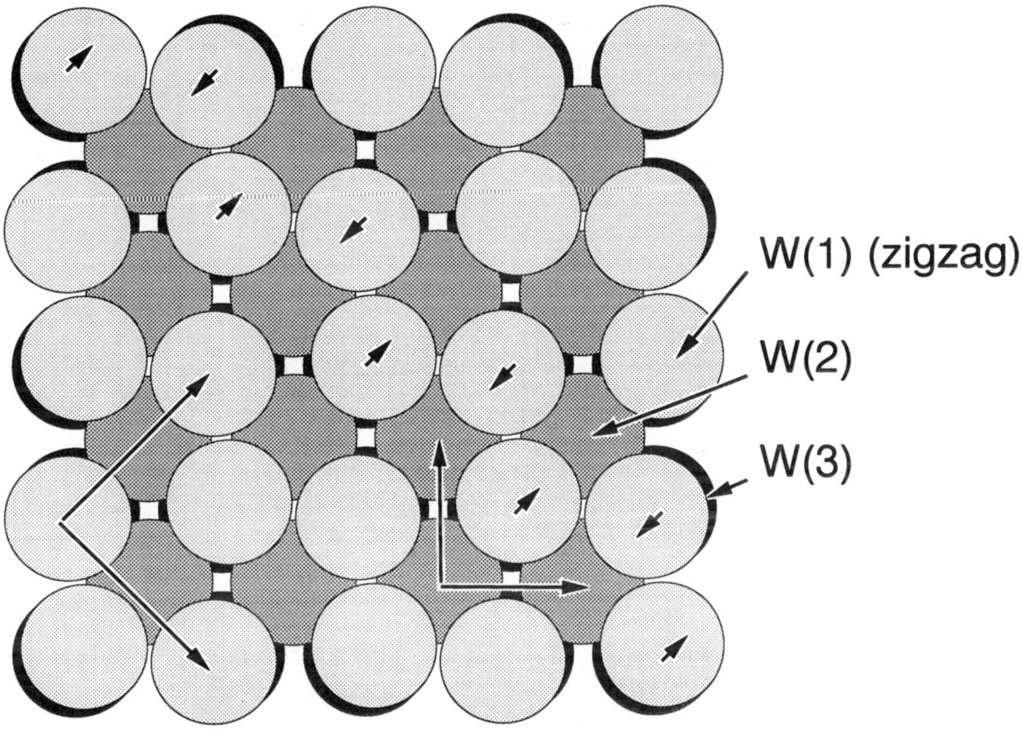

Fig. 13a : W(100)-c(2x2) (top view)

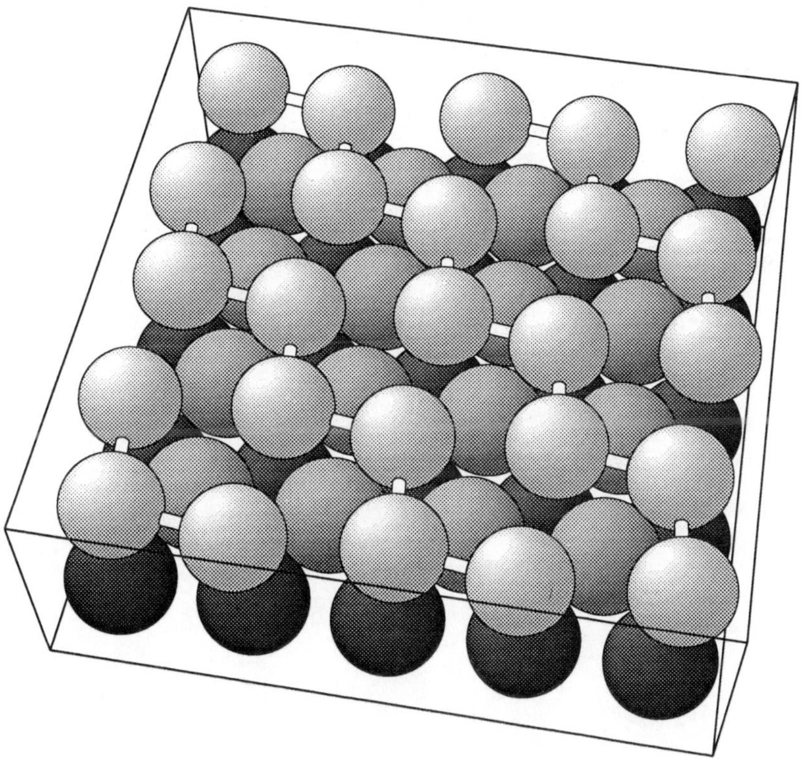

Fig. 13b : W(100)-c(2x2) (perspective)

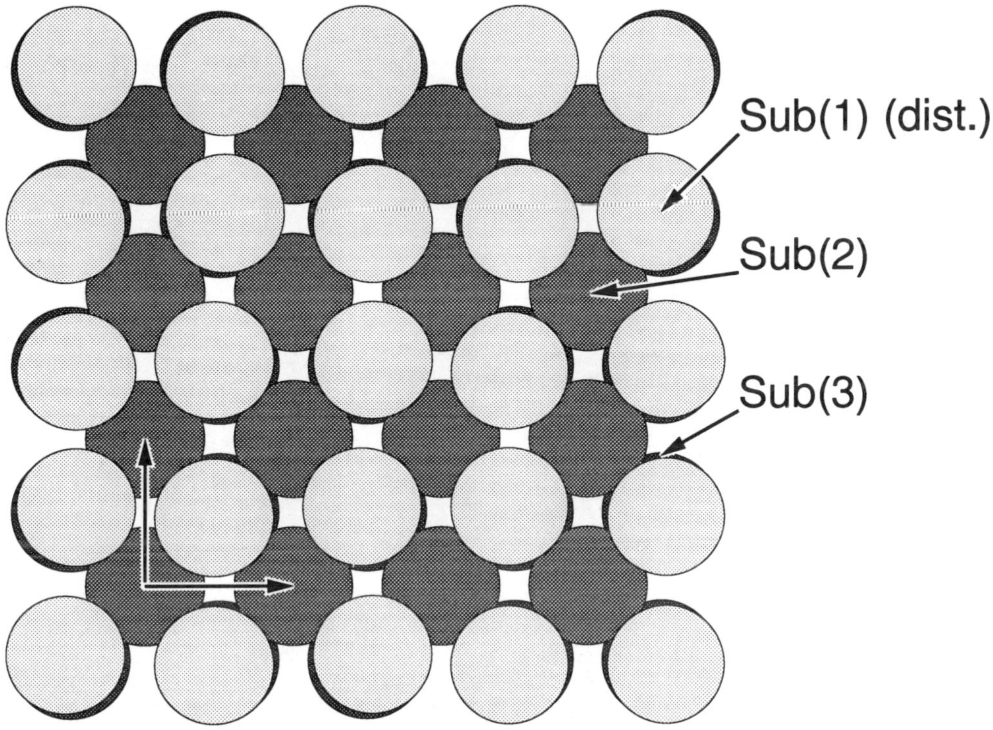

Fig. 14a : bcc(100)-(1x1) disordered (top view)

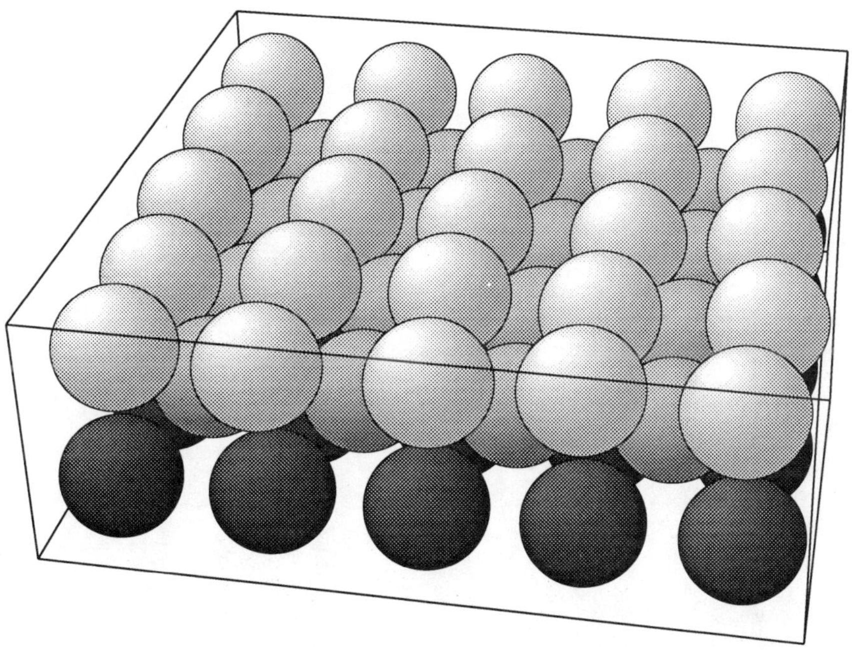

Fig. 14b : fcc(100)-(1x1) disordered (perspective)

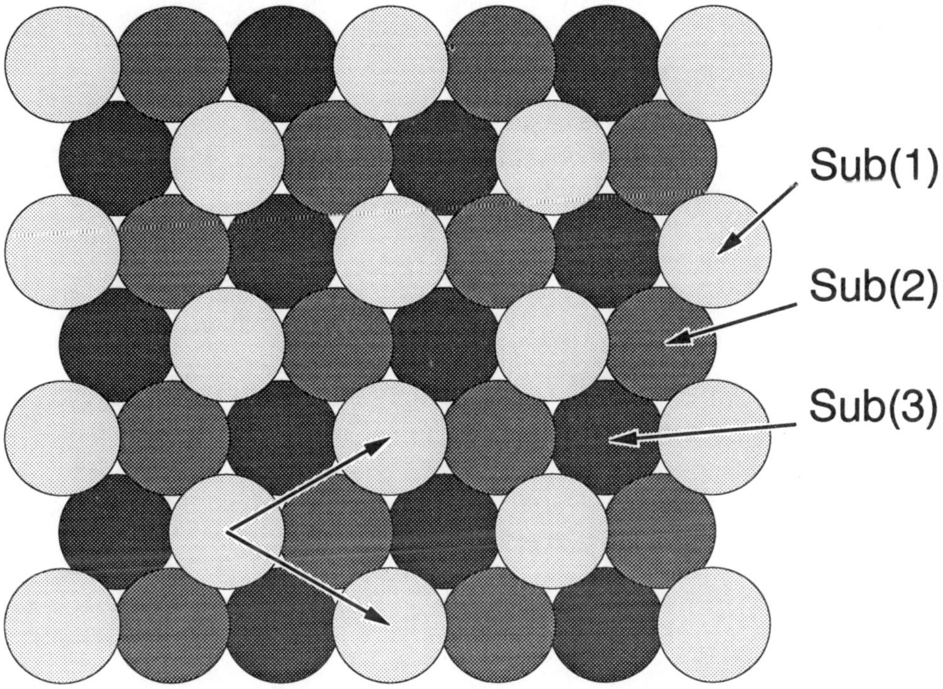

Fig. 15a : bcc(111)-(1x1) (top view)

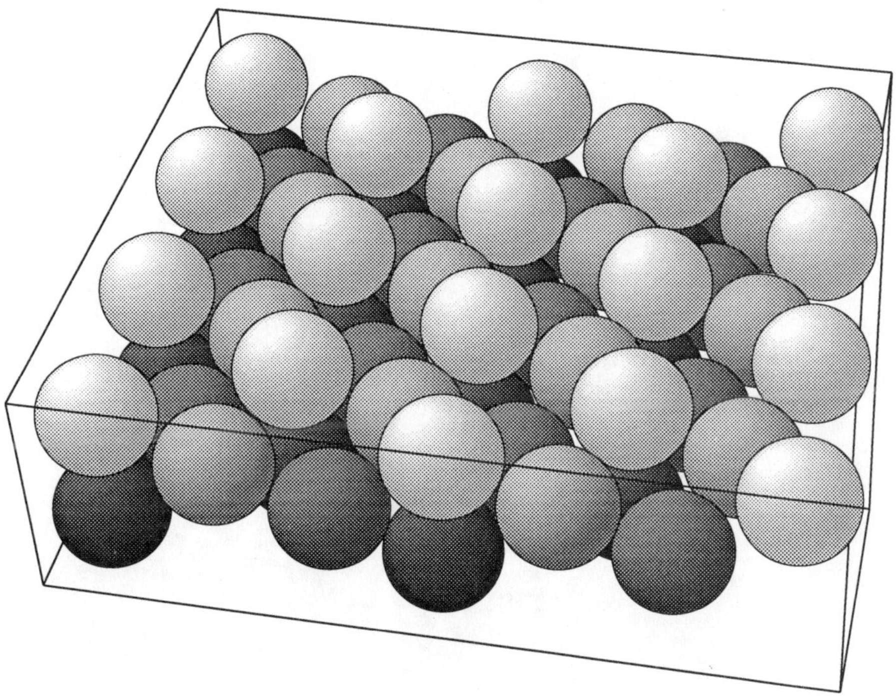

Fig. 15b : bcc(111)-(1x1) (perspective)

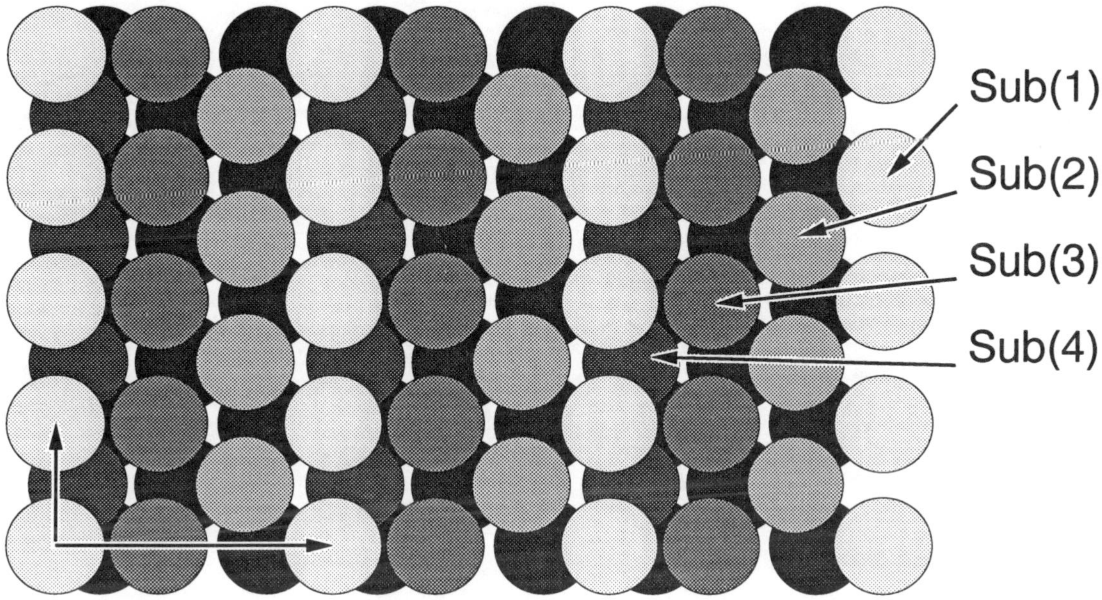

Fig. 16a : bcc(210)-(1x1) (top view)

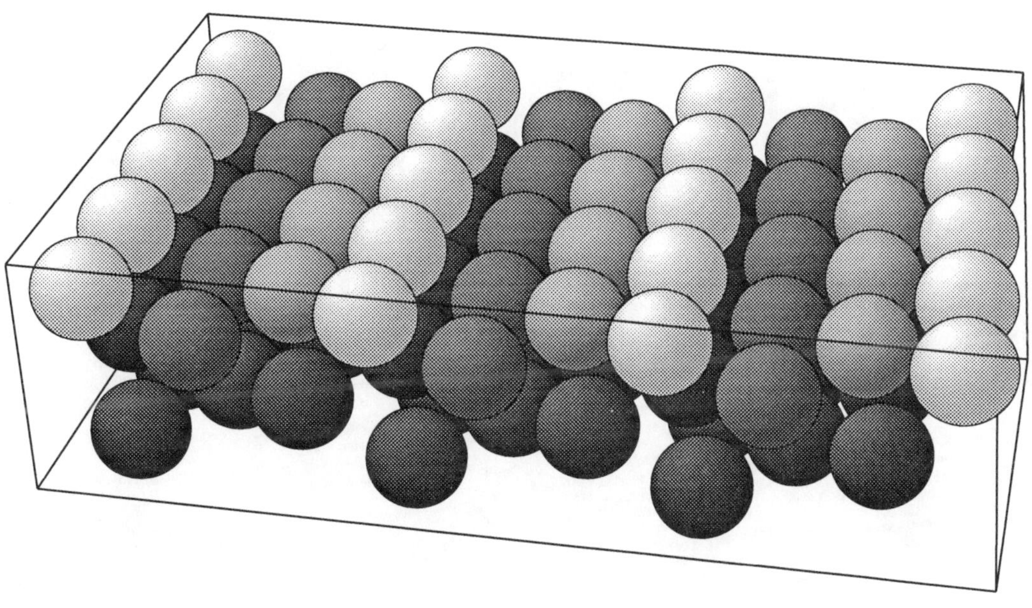

Fig. 16b : bcc(210)-(1x1) (perspective)

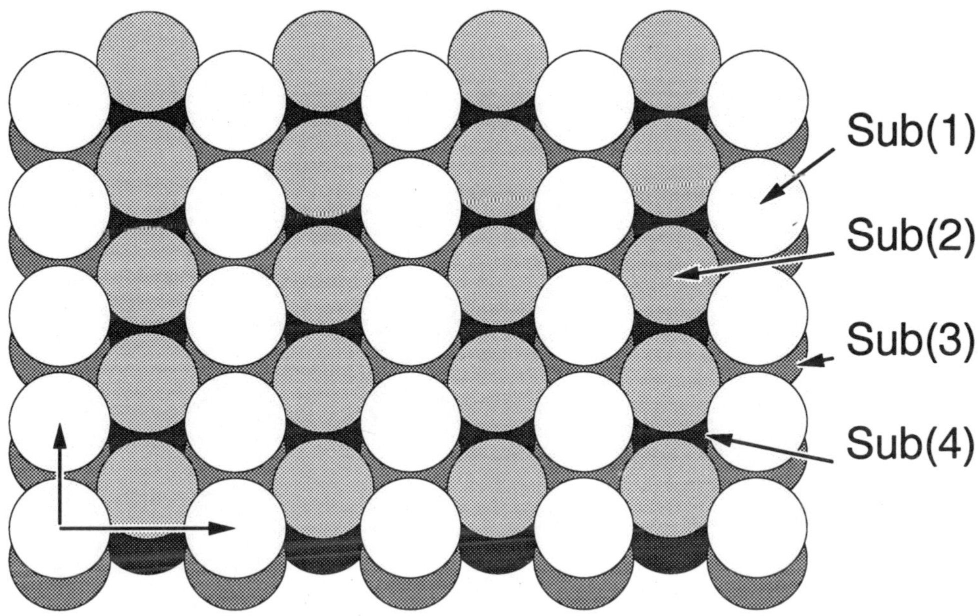

Fig. 17a : bcc(211)-(1x1) (top view)

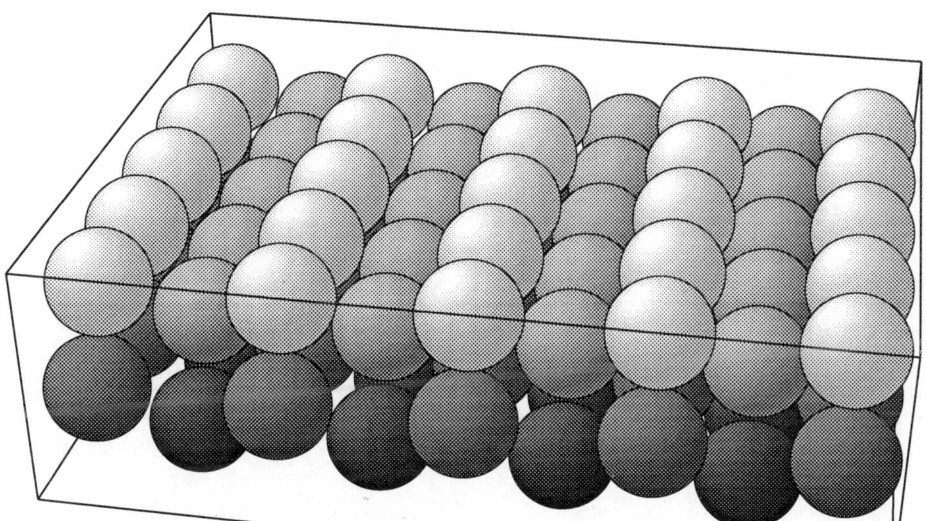

Fig. 17b : bcc(211)-(1x1) (perspective)

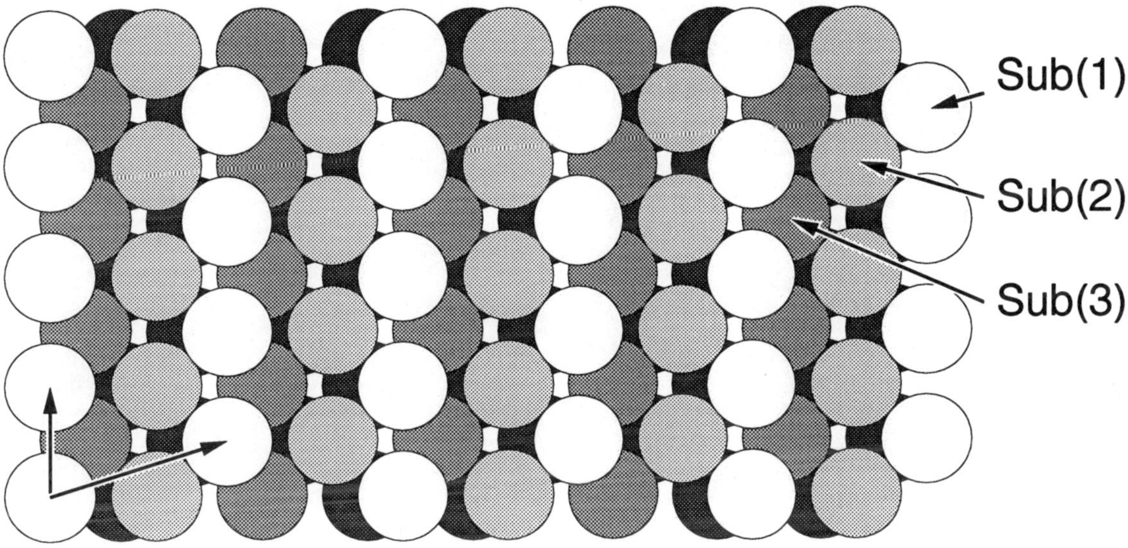

Fig. 18a : bcc(310)-(1x1) (top view)

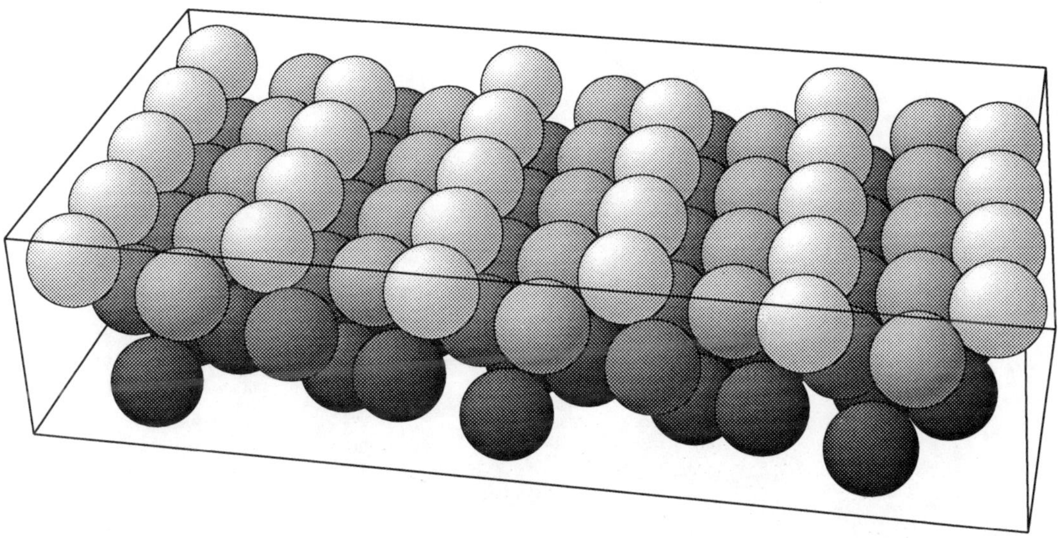

Fig. 18b : bcc(310)-(1x1) (perspective)

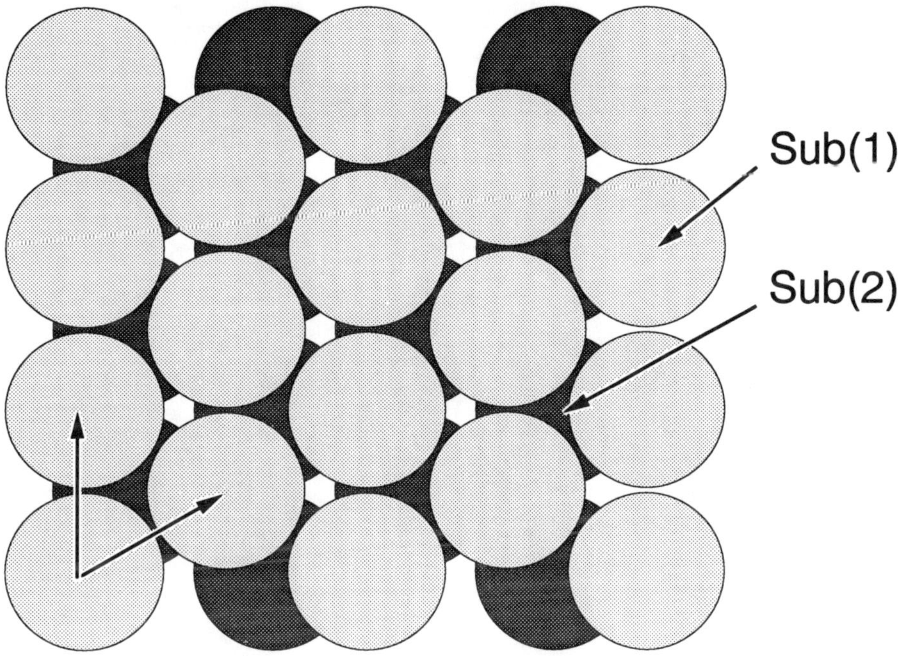

Fig. 19a : hcp(0001)-(1x1) (top view)

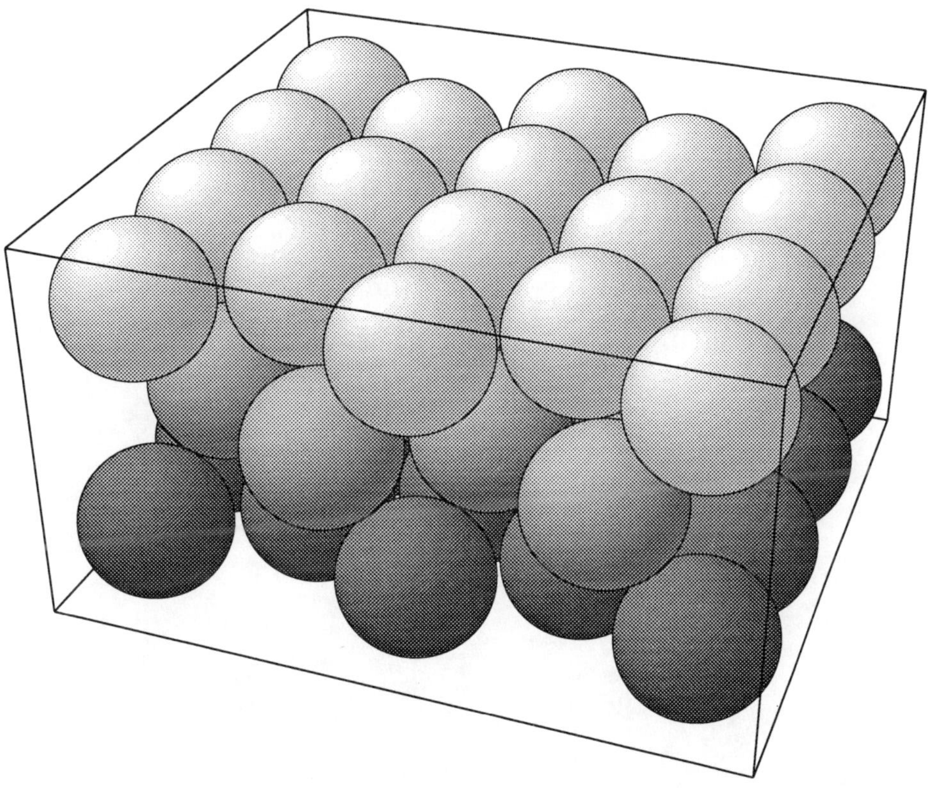

Fig. 19b : hcp(0001)-(1x1) (perspective)

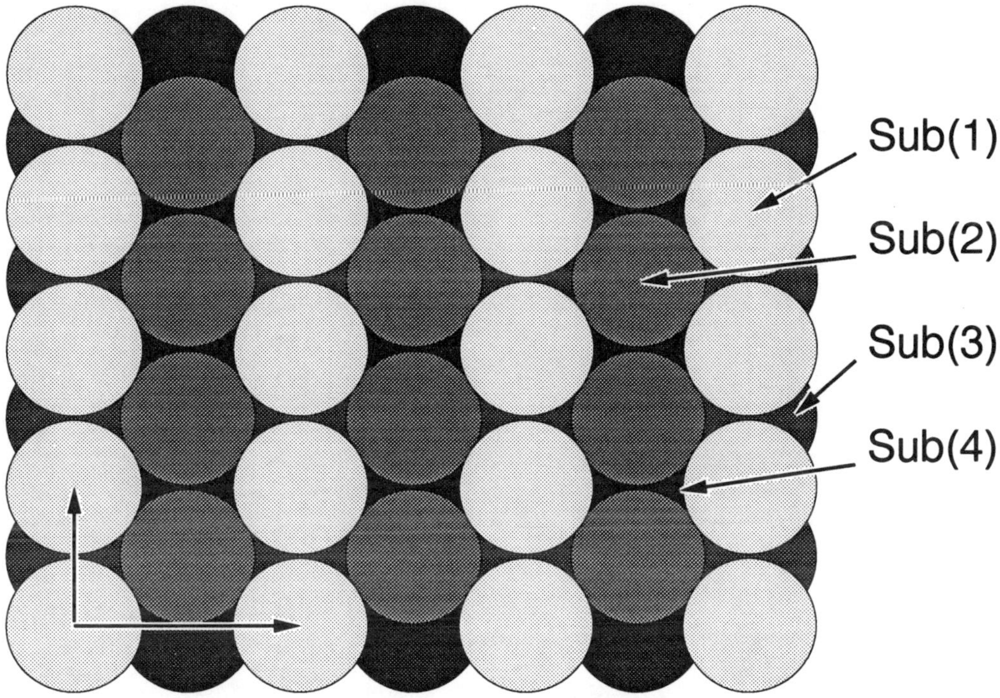

Fig. 20a : hcp(10-10)-(1x1) (top view)

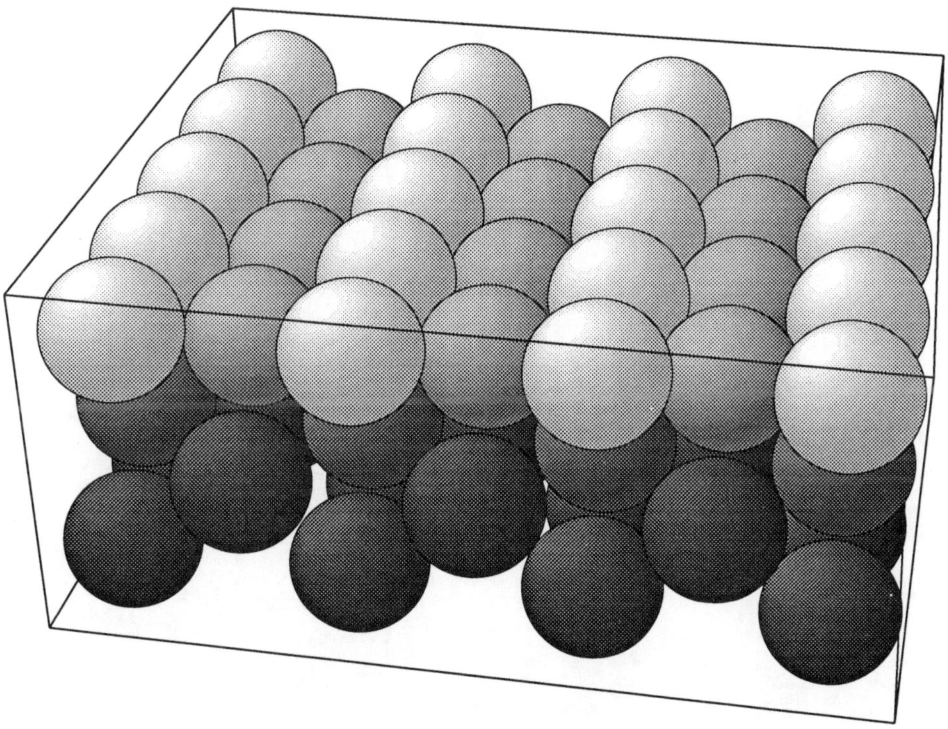

Fig. 20b : hcp(10-10)-(1x1) (perspective)

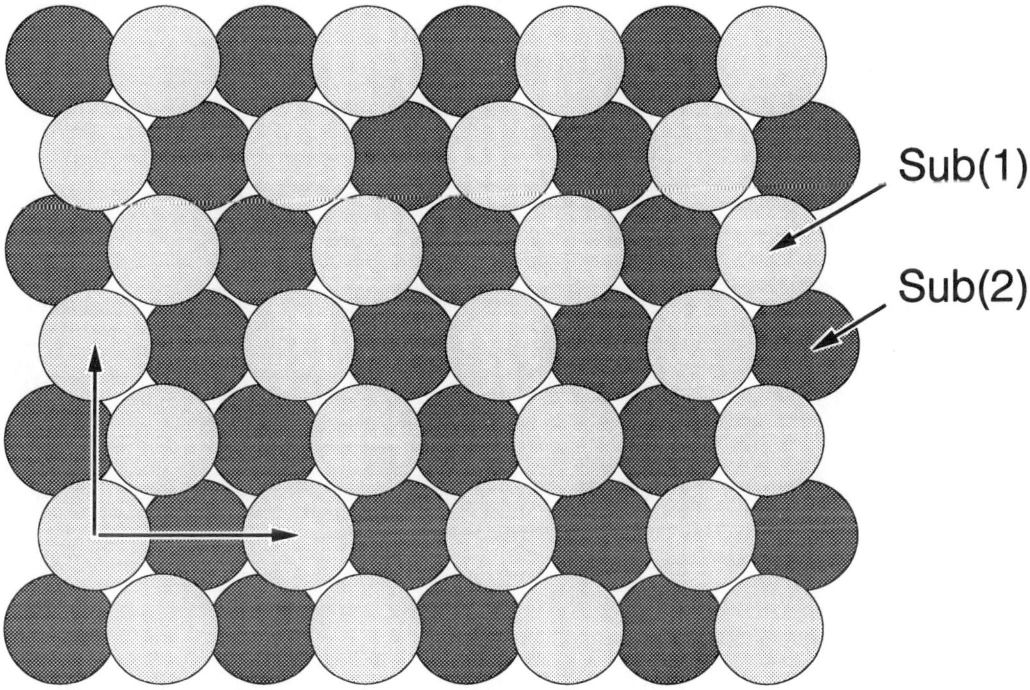

Fig. 21a : hcp(11-20)-(1x1) (top view)

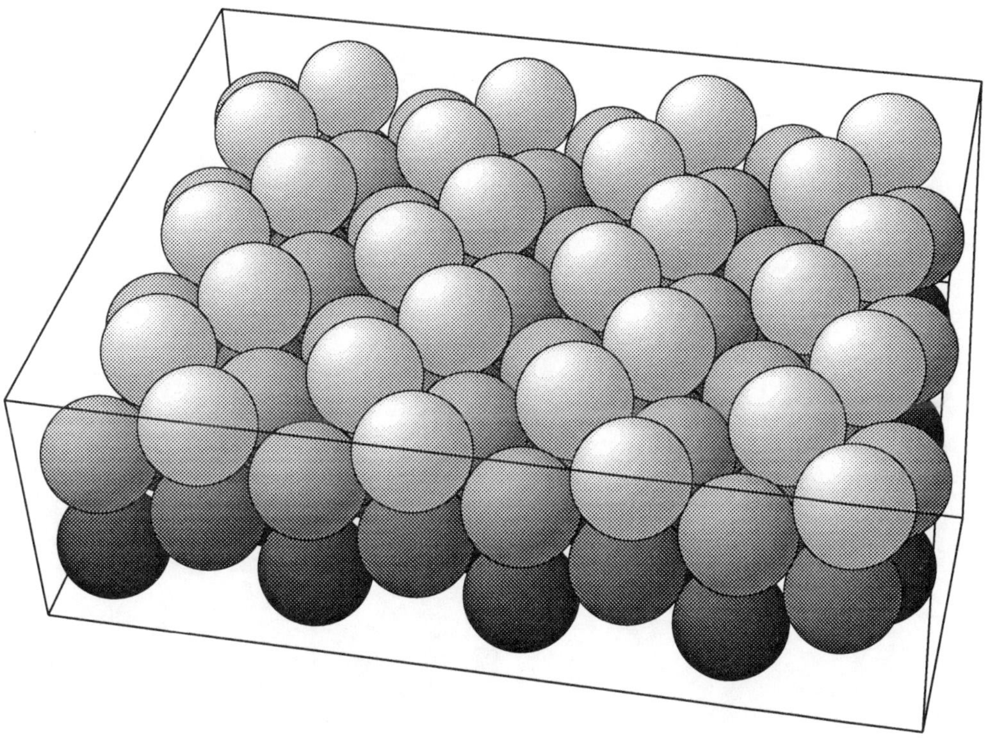

Fig. 21b : hcp(11-20)-(1x1) (perspective)

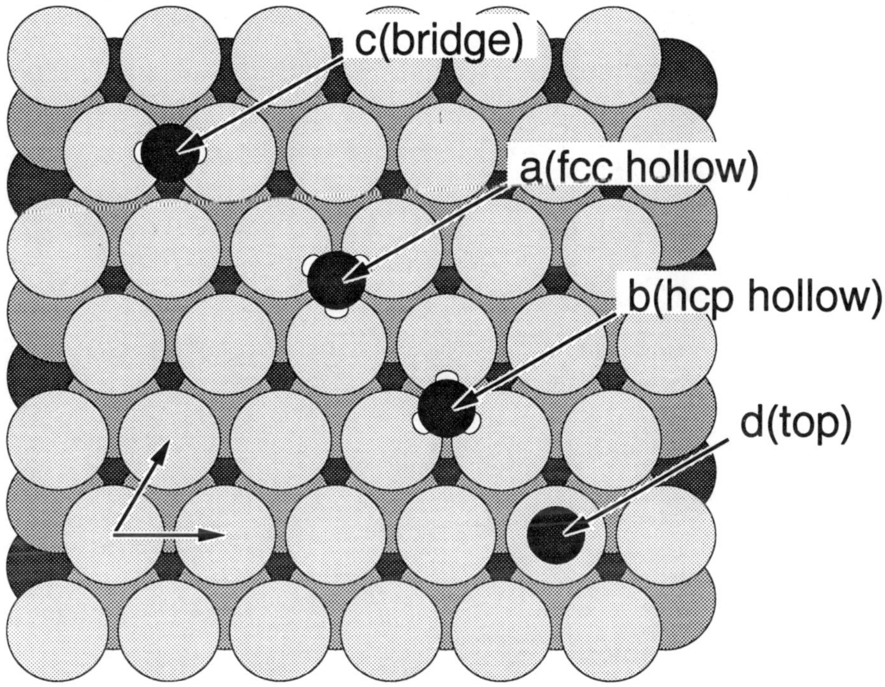

Fig. 22a : fcc(111) high symmetry adsorbate sites (top view)

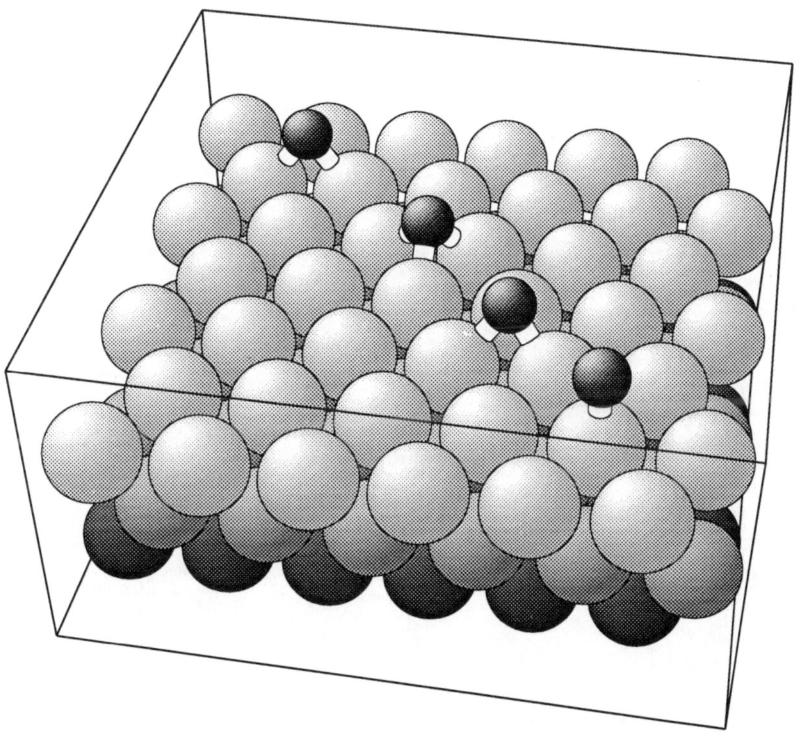

Fig. 22b : fcc(111) high symmetry adsorbate sites (perspective)

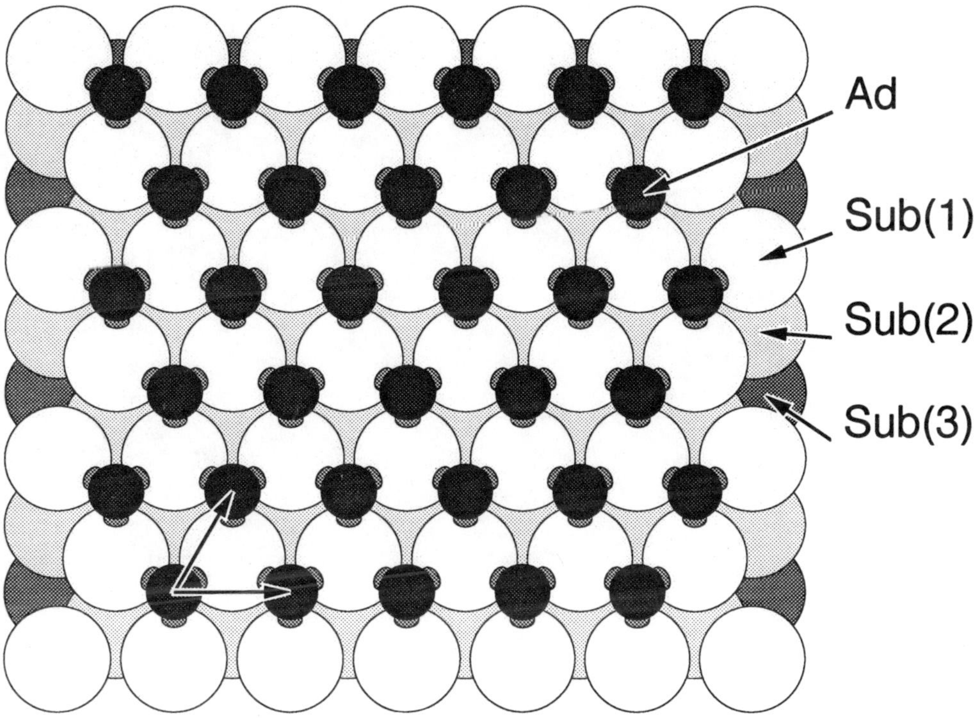

Fig. 23a : fcc(111)-(1x1)Ad (top view)

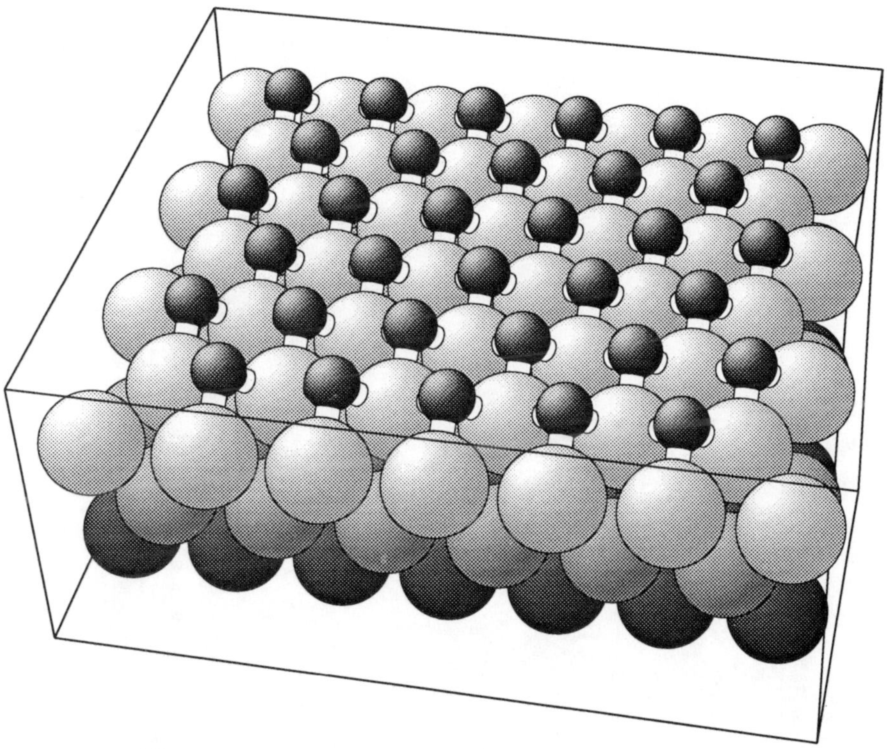

Fig. 23b : fcc(111)-(1x1)Ad (perspective)

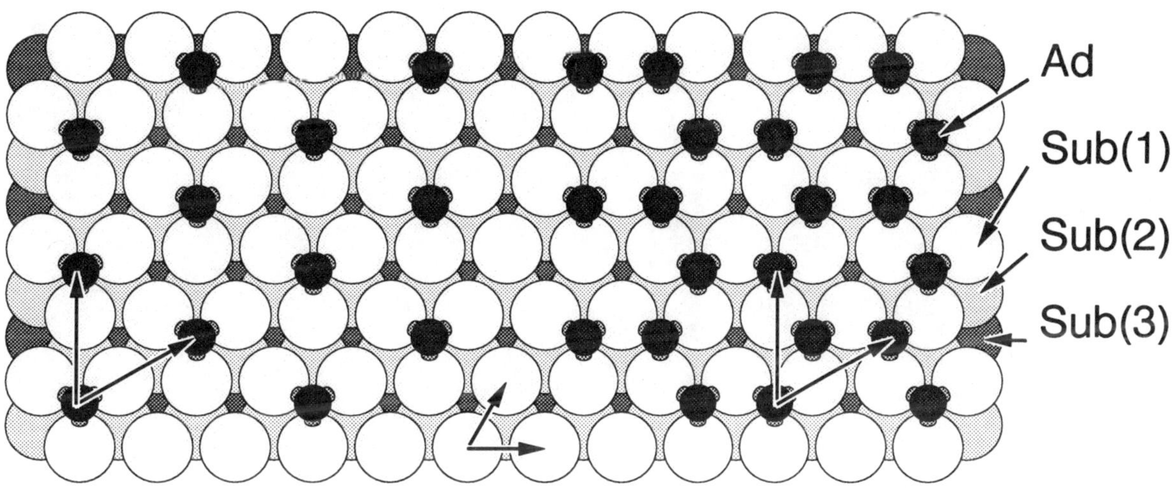

Fig. 24a : fcc(111)-(√3x√3)R30°-Ad; ~-2Ad (top view)

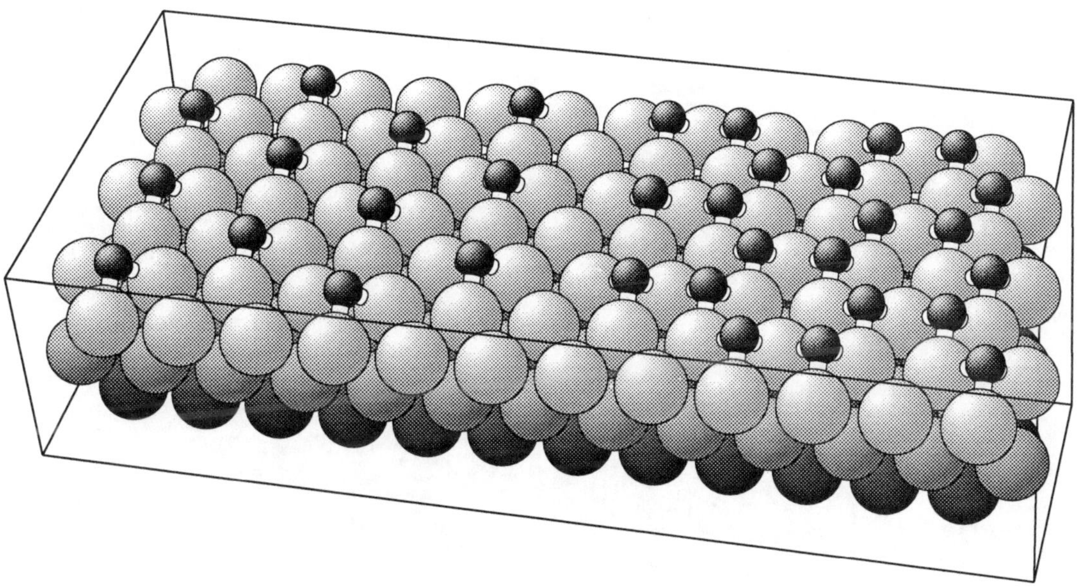

Fig. 24b : fcc(111)-(√3x√3)R30°-Ad; ~-2Ad (perspective)

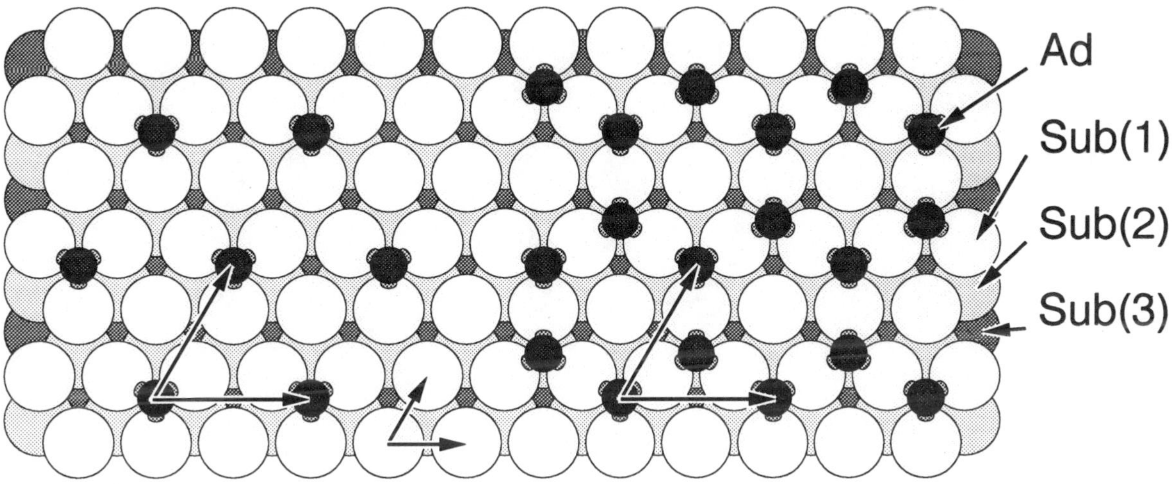

Fig. 25a : fcc(111)-p(2x2)-Ad; ~-2Ad (top view)

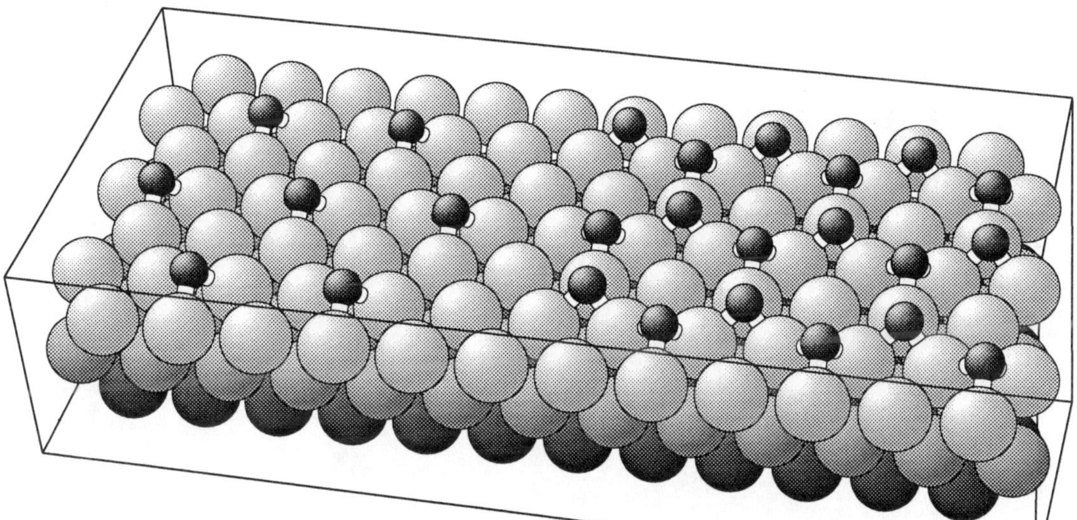

Fig. 25b : fcc(111)-p(2x2)-Ad; ~-2Ad (perspective)

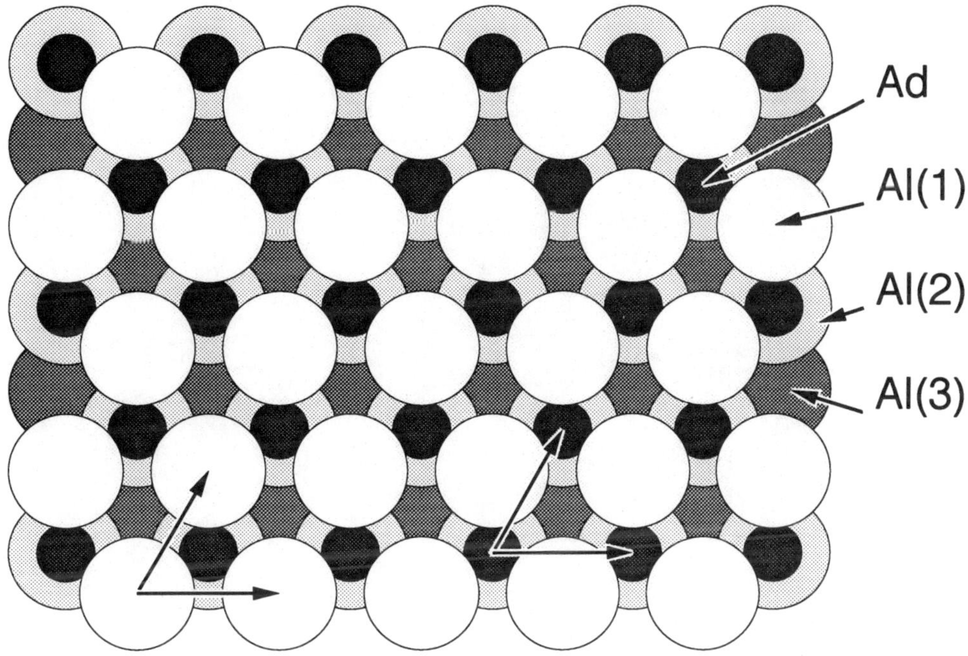

Fig. 26a : Al(111)-(1x1)-O (subsurface) (top view)

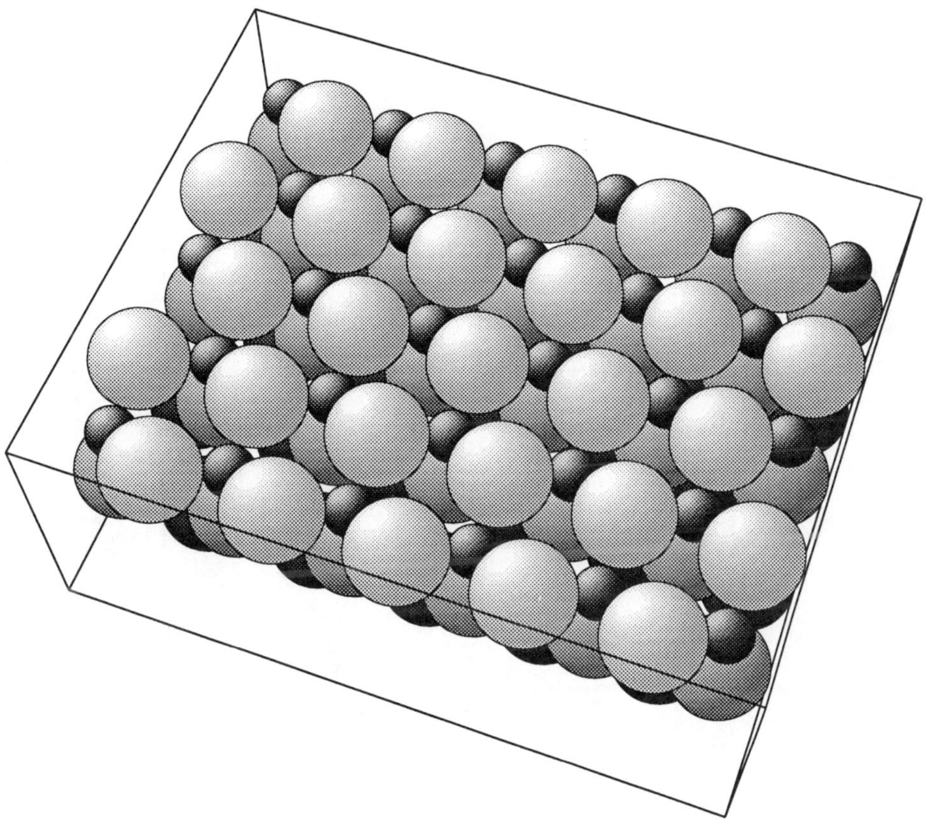

Fig. 26b : Al(111)-(1x1)-O (subsurface) (perspective)

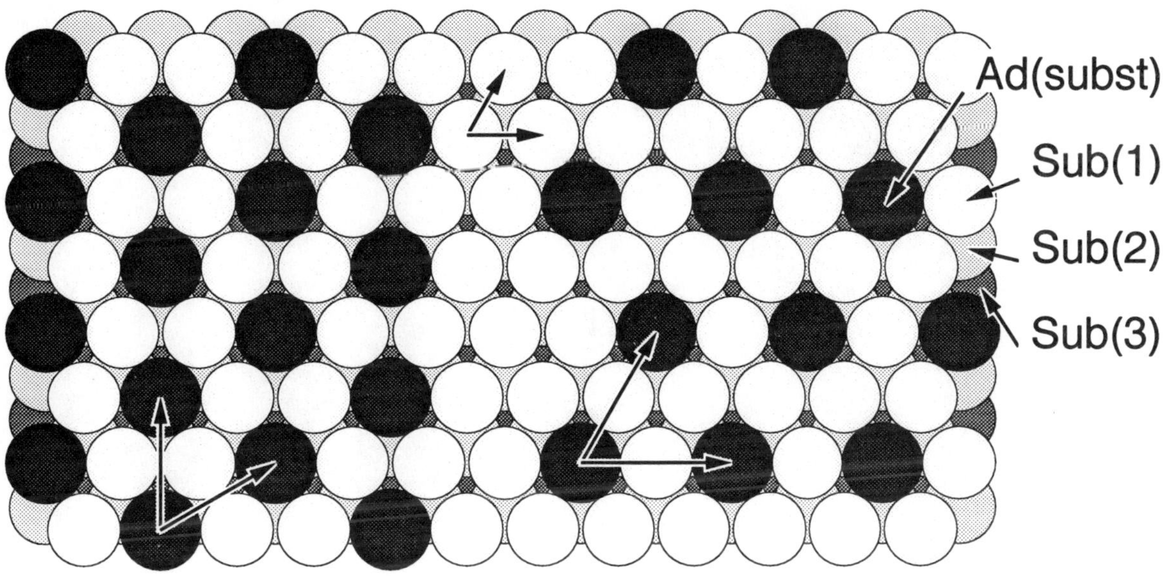

Fig. 27a : fcc(111)-(√3x√3)R30°-Ad; ~-(2x2)Ad (substitutional) (top view)

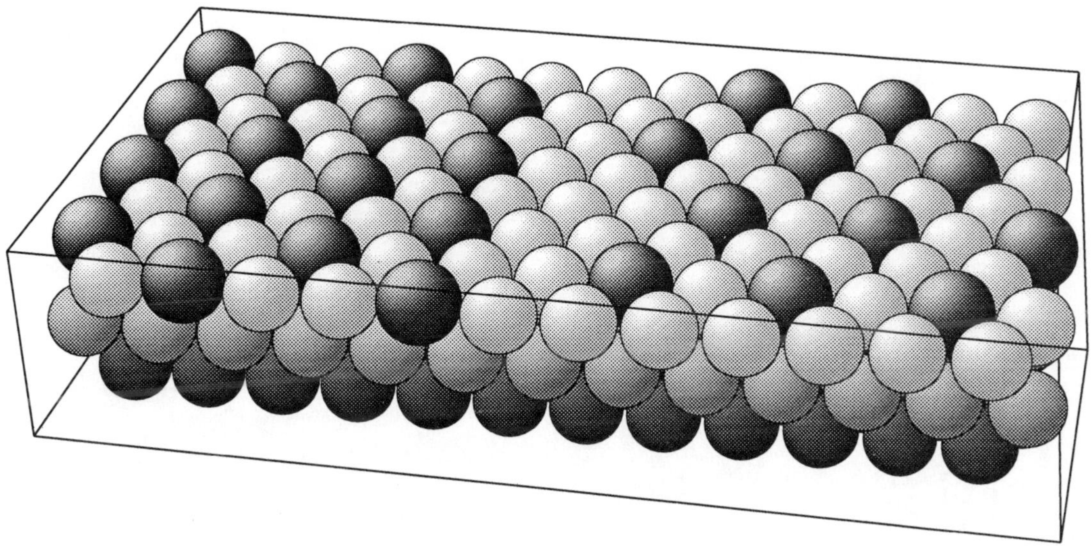

Fig. 27b : fcc(111)-(√3x√3)R30°-Ad; ~-(2x2)Ad (substitutional) (perspective)

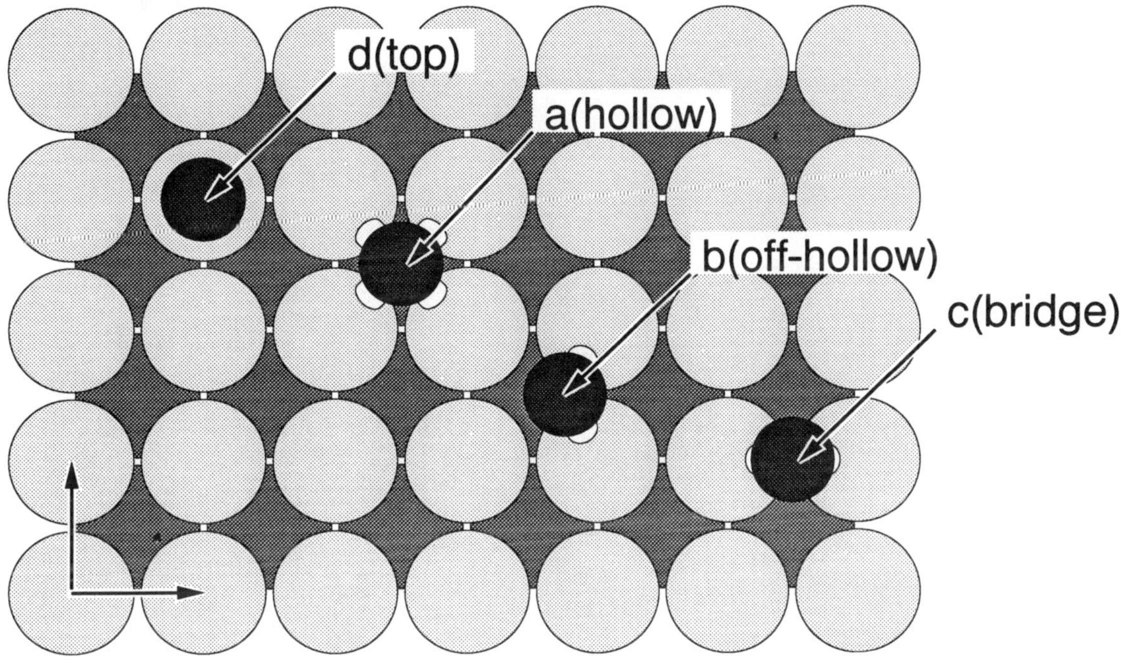

Fig. 28a : fcc(100) adsorbate sites (top view)

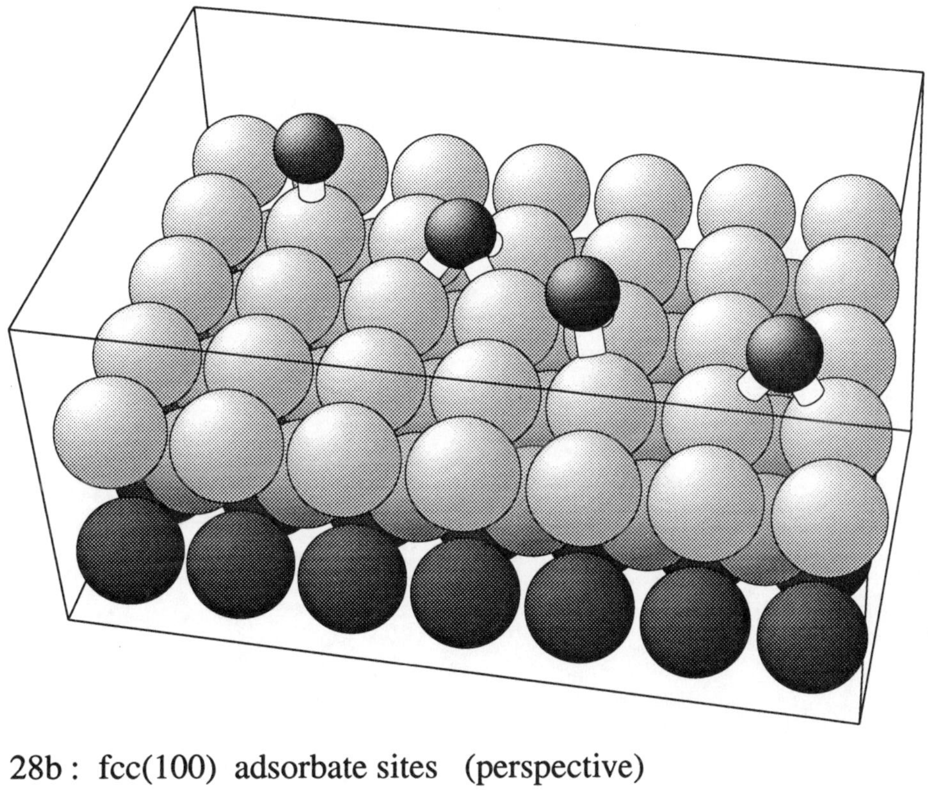

Fig. 28b : fcc(100) adsorbate sites (perspective)

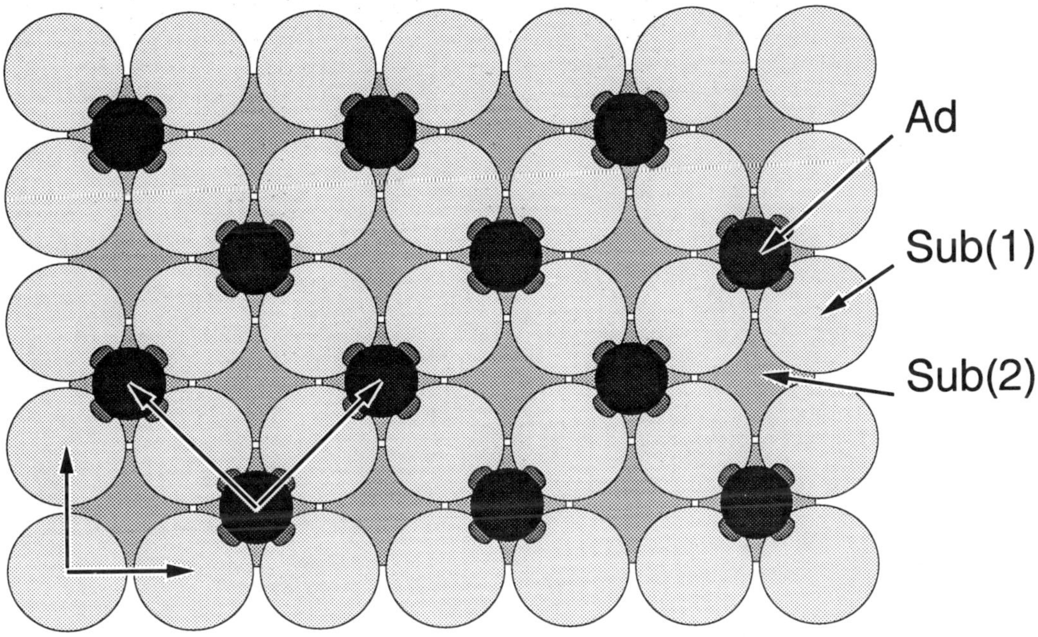

Fig. 29a : fcc(100)-c(2x2)-Ad (top view)

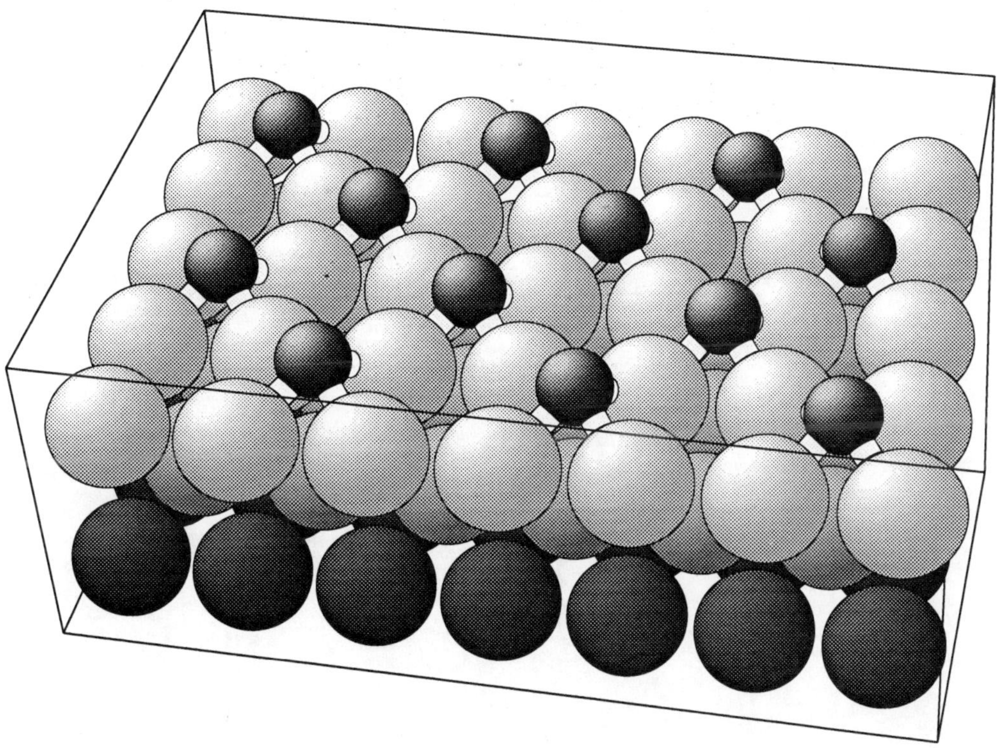

Fig. 29b : fcc(100)-c(2x2)-Ad (perspective)

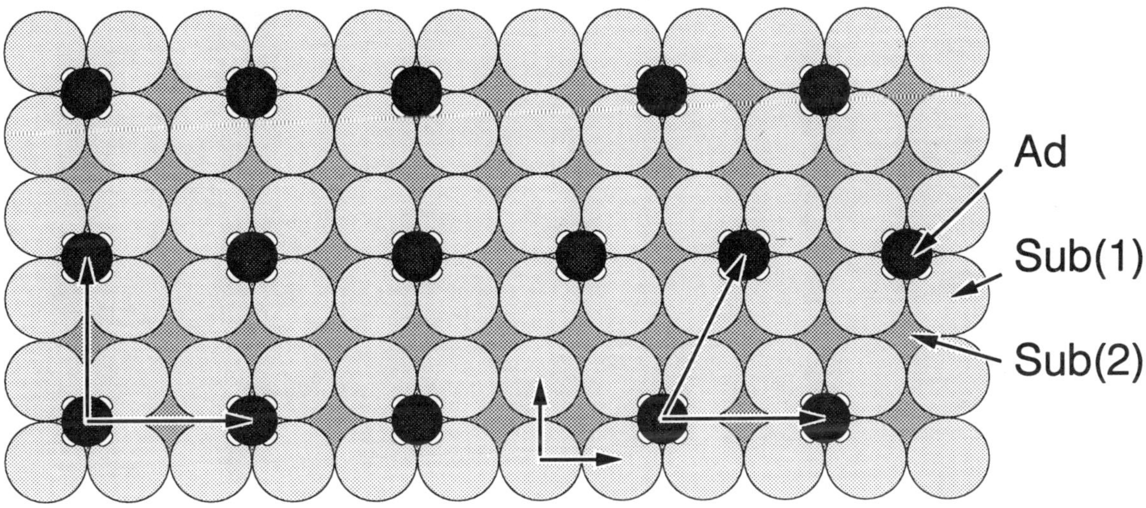

Fig. 30a : fcc(100)-p(2x2)-Ad; ~-c(4x2)-Ad (top view)

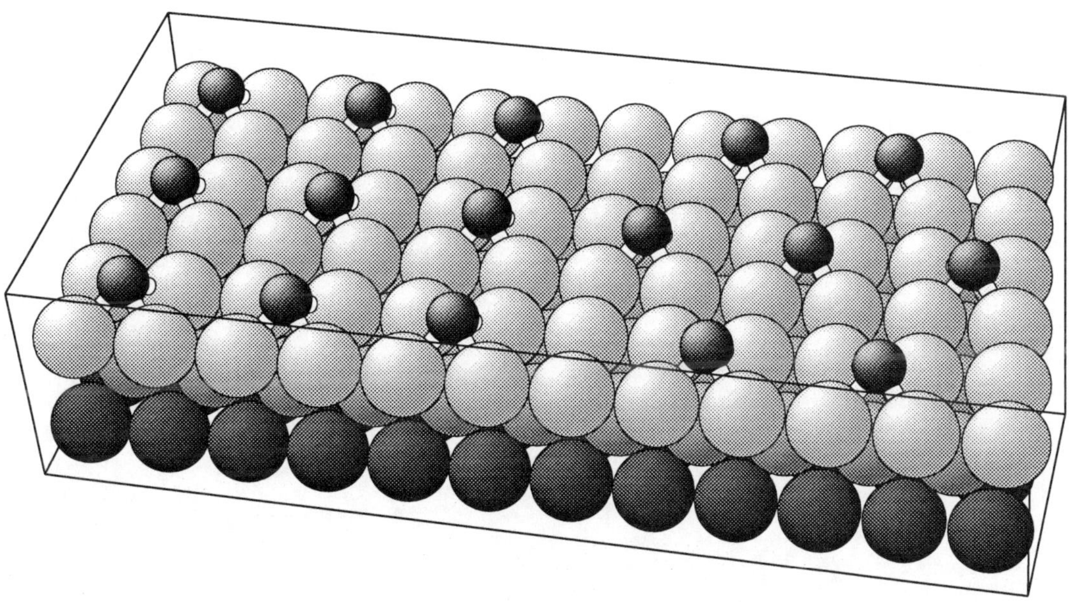

Fig. 30b : fcc(100)-p(2x2)-Ad; ~-c(4x2)-Ad (perspective)

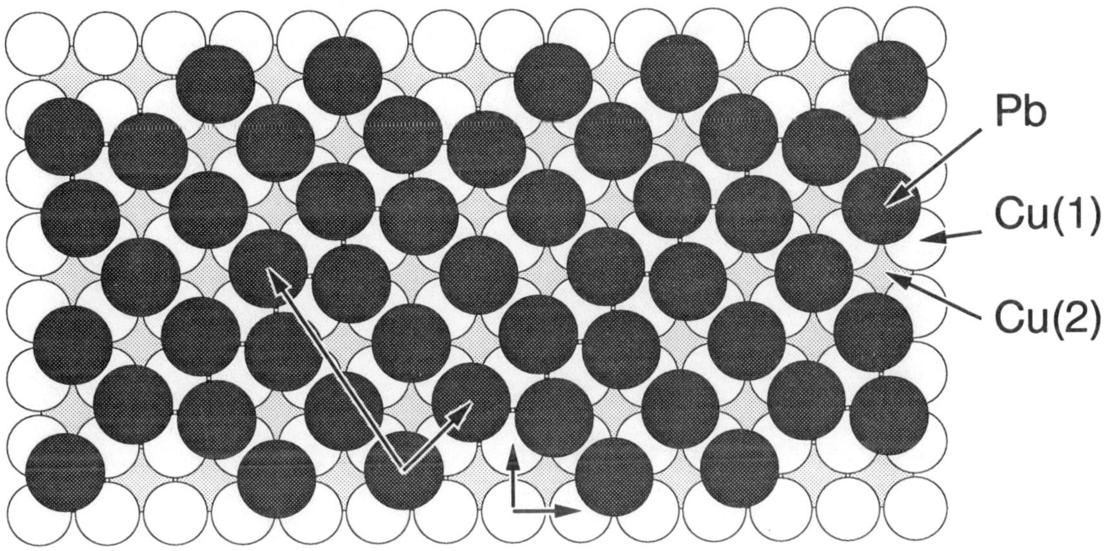

Fig. 31a : Cu(100)-c(5√2x√2)R45°-3Pb (top view)

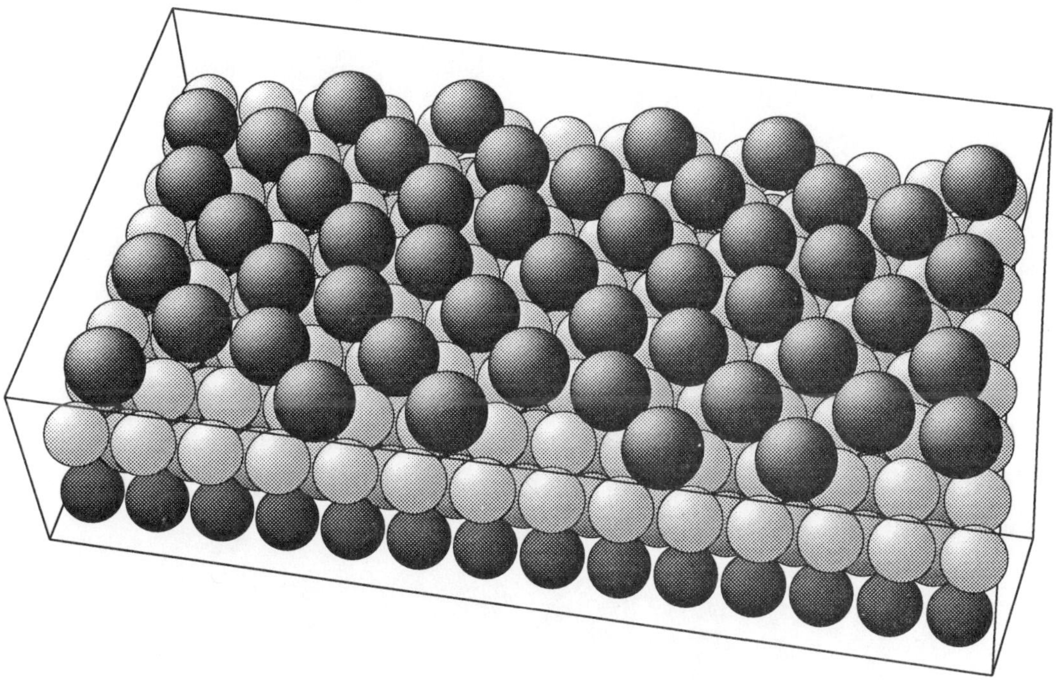

Fig. 31b : Cu(100)-c(5√2x√2)R45°-3Pb (perspective)

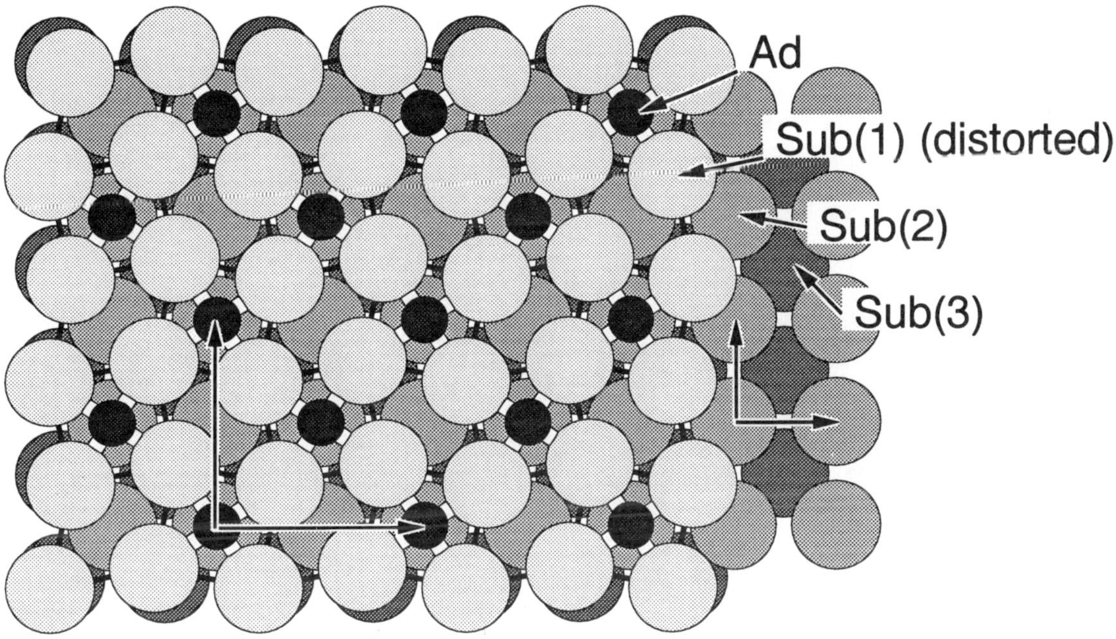

Fig. 32a : fcc(100)-p4g(2x2)-2Ad (top view)

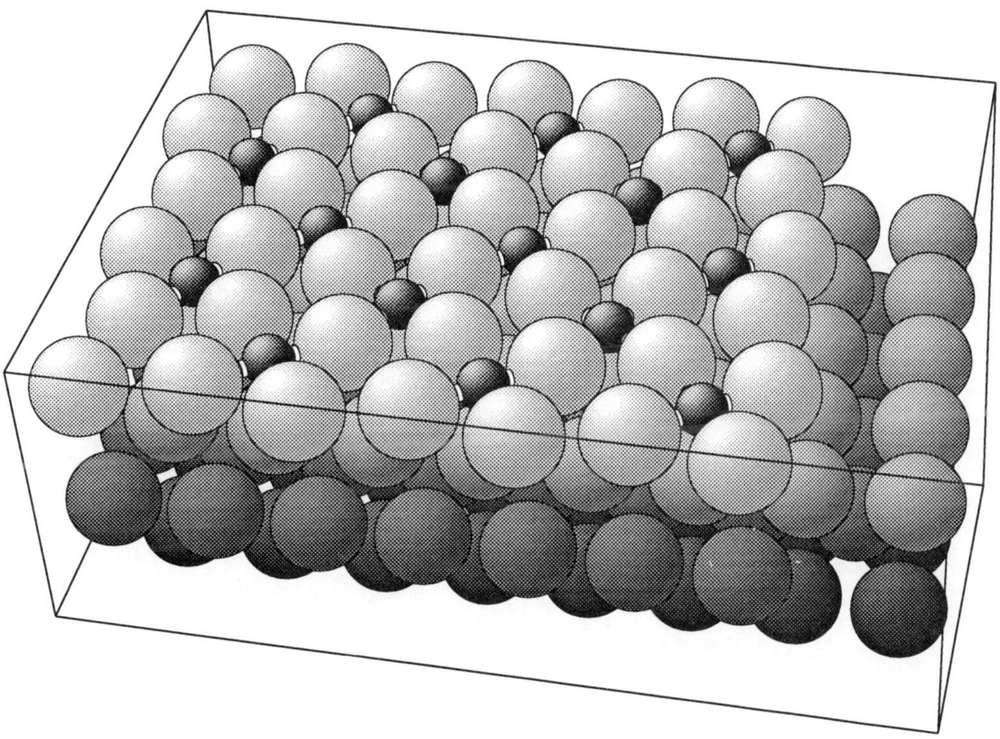

Fig. 32b : fcc(100)-p4g(2x2)-2Ad (perspective)

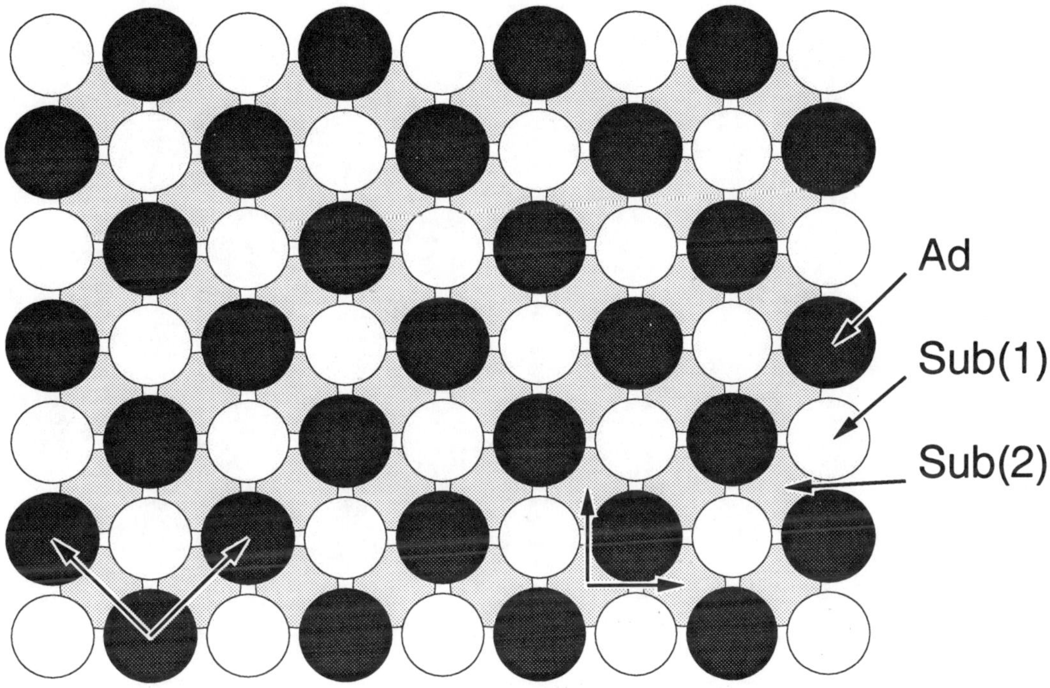

Fig. 33a : fcc(100)-c(2x2)-Ad (substitutional) (top view)

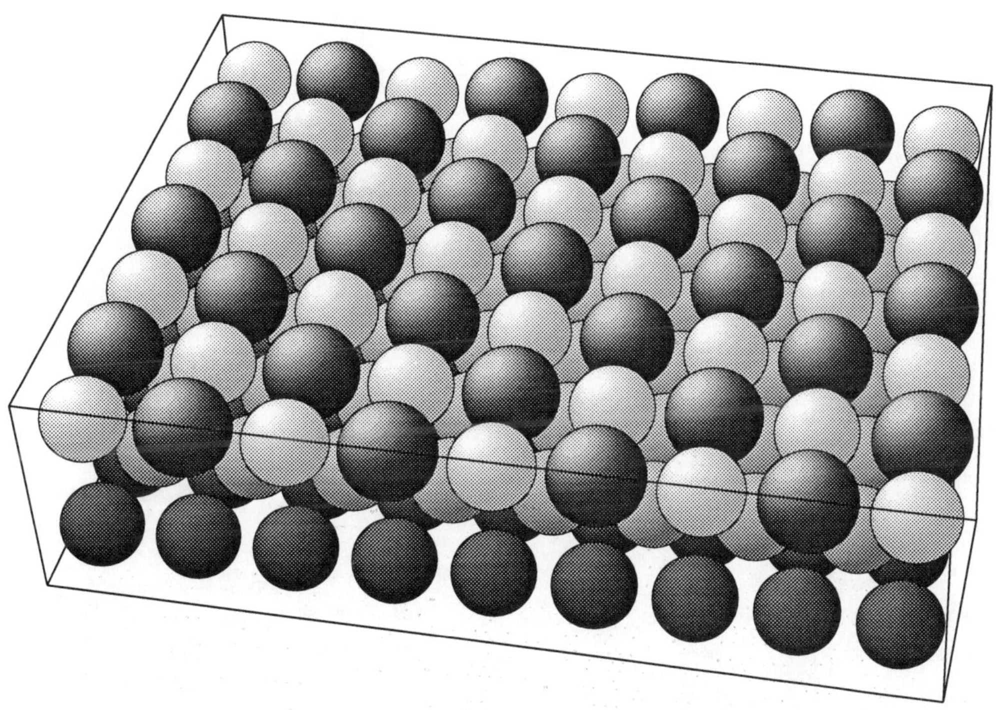

Fig. 33b : fcc(100)-c(2x2)-Ad (substitutional) (perspective)

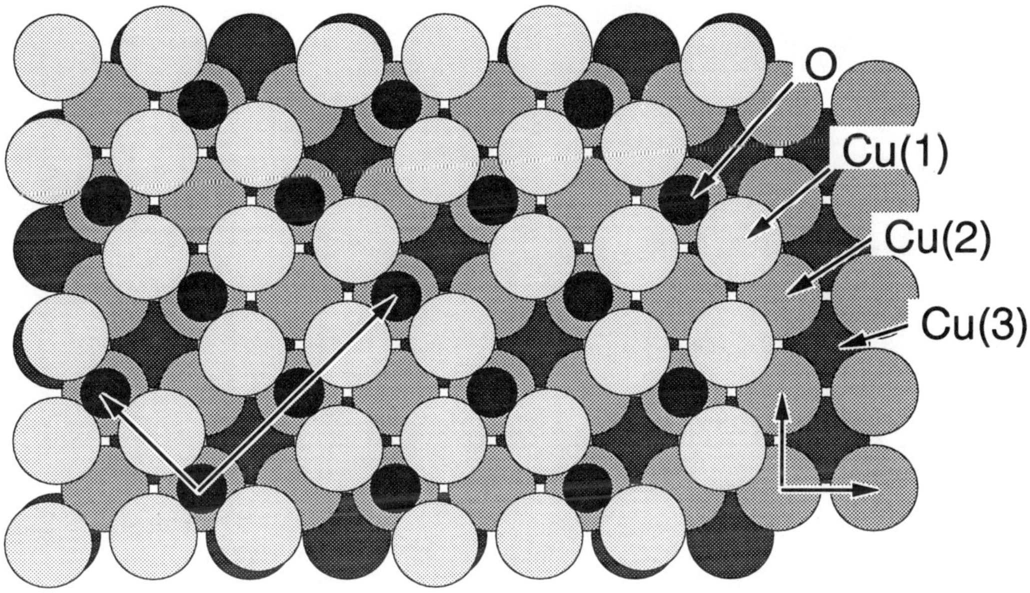

Fig. 34a : Cu(100)-(2√2x√2)R45°-2O (top view)

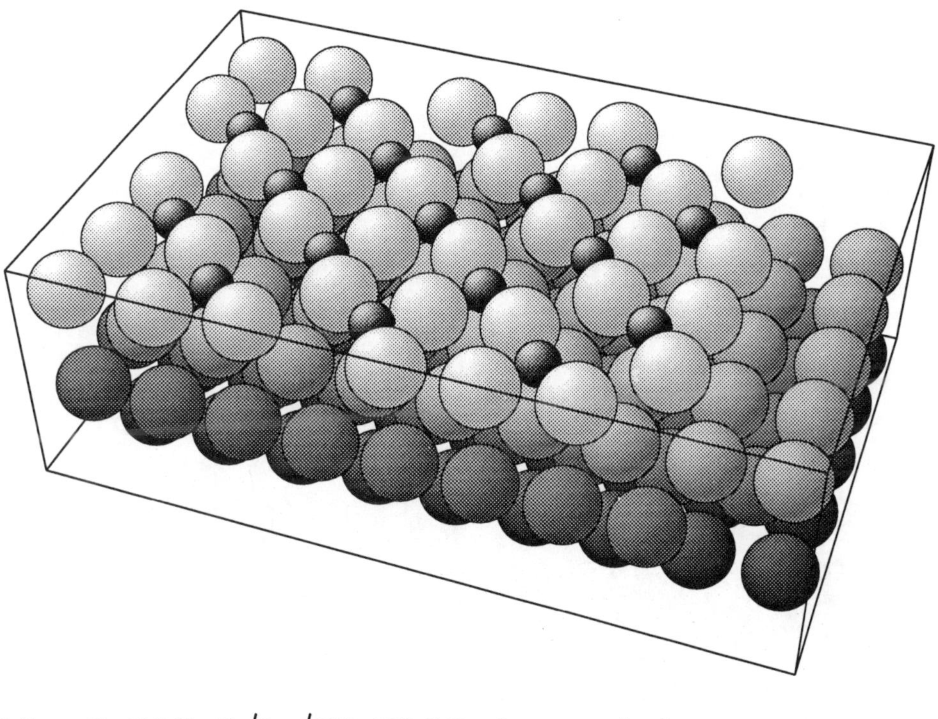

Fig. 34b : Cu(100)-(2√2x√2)R45°-2O (perspective)

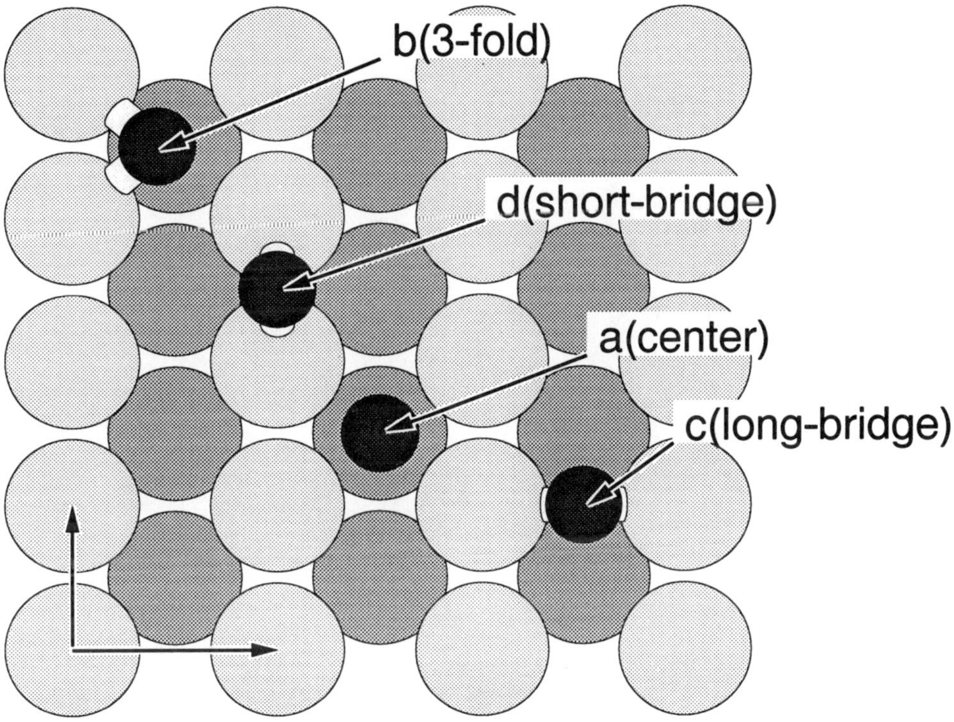

Fig. 35a : fcc(110) high symmetry adsorbate sites (top view)

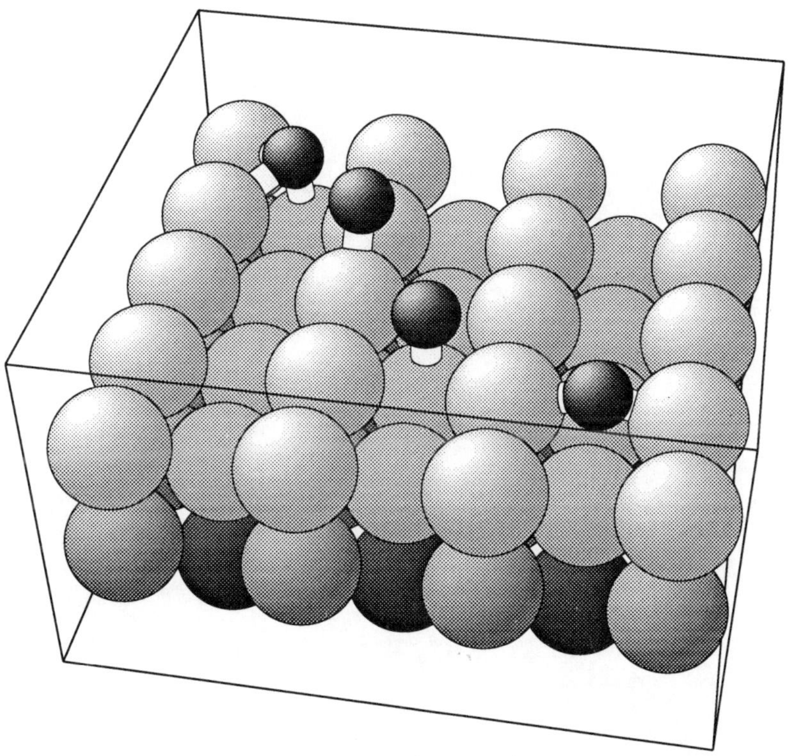

Fig. 35b : fcc(110) high symmetry adsorbate sites (perspective)

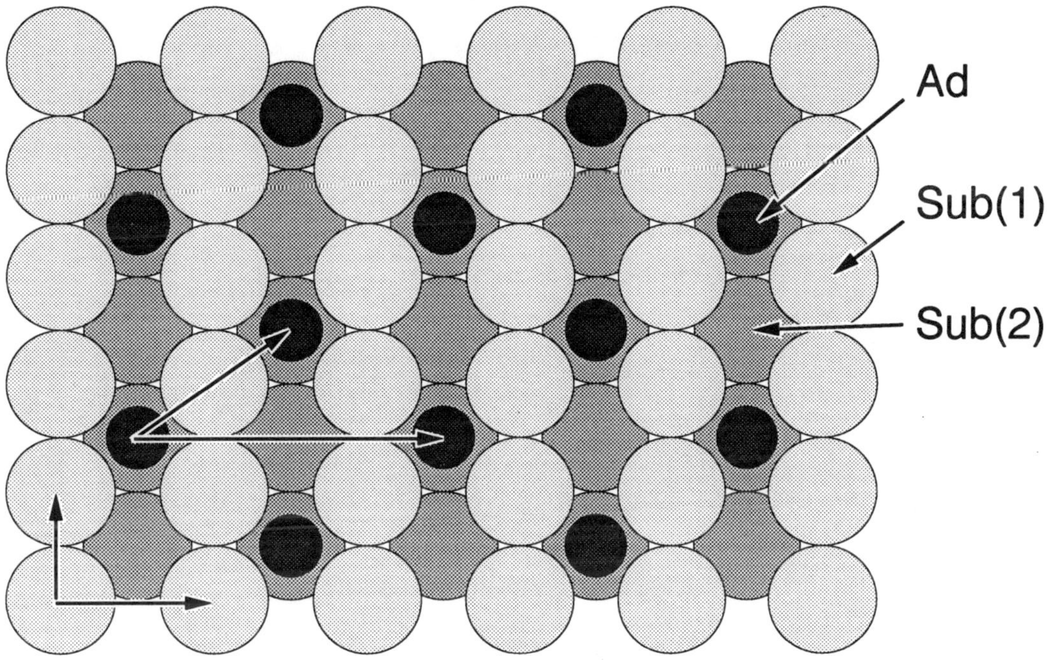

Fig. 36a : fcc(110)-c(2x2)-Ad (top view)

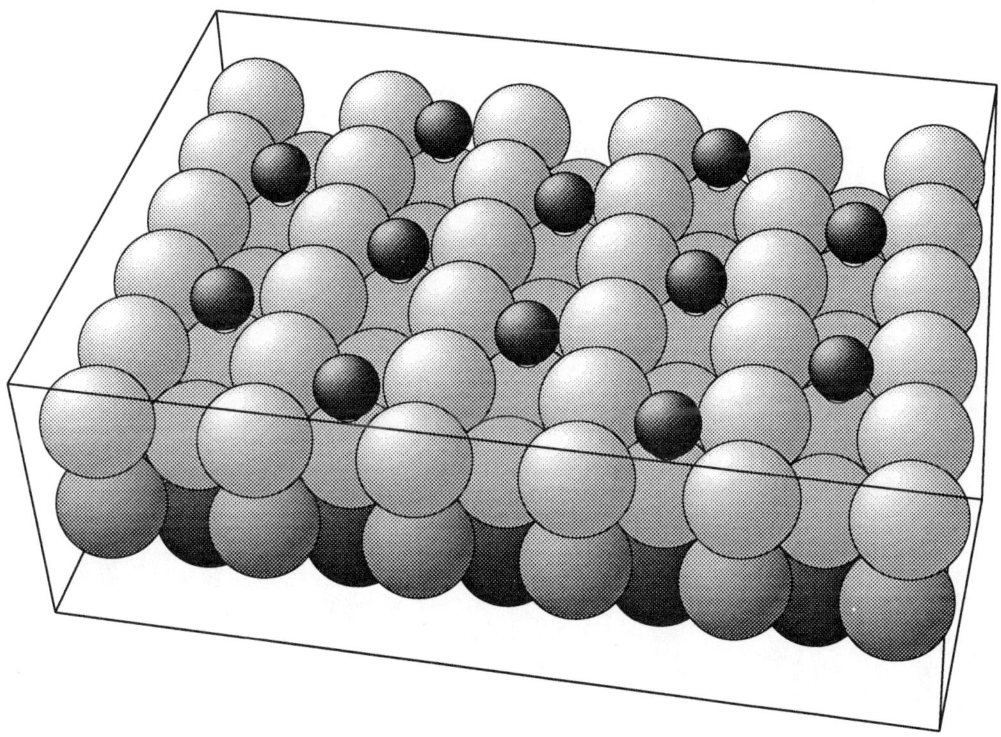

Fig. 36b : fcc(110)-c(2x2)-Ad (perspective)

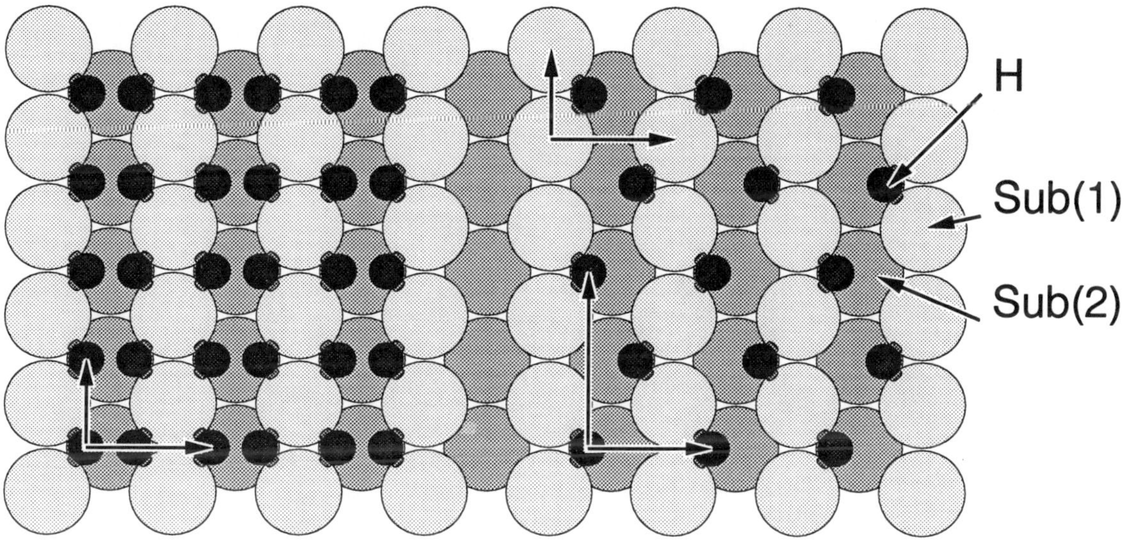

Fig. 37a : Rh(110)-(1x1)-2H; Pd(110)-(2x1)-2H (top view)

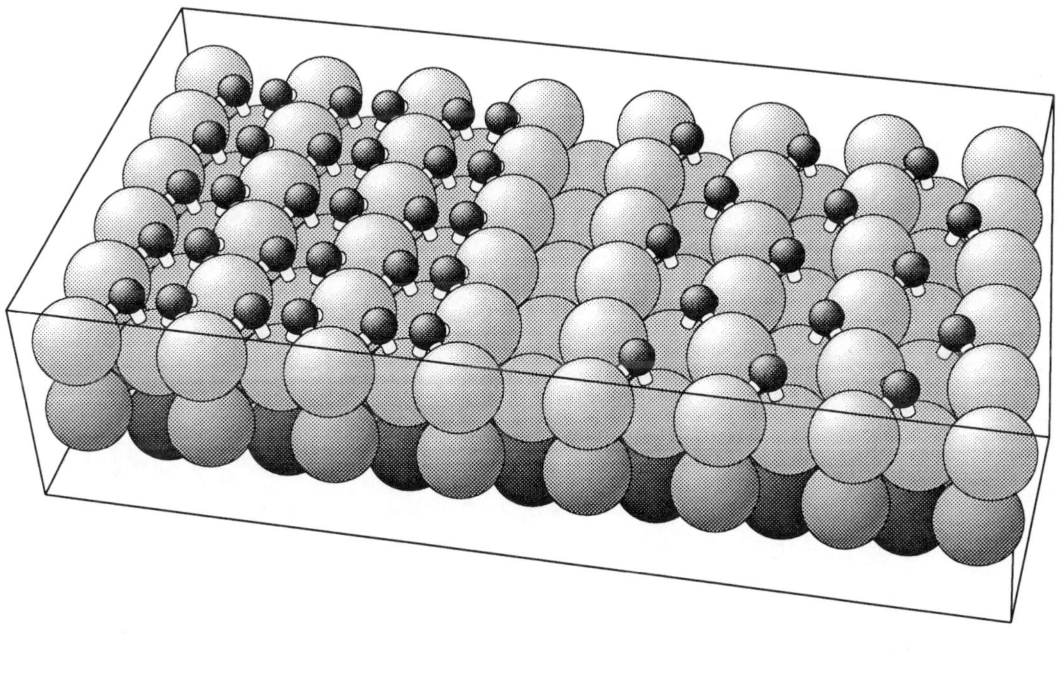

Fig. 37b : Rh(110)-(1x1)-2H; Pd(110)-(2x1)-2H (perspective)

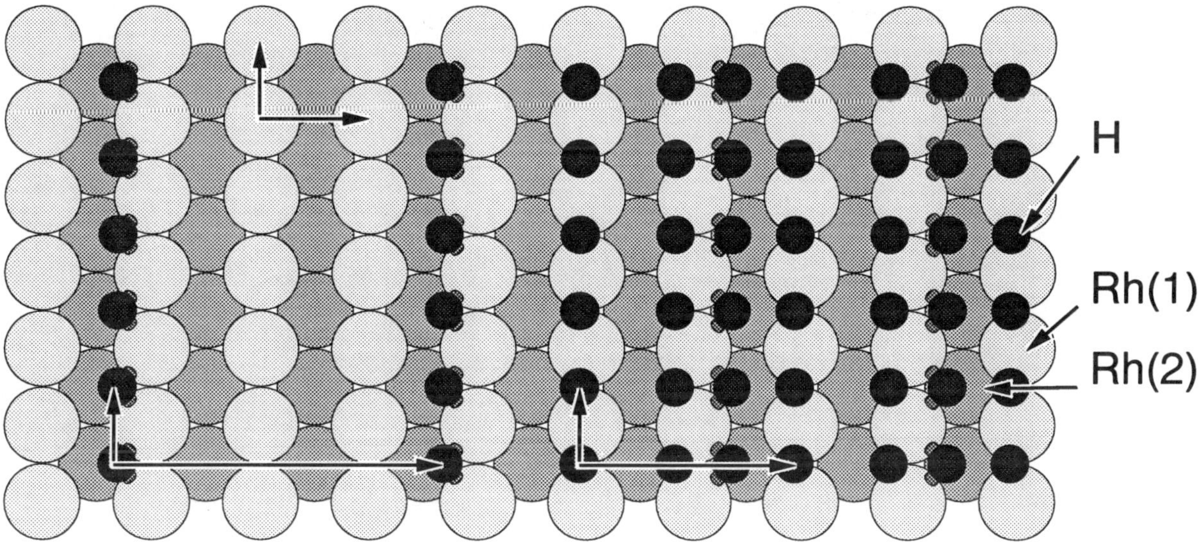

Fig. 38a : Rh(110)-(1x3)-H; ~-(1x2)-3H (top view)

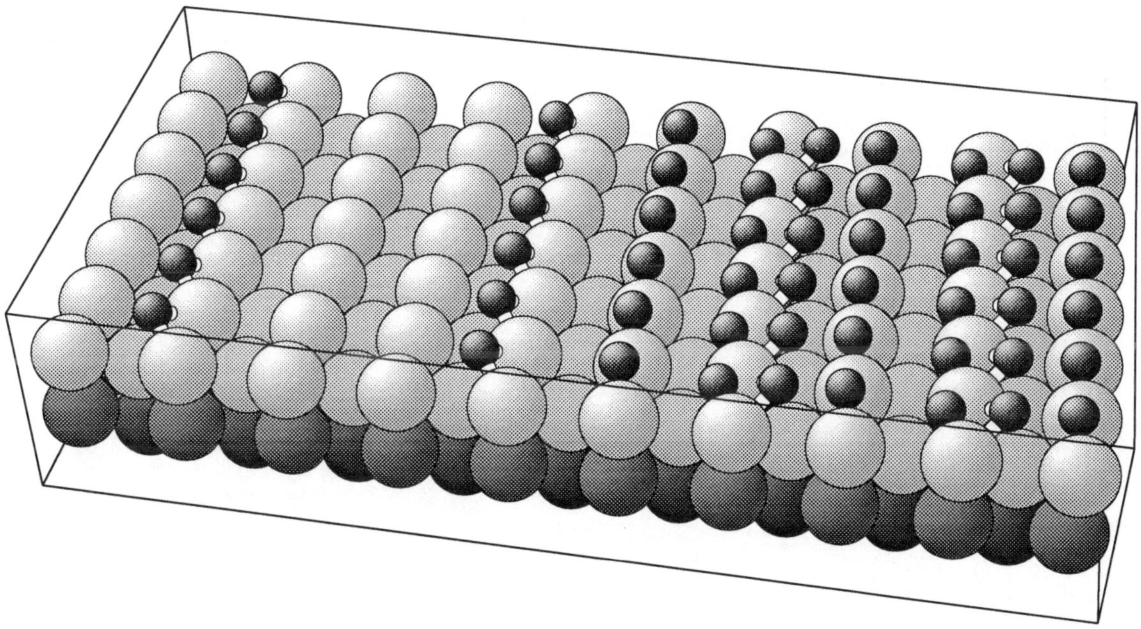

Fig. 38b : Rh(110)-(1x3)-H; ~-(1x2)-3H (perspective)

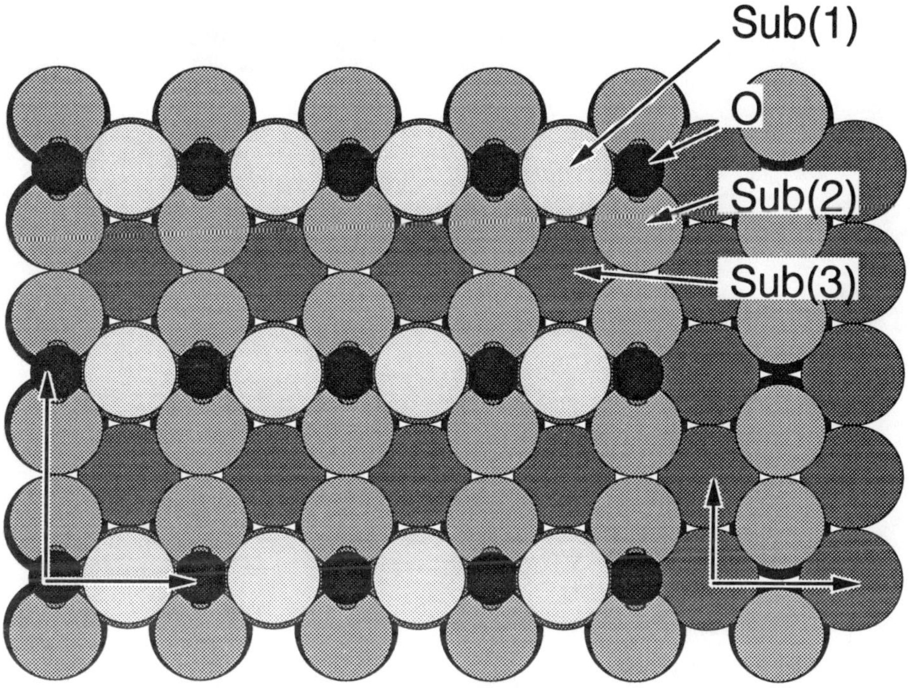

Fig. 39a : fcc(110)-(2x1)-O missing/added row (top view)

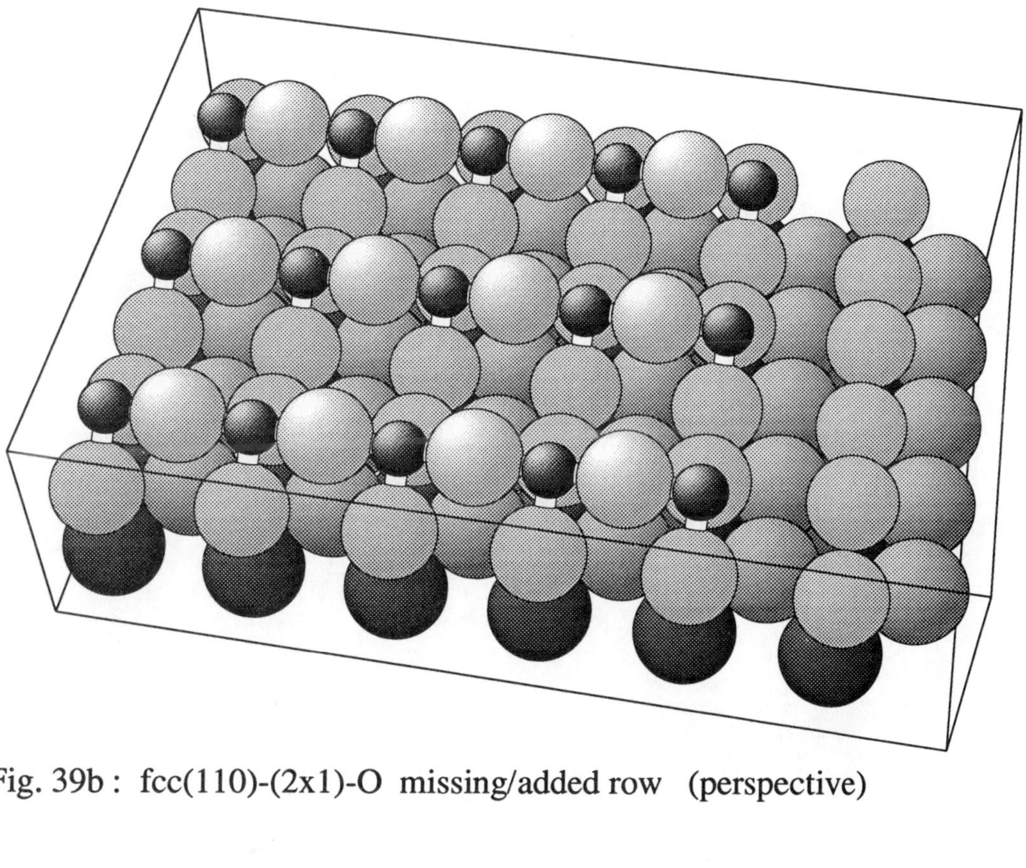

Fig. 39b : fcc(110)-(2x1)-O missing/added row (perspective)

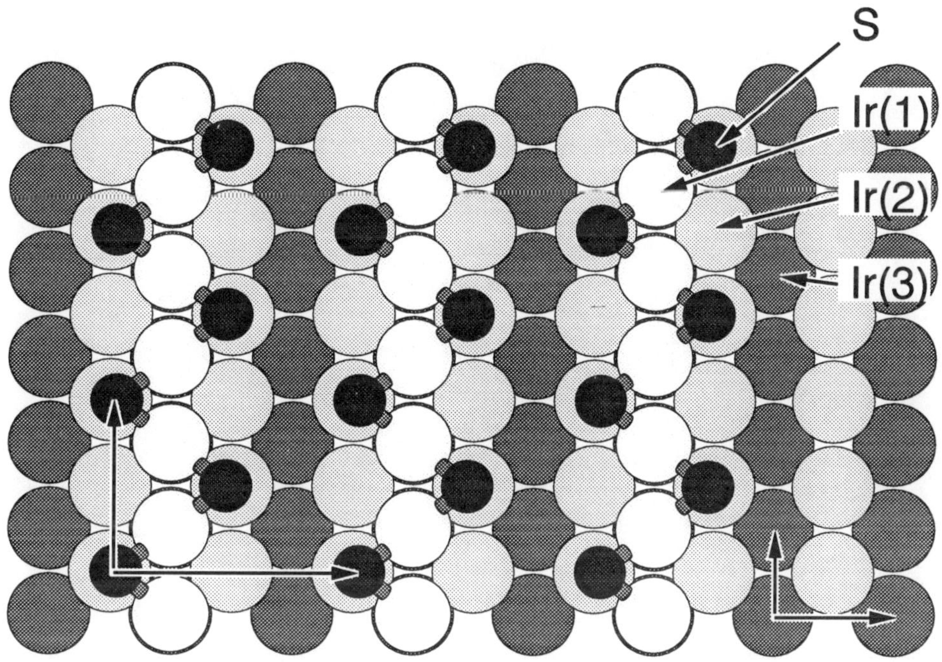

Fig. 40a : Ir(110)-(2x2)-2S (top view)

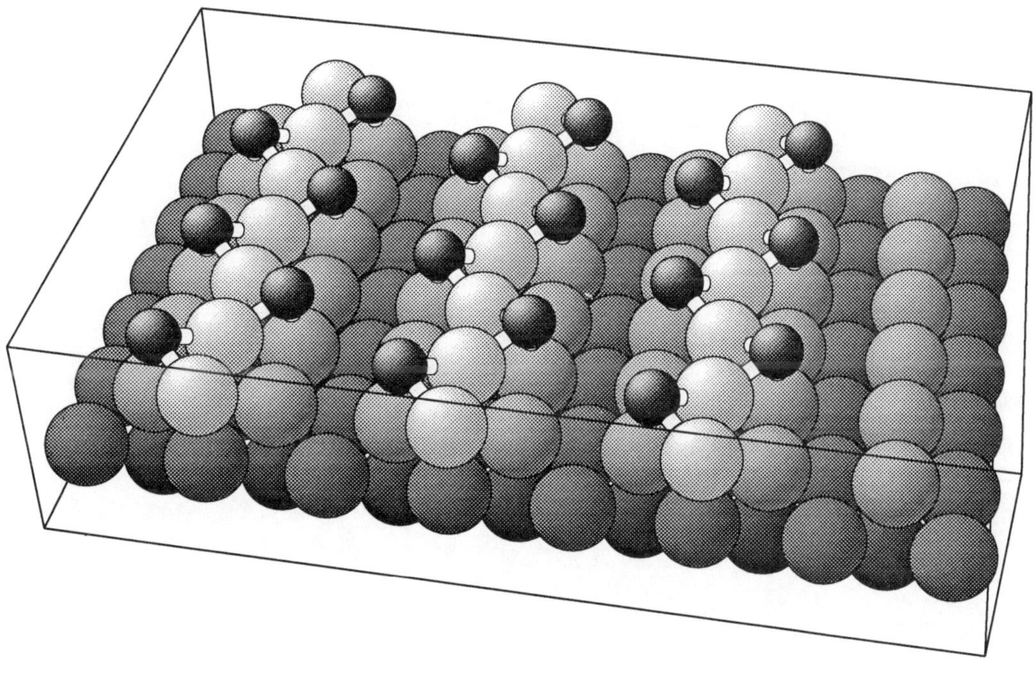

Fig. 40b : Ir(110)-(2x2)-2S (perspective)

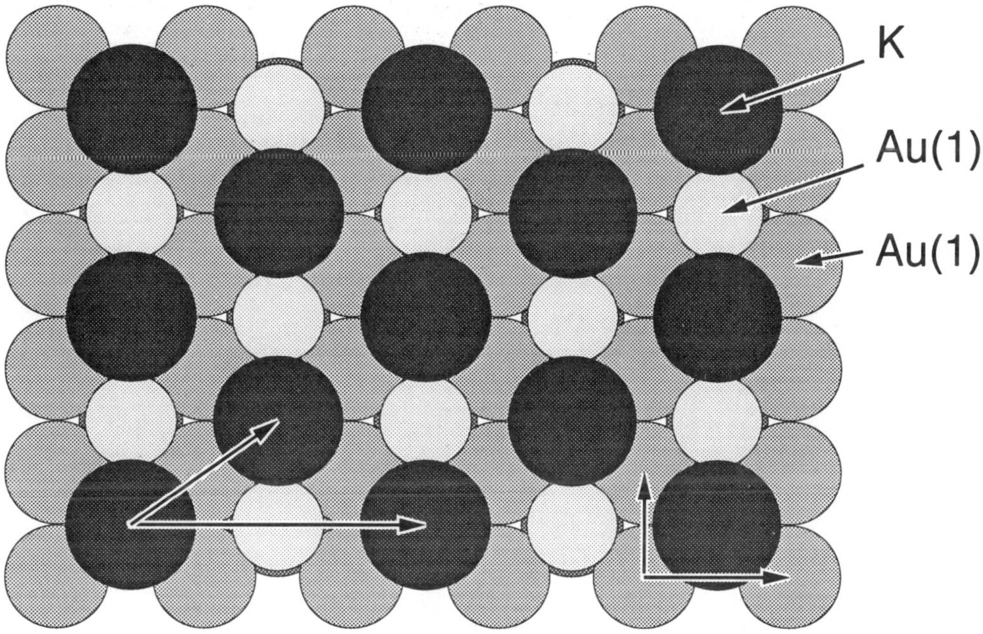

Fig. 41a : Au(110)-c(2x2)-K (substitutional) (top view)

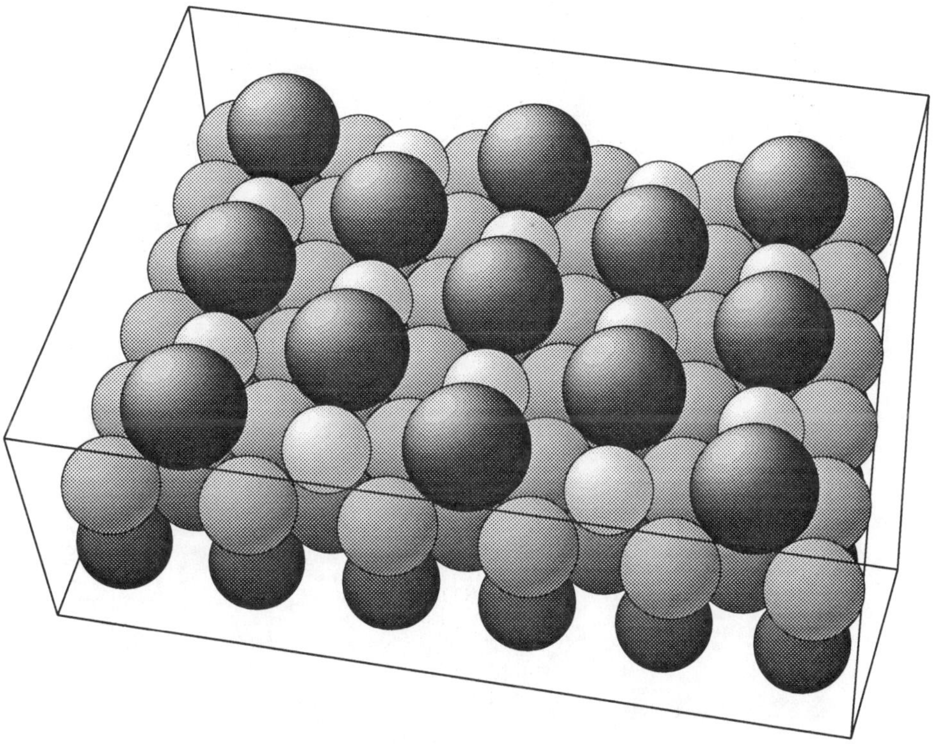

Fig. 41b : Au(110)-c(2x2)-K (substitutional) (perspective)

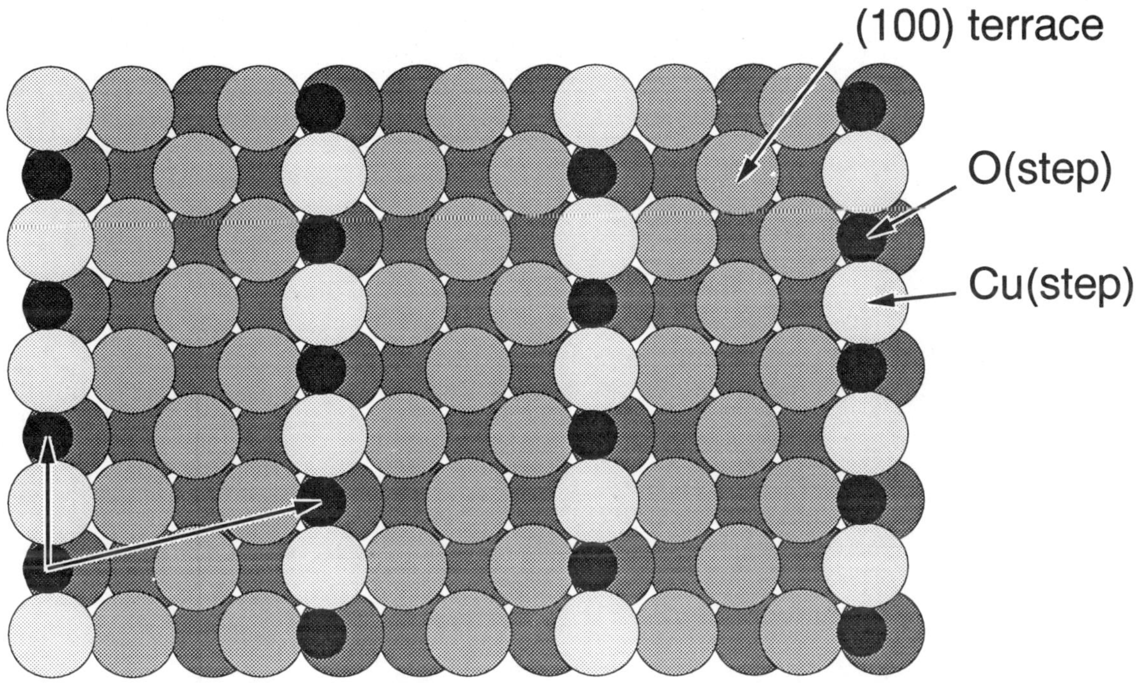

Fig. 42a : Cu(410)-(1x1)-O (top view)

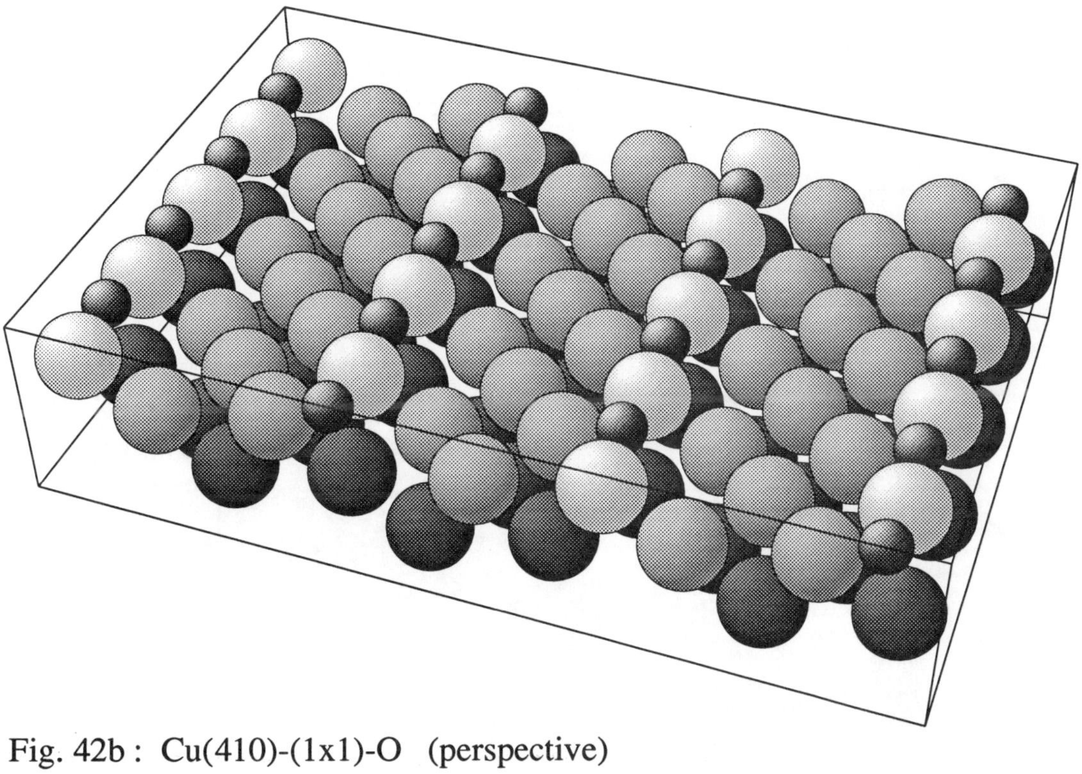

Fig. 42b : Cu(410)-(1x1)-O (perspective)

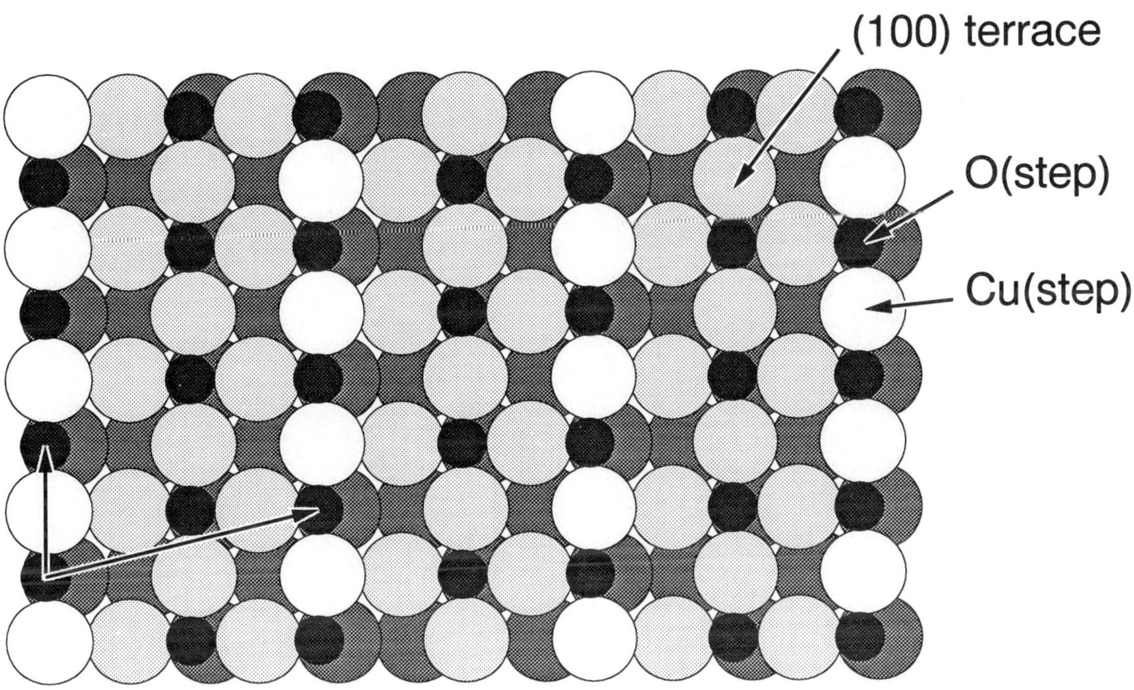

Fig. 43a : Cu(410)-(1x1)-2O (top view)

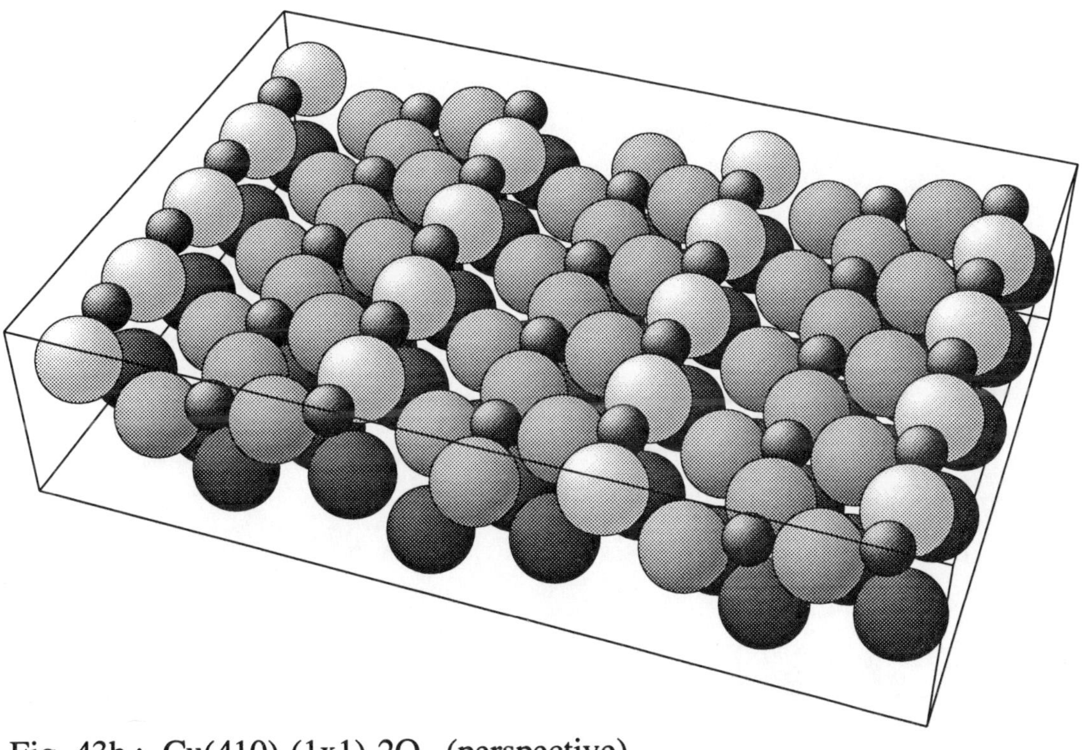

Fig. 43b : Cu(410)-(1x1)-2O (perspective)

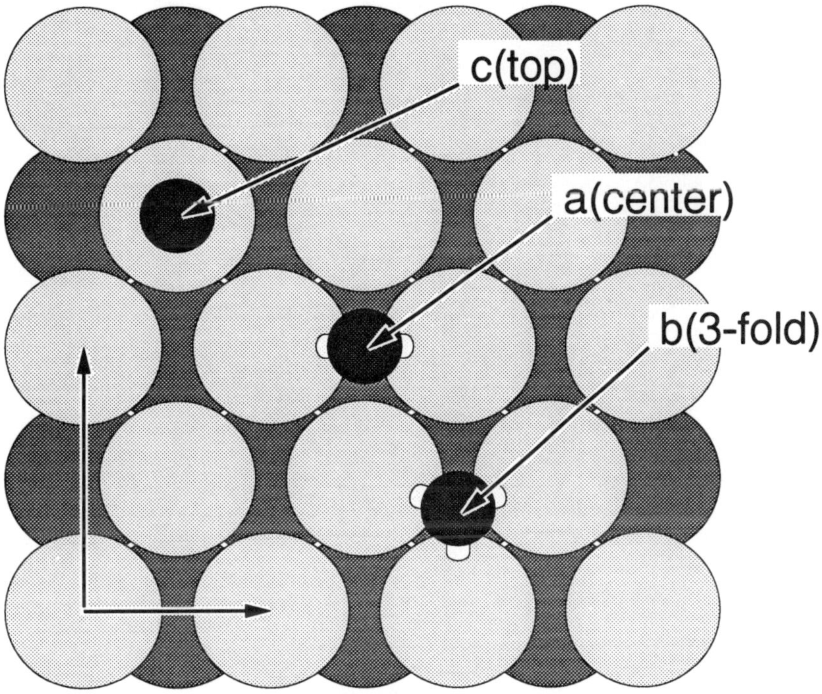

Fig. 44a : bcc(110) high symmetry adsorbate sites (top view)

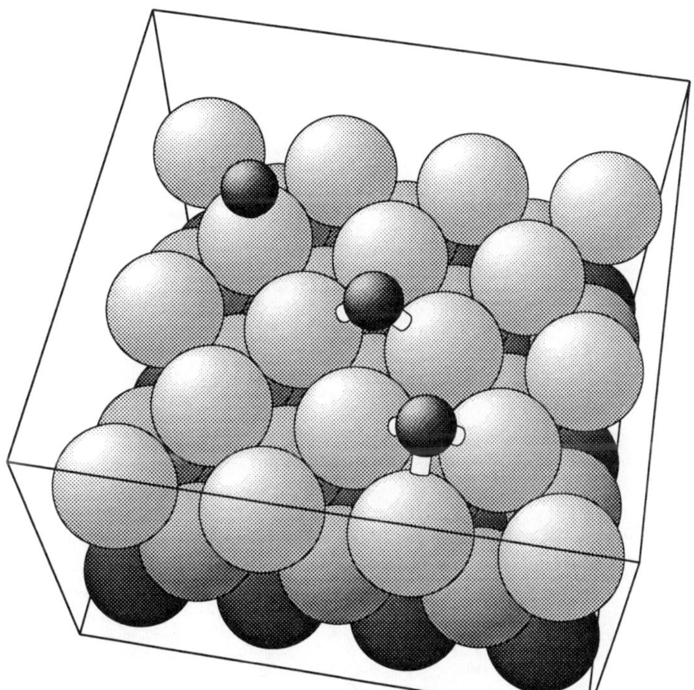

Fig. 44b : bcc(110) high symmetry adsorbate sites (perspective)

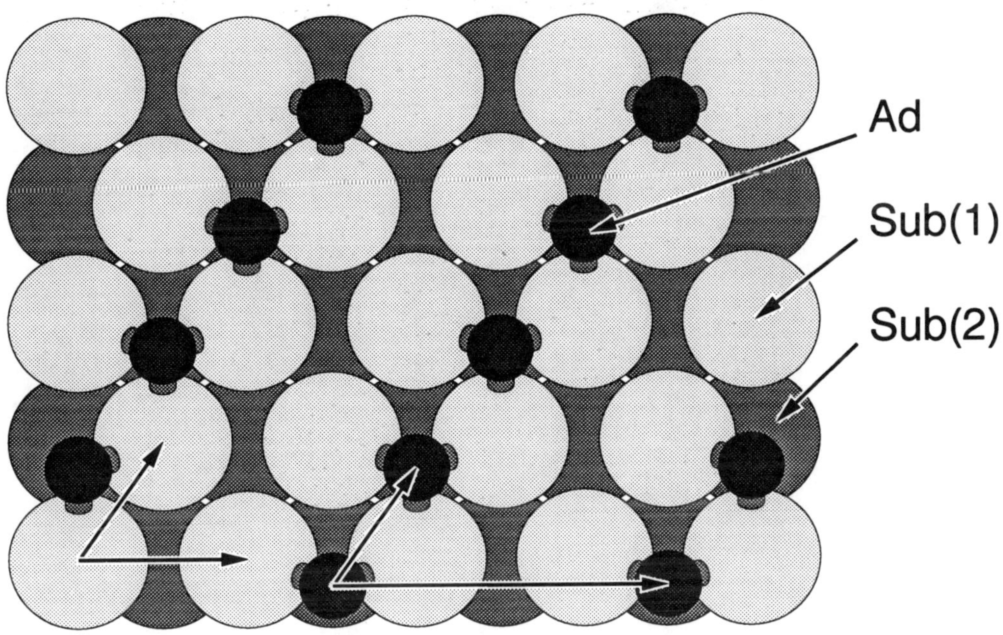

Fig. 45a : bcc(110)-(2x1)-Ad (top view)

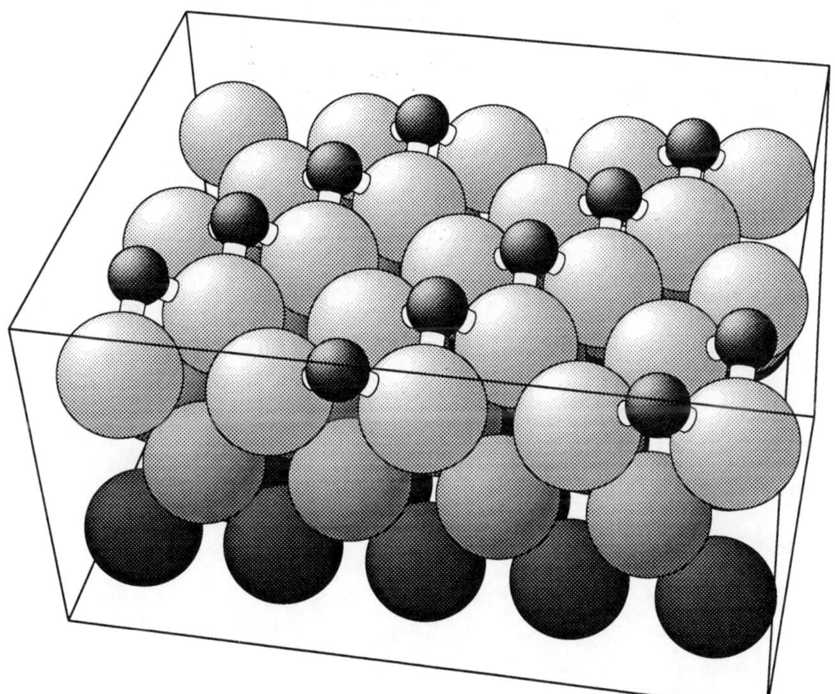

Fig. 45b : bcc(110)-(2x1)-Ad (perspective)

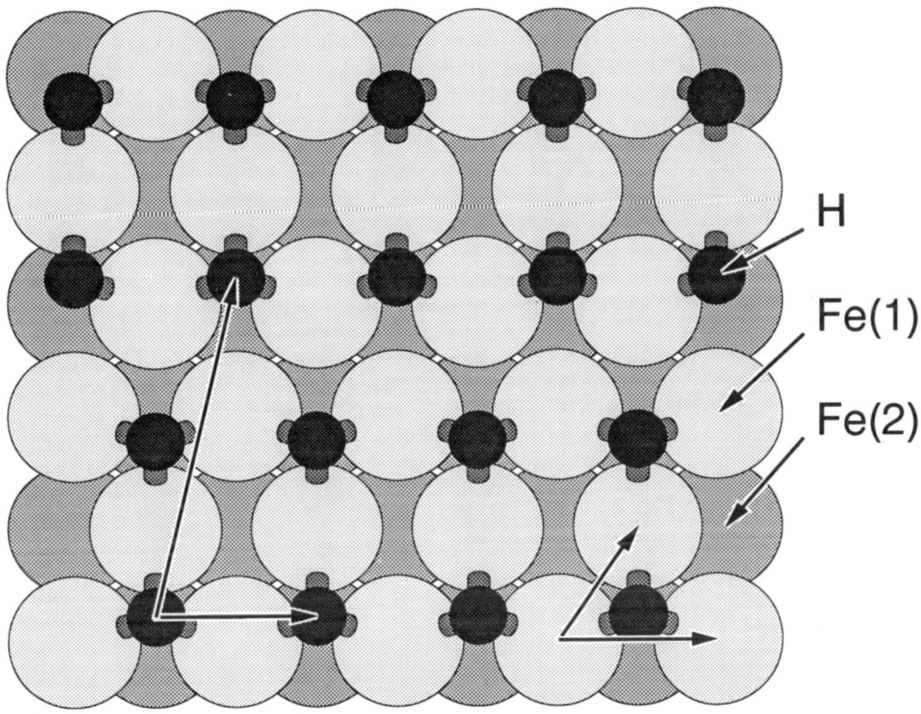

Fig. 46a : Fe(110)-(3x1)-2H (top view)

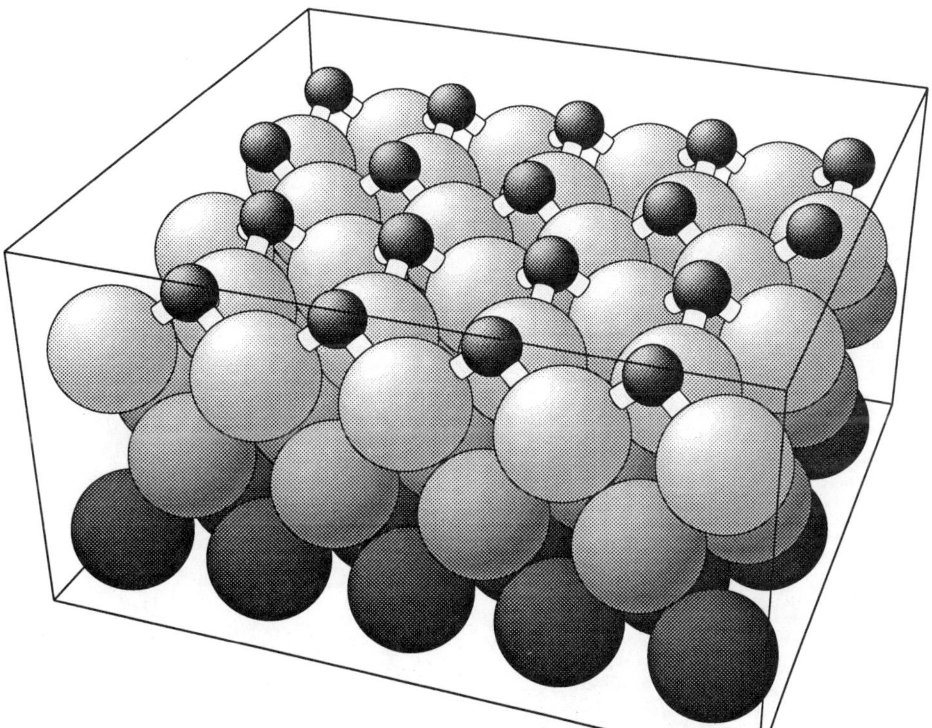

Fig. 46b : Fe(110)-(3x1)-2H (perspective)

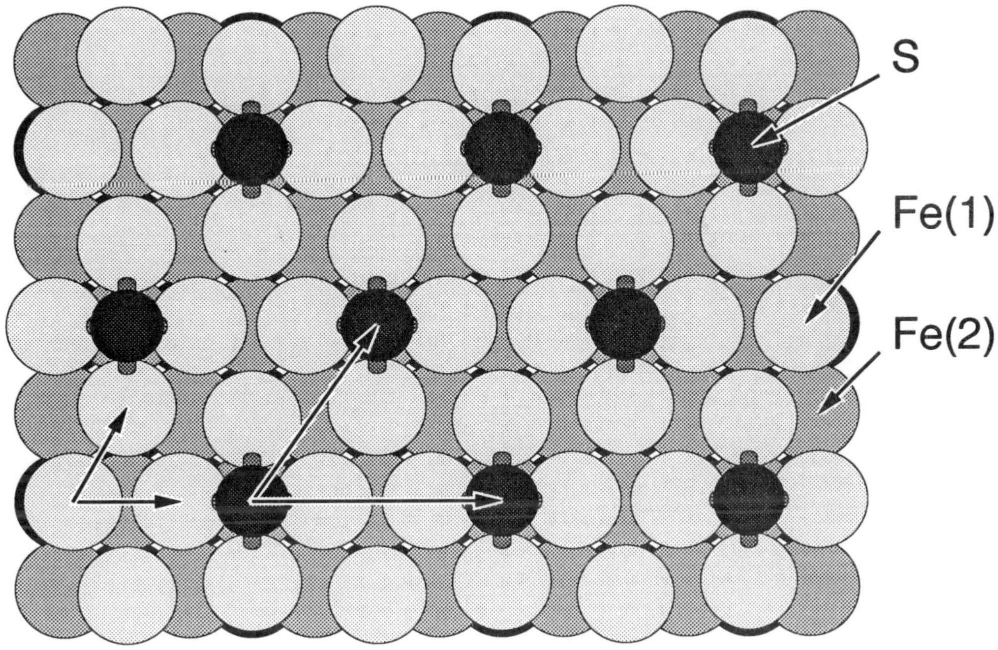

Fig. 47a : Fe(110)-p(2x2)-S (top view)

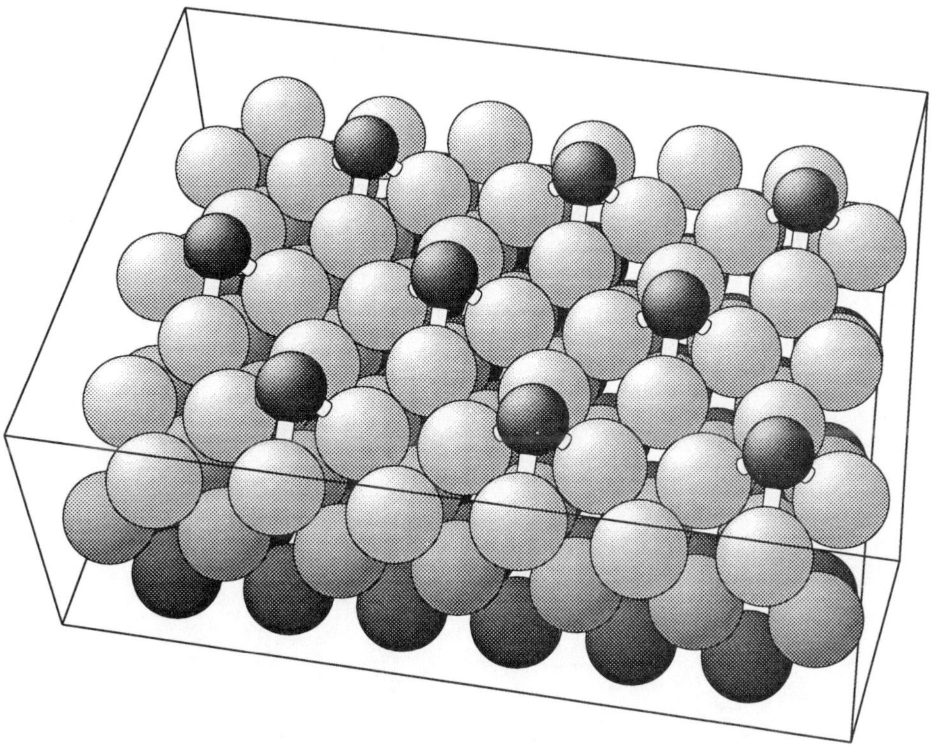

Fig. 47b : Fe(110)-p(2x2)-S (perspective)

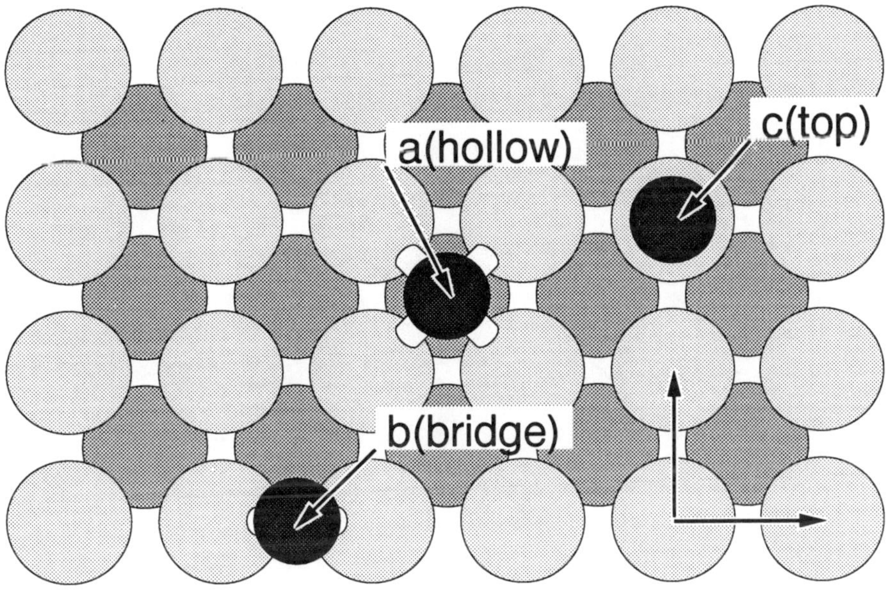

Fig. 48a : bcc(100) high symmetry adsorbate sites (top view)

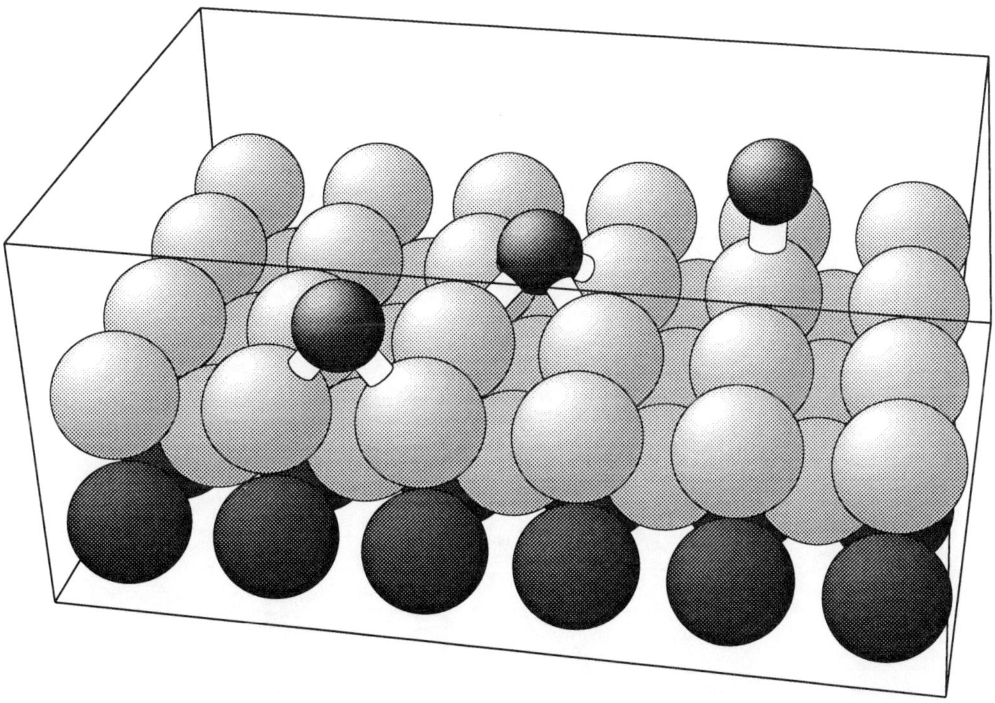

Fig. 48b : bcc(100) high symmetry adsorbate sites (perspective)

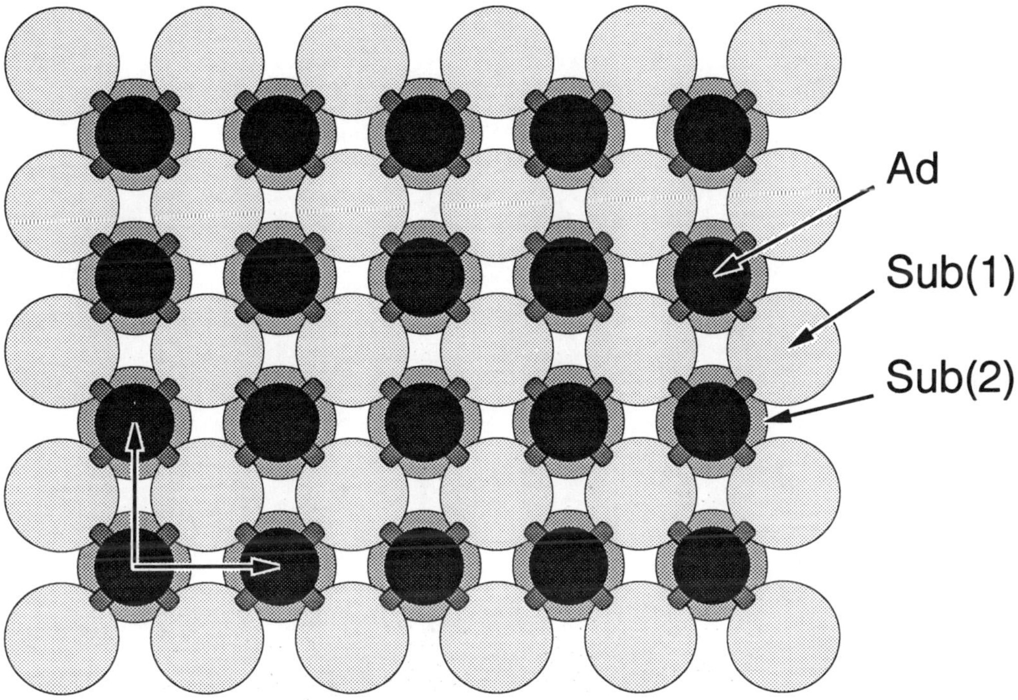

Fig. 49a : bcc(100)-(1x1)-Ad (top view)

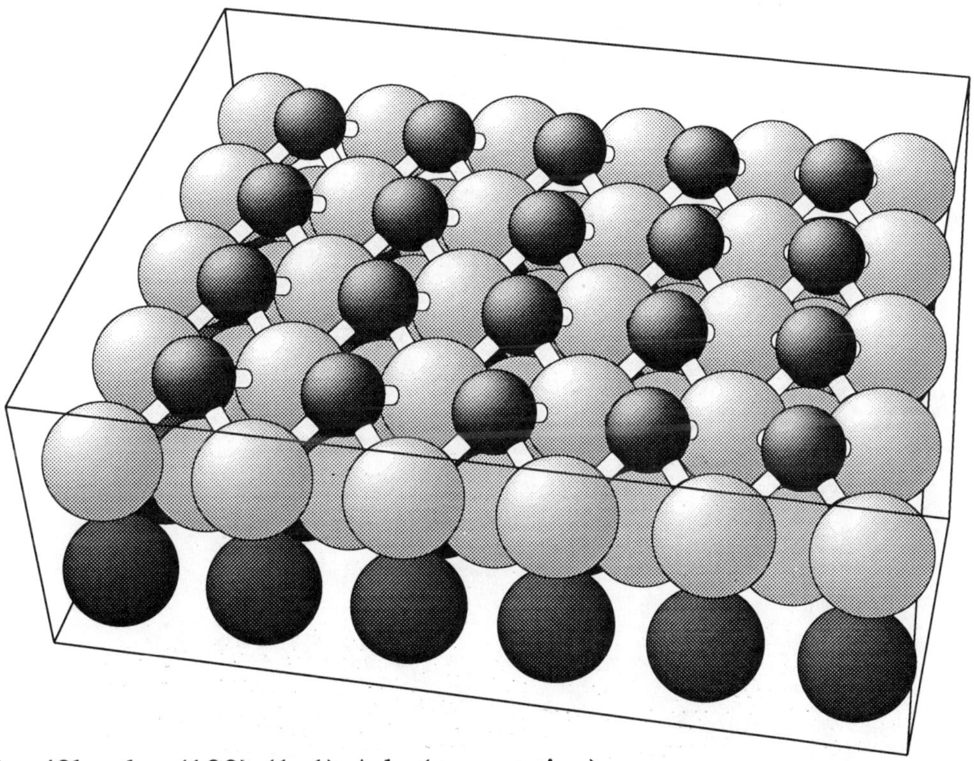

Fig. 49b : bcc(100)-(1x1)-Ad (perspective)

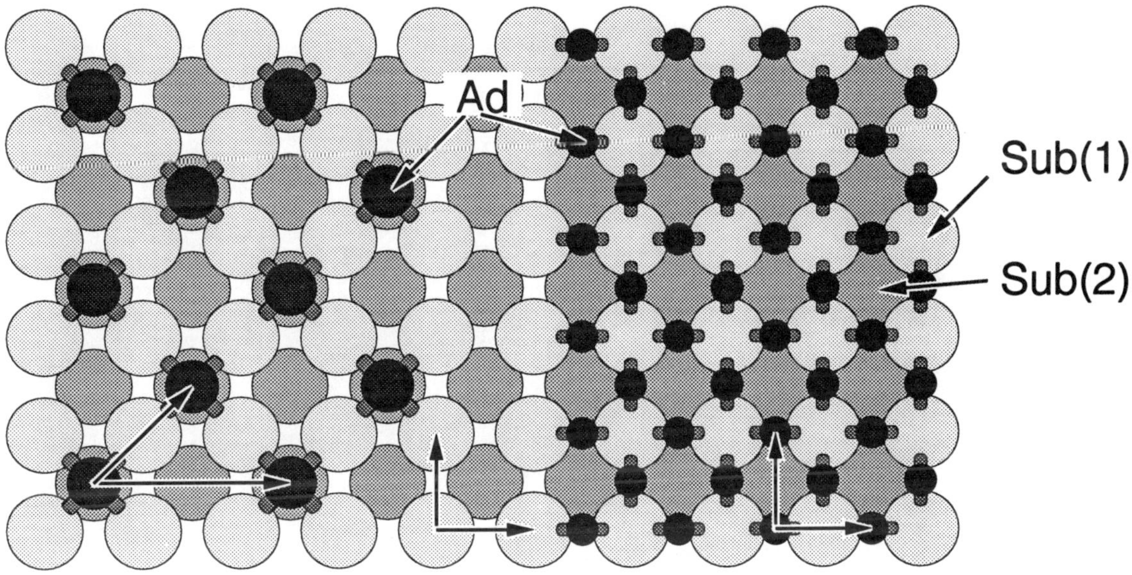

Fig. 50a : bcc(100)-c(2x2)-Ad; ~-(1x1)-2Ad (top view)

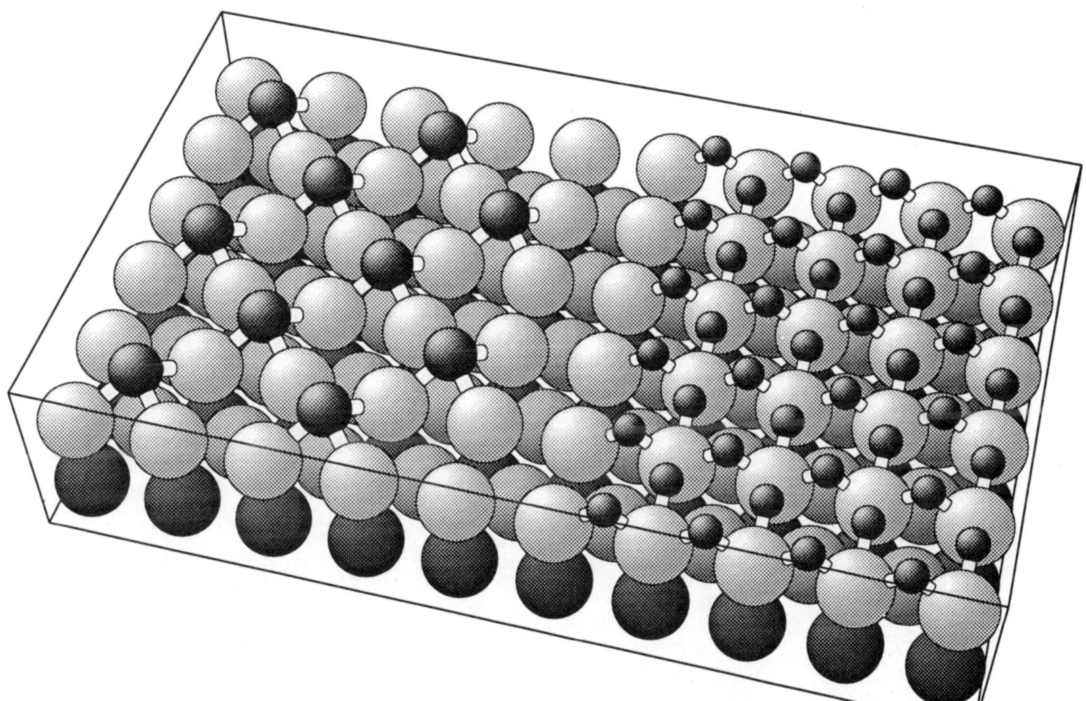

Fig. 50b : bcc(100)-c(2x2)-Ad; ~-(1x1)-2Ad (perspective)

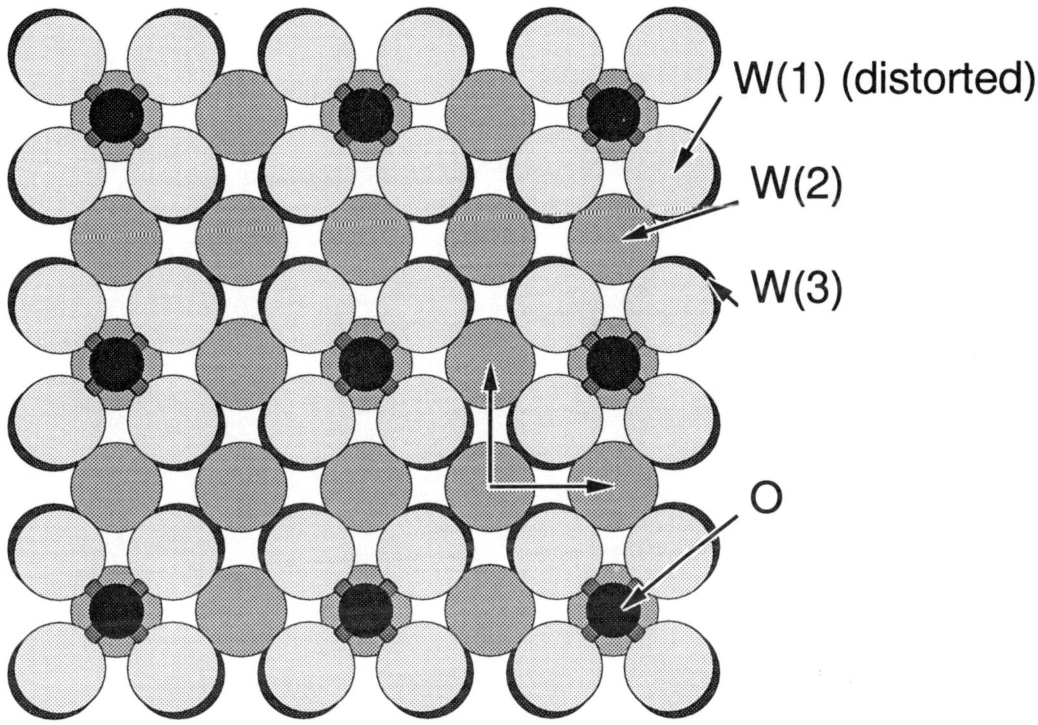

Fig. 51a : W(100)-O disordered (simulated as 2x2) (top view)

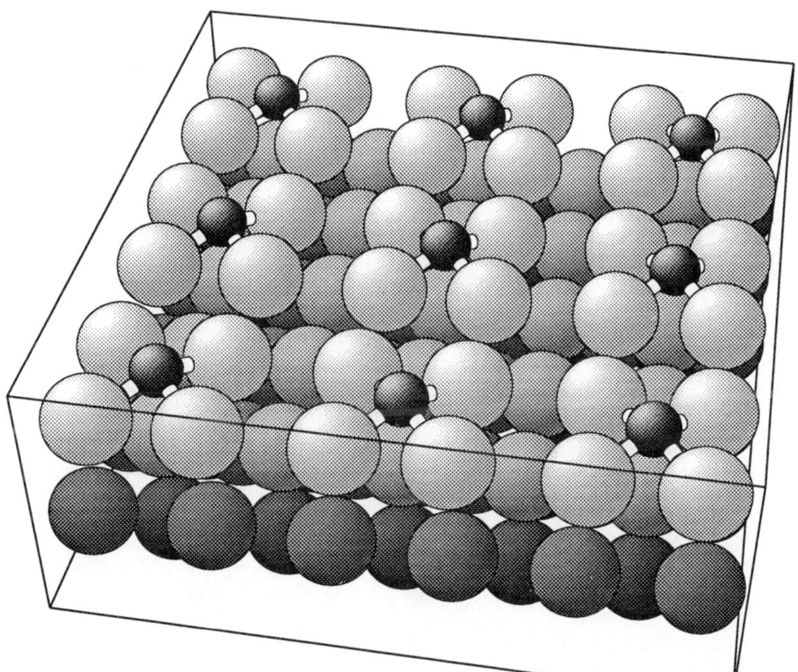

Fig. 51b : W(100)-O disordered (simulated as 2x2) (perspective)

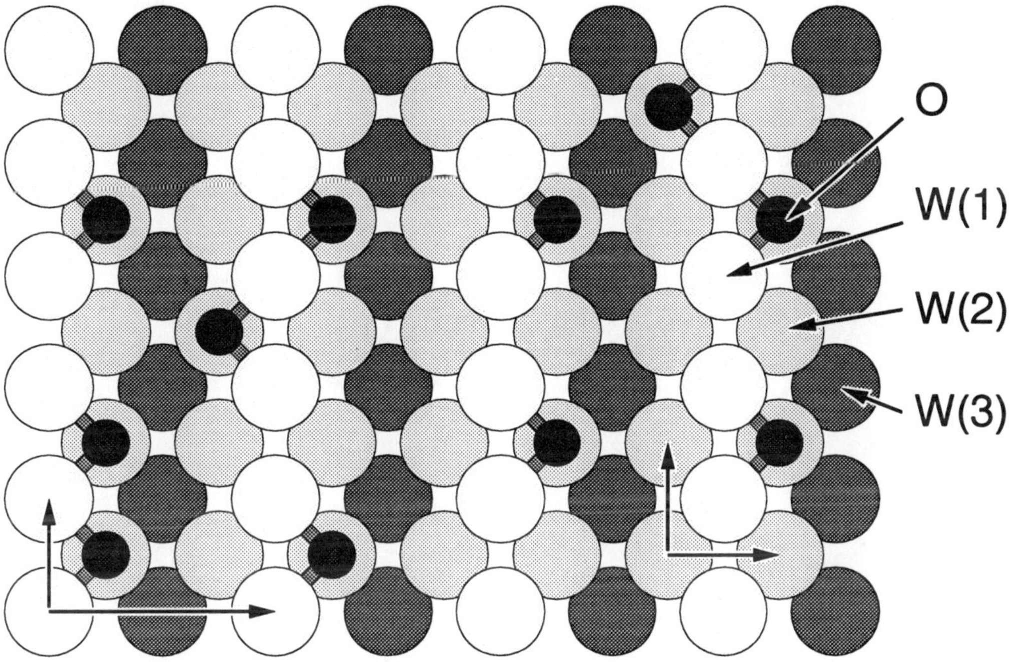

Fig. 52a : W(100)-p(2x1)-disordered O (top view)

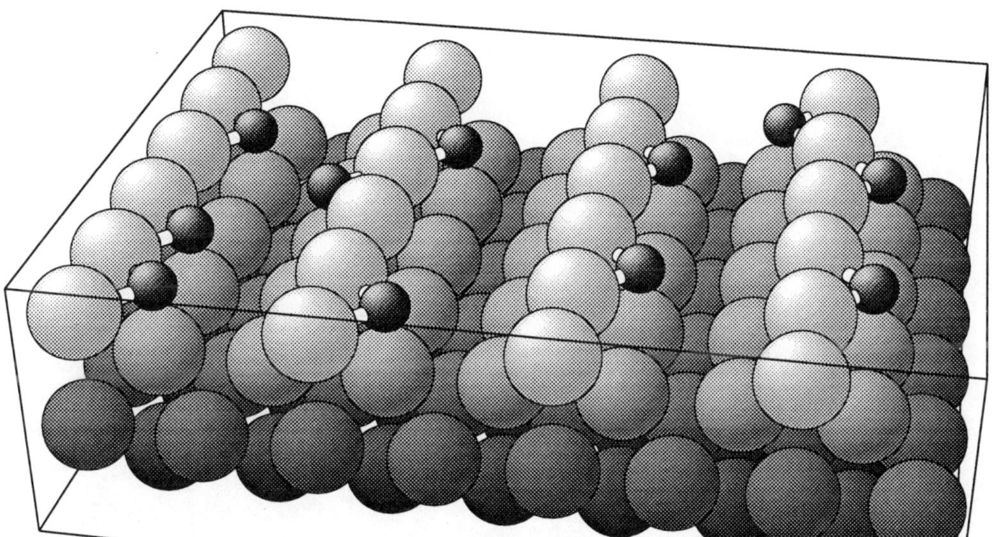

Fig. 52b : W(100)-p(2x1)-disordered O (perspective)

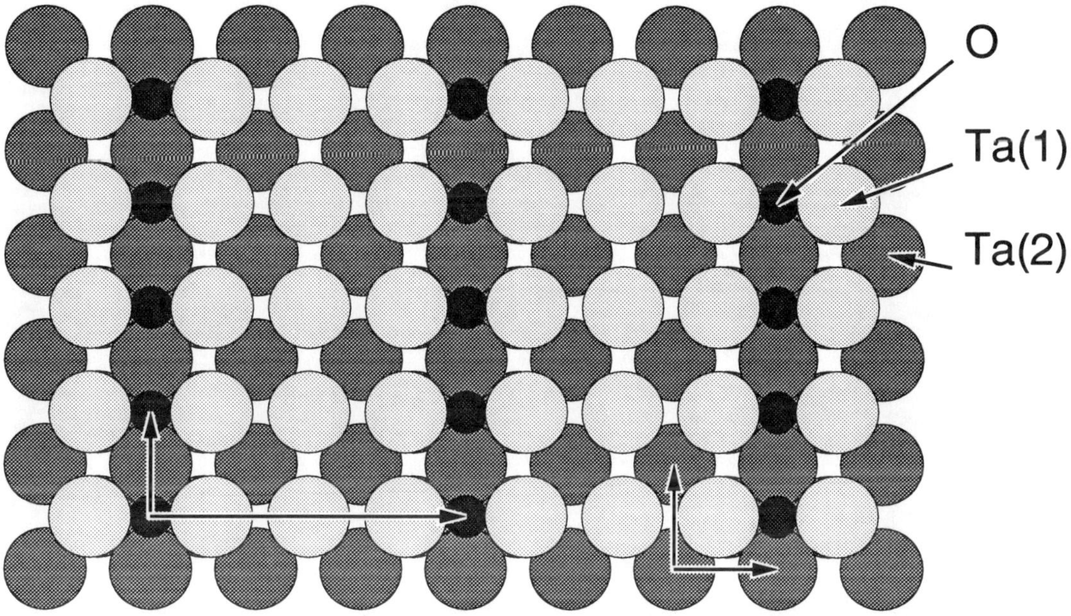

Fig. 53a : Ta(100)-(1x3)-O (top view)

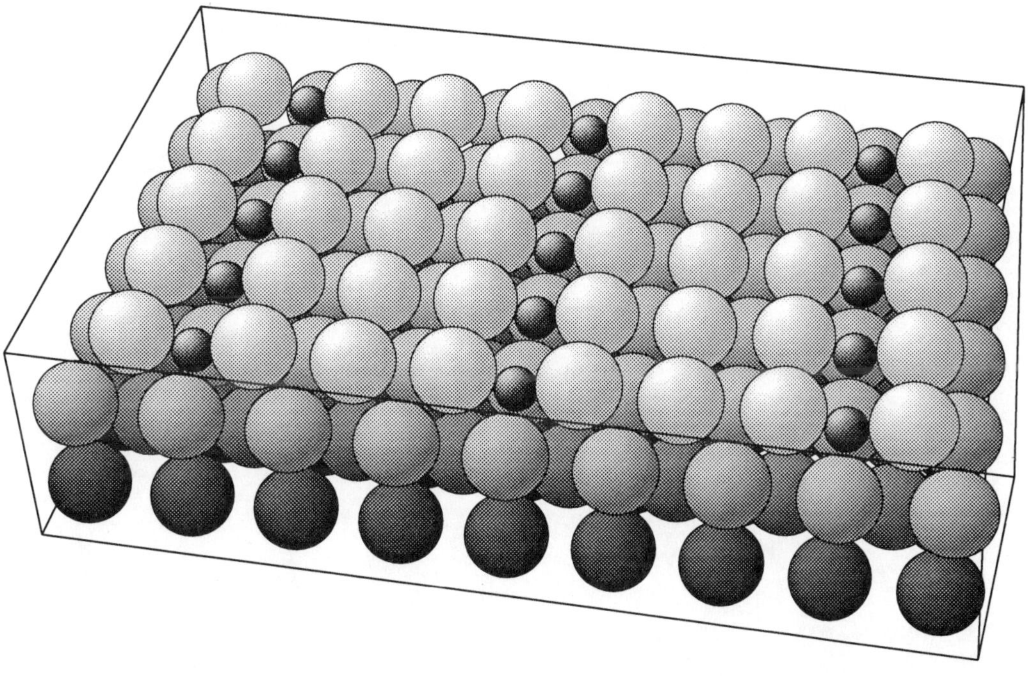

Fig. 53b : Ta(100)-(1x3)-O (perspective)

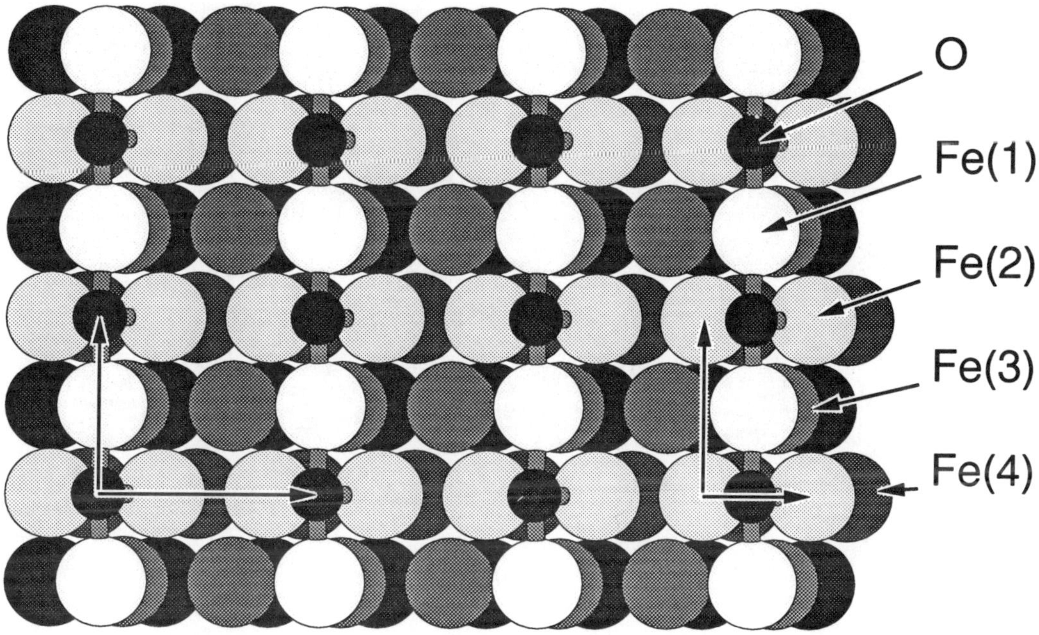

Fig. 54a : Fe(211)-(2x1)-O (top view)

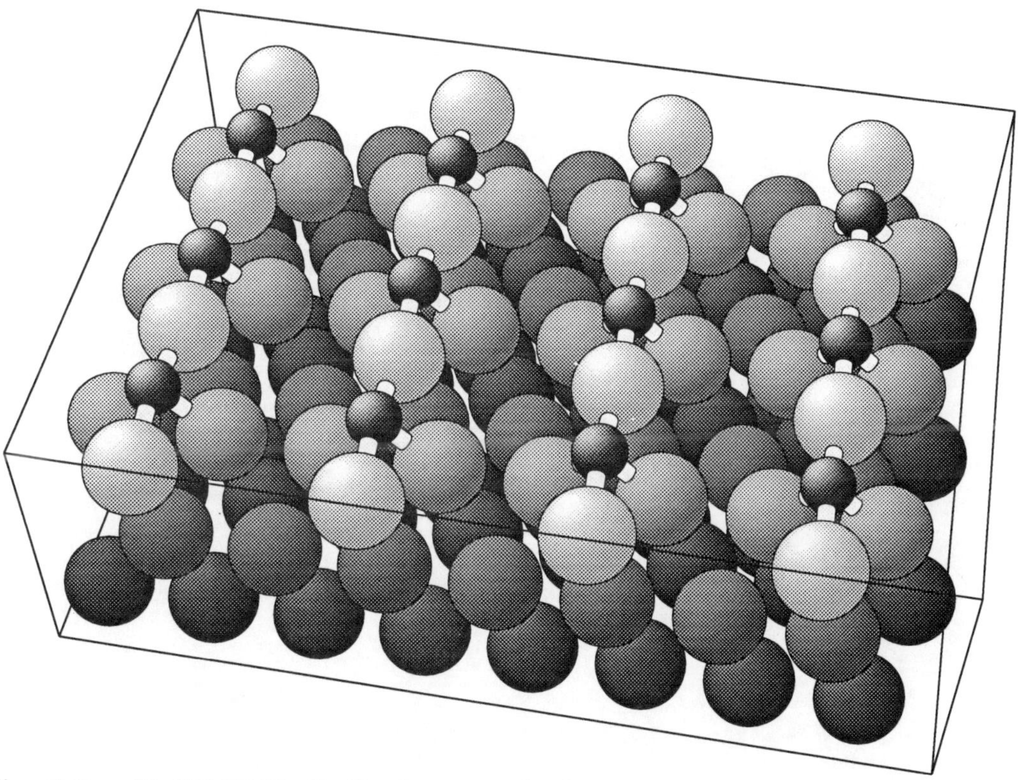

Fig. 54b : Fe(211)-(2x1)-O (perspective)

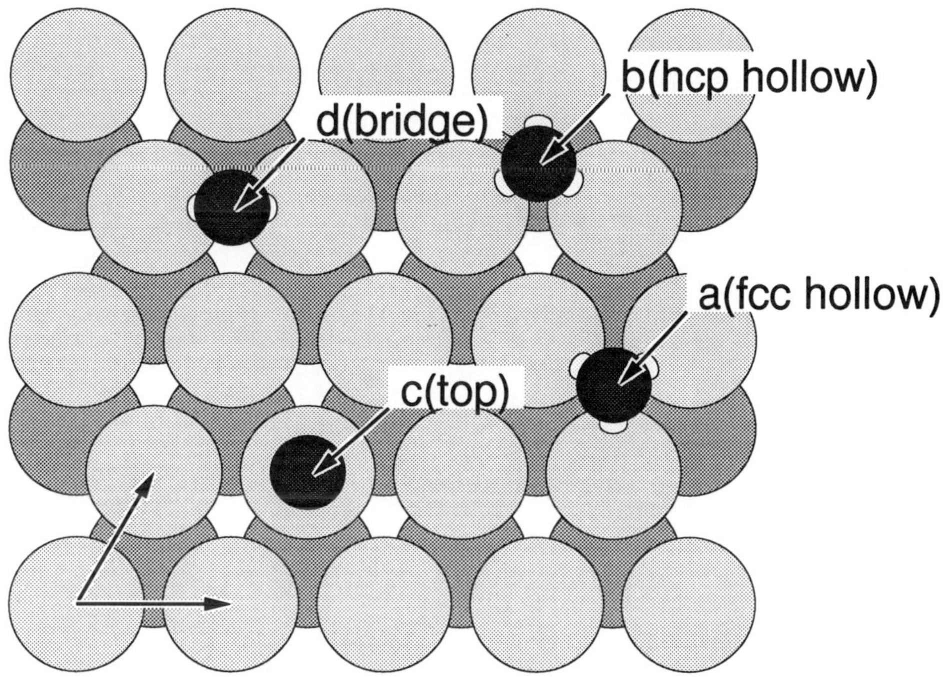

Fig. 55a : hcp(0001) high symmetry adsorbate sites (top view)

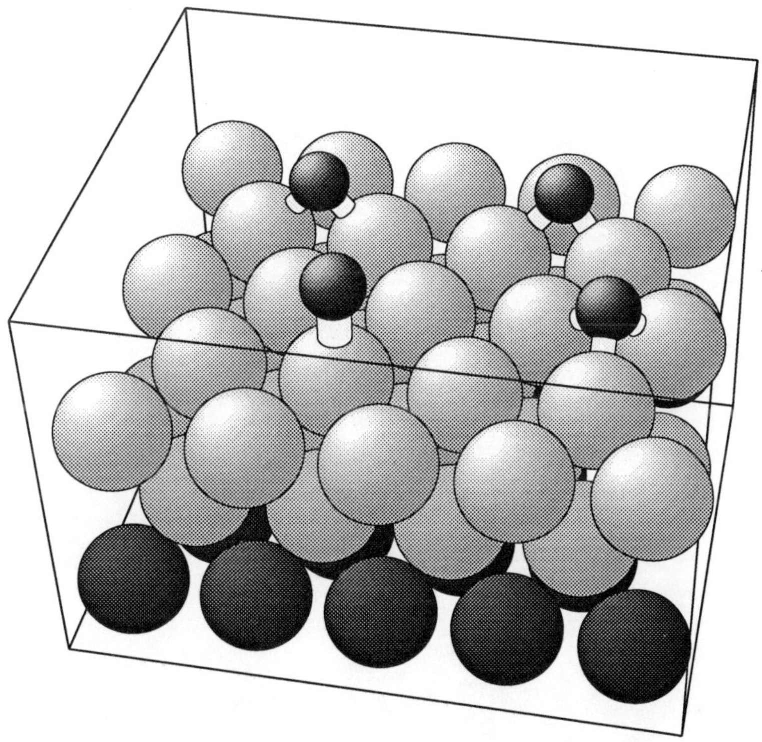

Fig. 55b : hcp(0001) high symmetry adsorbate sites (perspective)

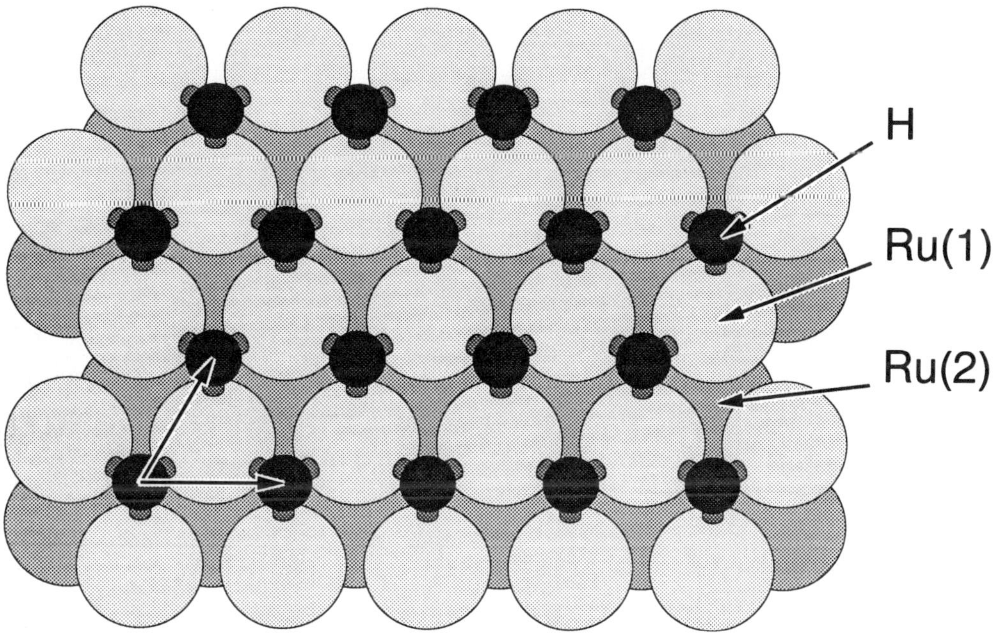

Fig. 56a : Ru(0001)-(1x1)-H (top view)

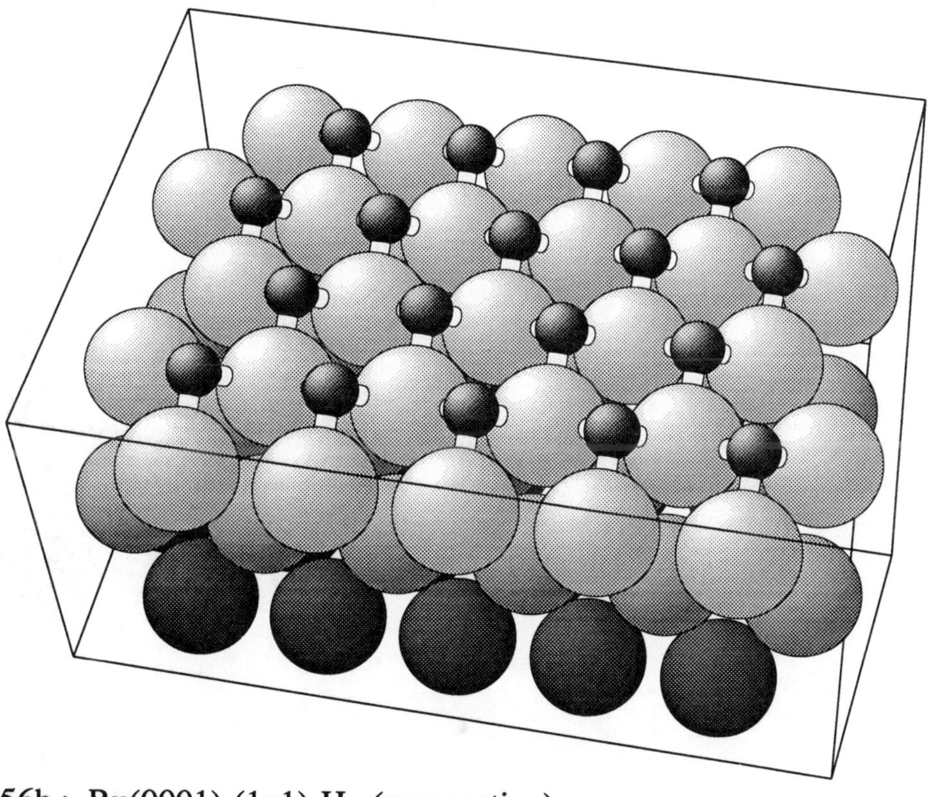

Fig. 56b : Ru(0001)-(1x1)-H (perspective)

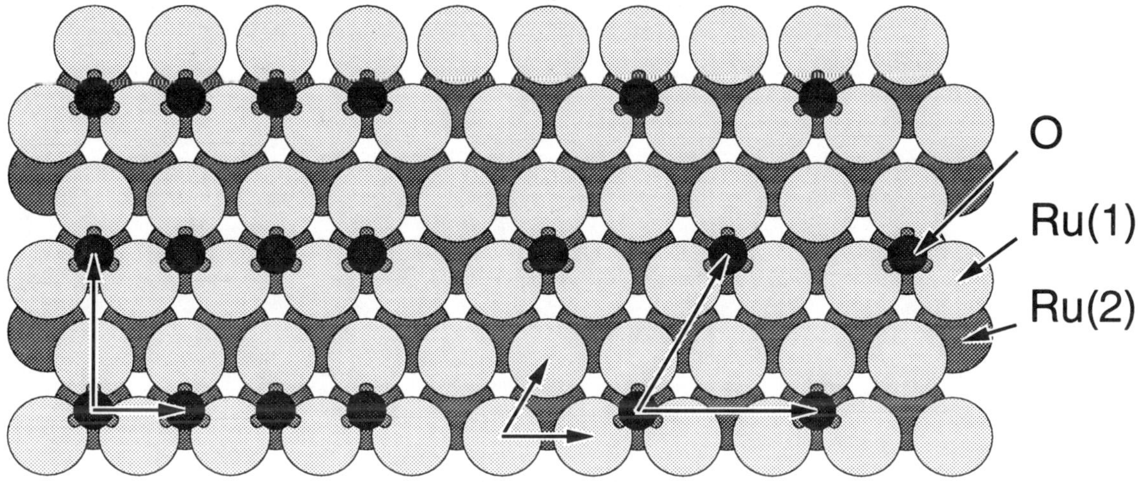

Fig. 57a : Ru(0001)-p(2x1)-O; ~-p(2x2)-O (top view)

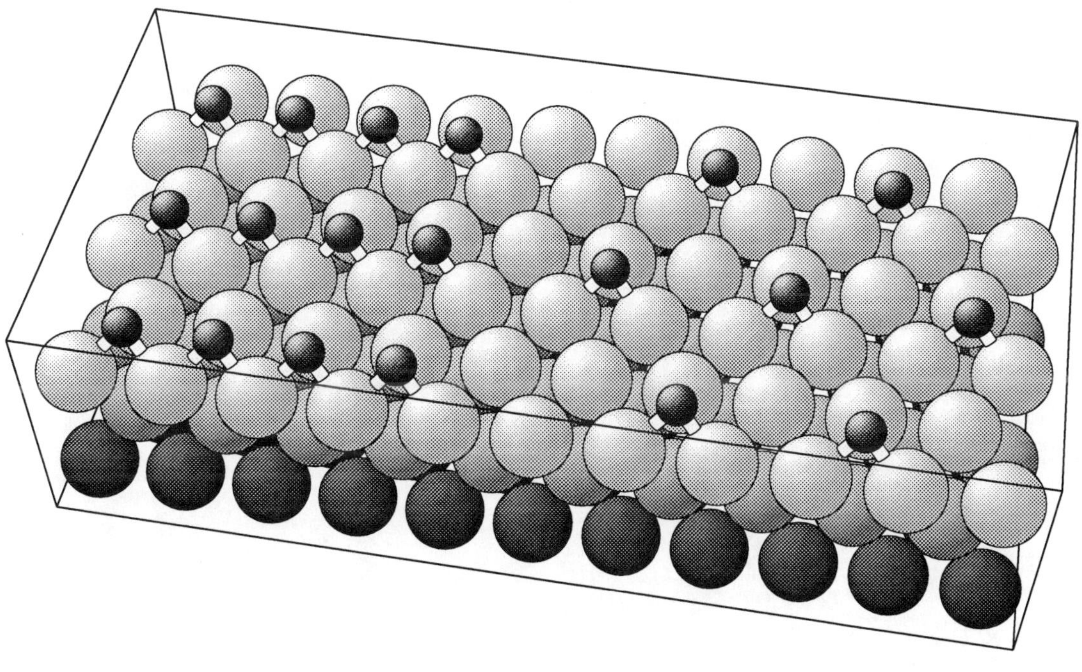

Fig. 57b : Ru(0001)-p(2x1)-O; ~-p(2x2)-O (perspective)

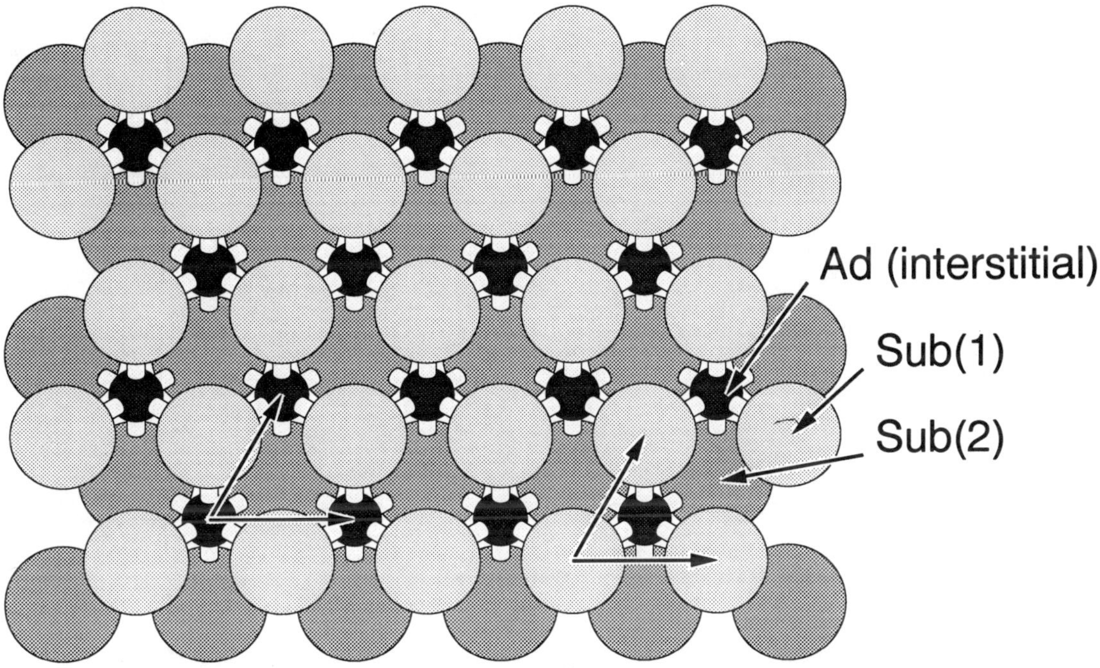

Fig. 58a : hcp(0001)-(1x1)-Ad (interstitial) (top view)

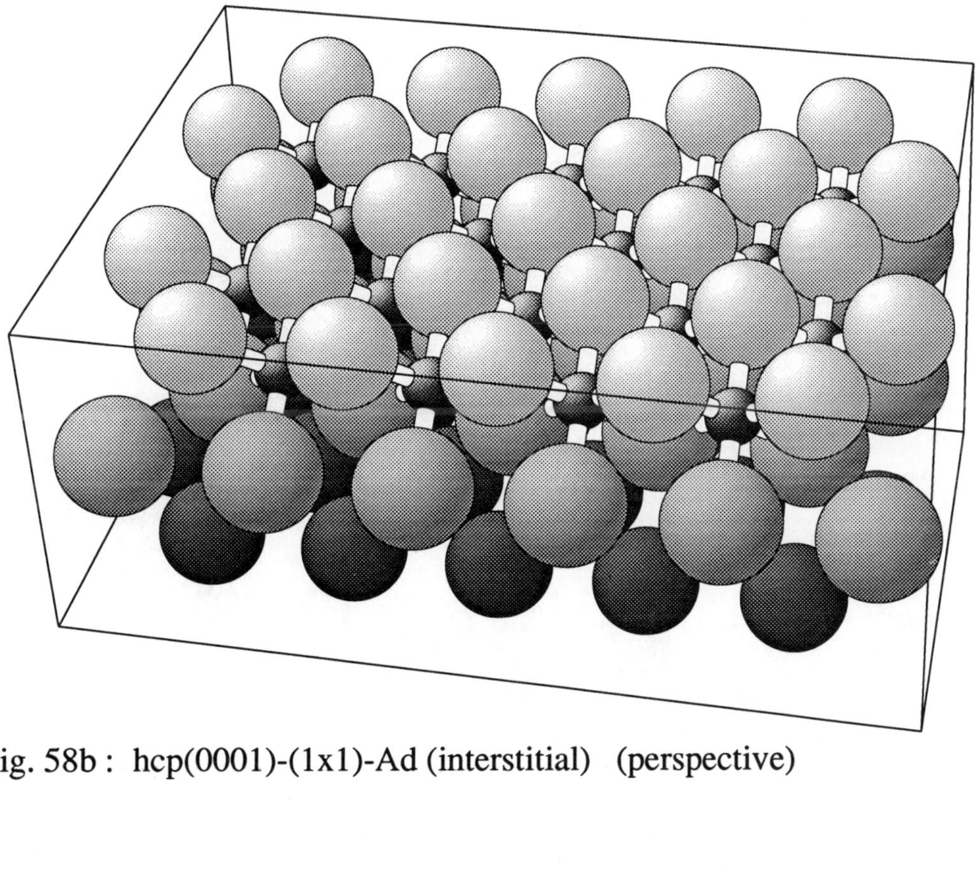

Fig. 58b : hcp(0001)-(1x1)-Ad (interstitial) (perspective)

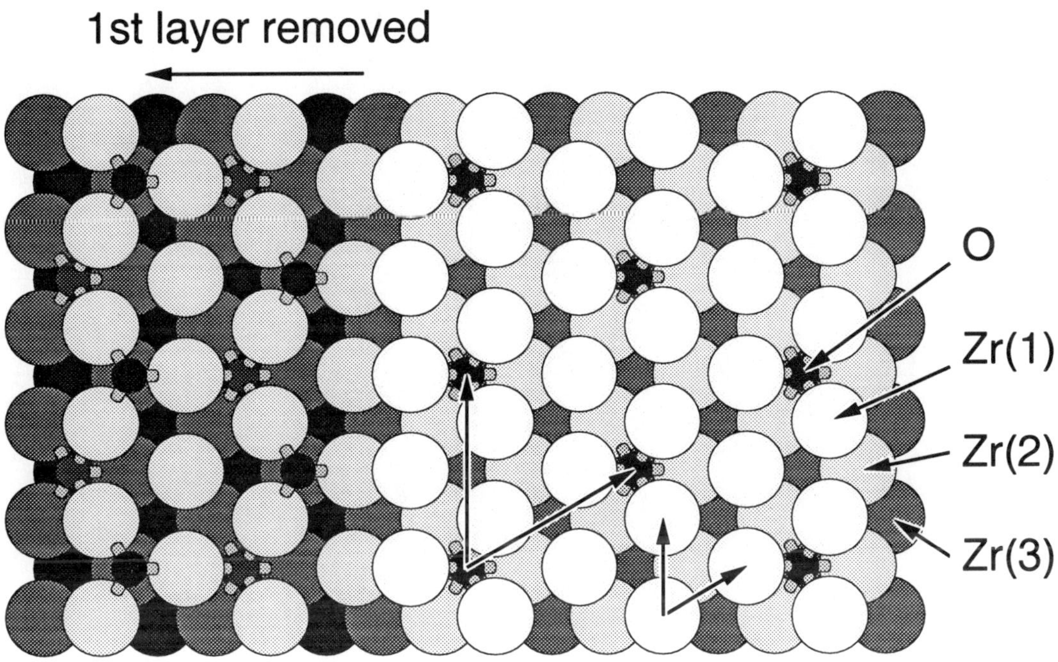

Fig. 59a : Zr(0001)-(2x2)-O (fcc, interstitial) (top view)

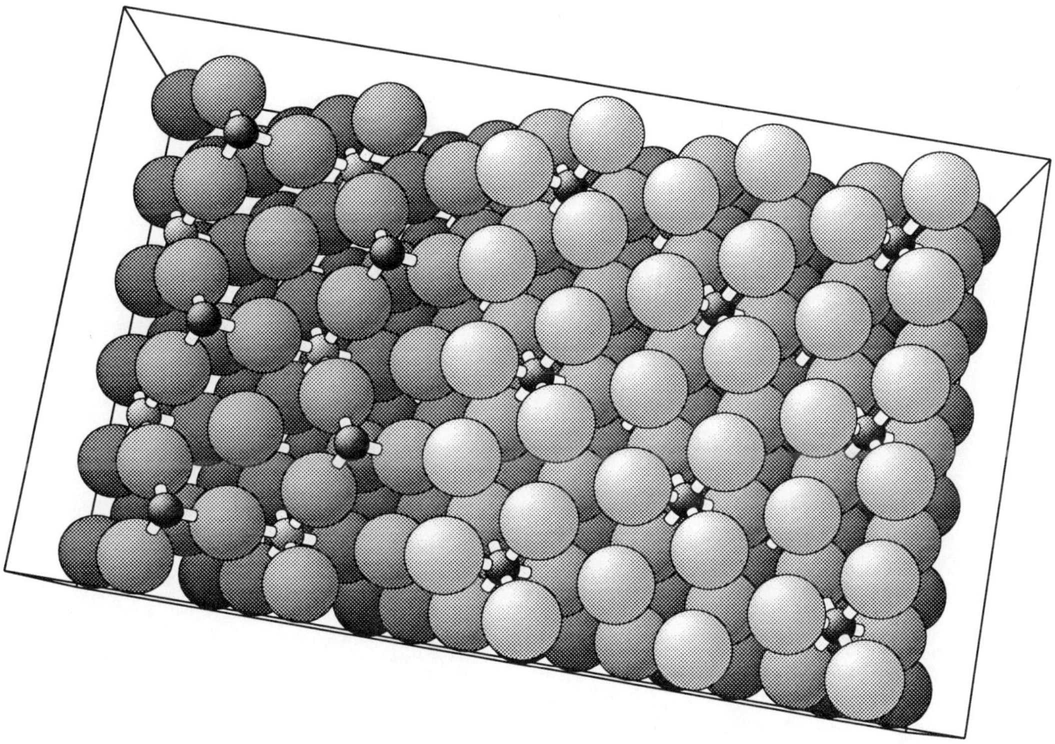

Fig. 59b : Zr(0001)-(2x2)-O (fcc, interstitial) (perspective)

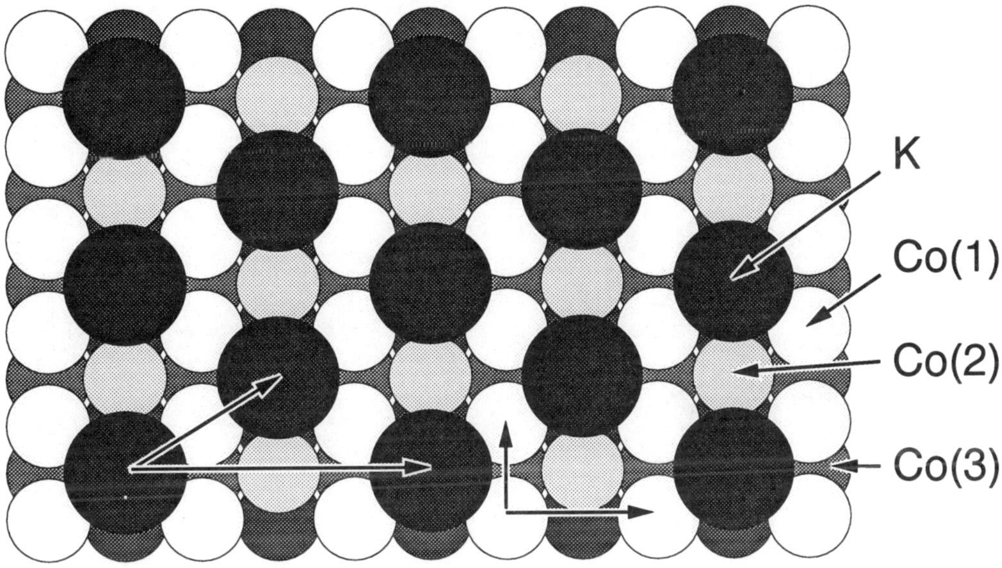

Fig. 60a : Co(10-10)-c(2x2)-K (top view)

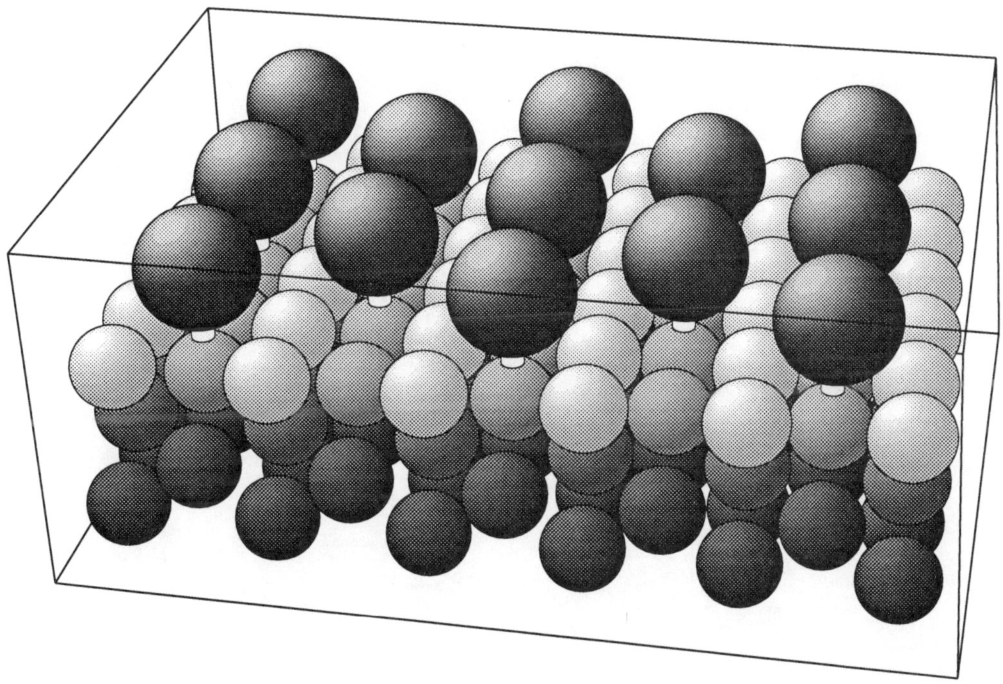

Fig. 60b : Co(10-10)-c(2x2)-K (perspective)

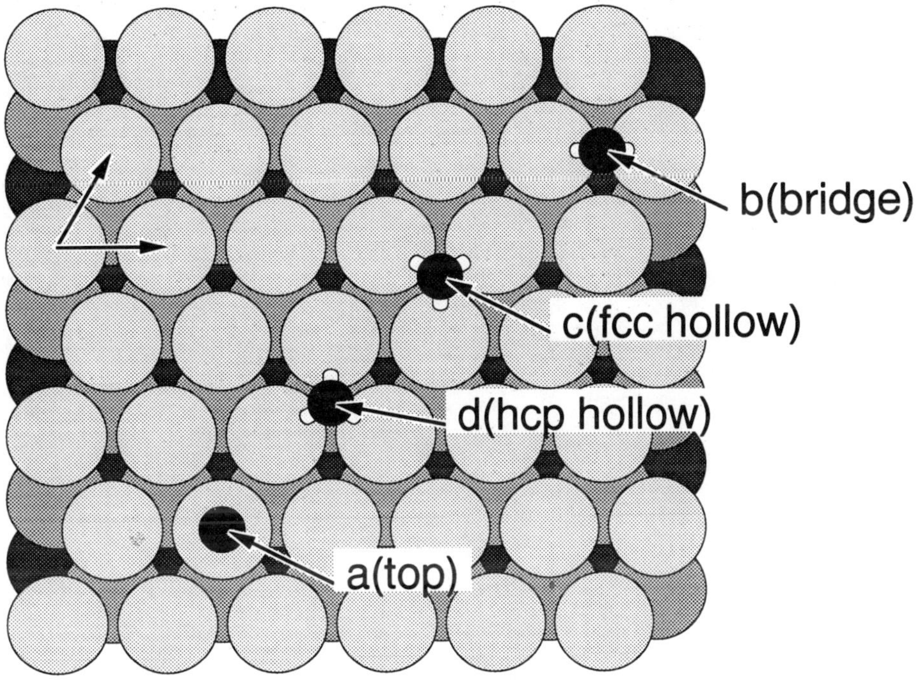

Fig. 61a : fcc(111)-CO high symmetry sites (top view)

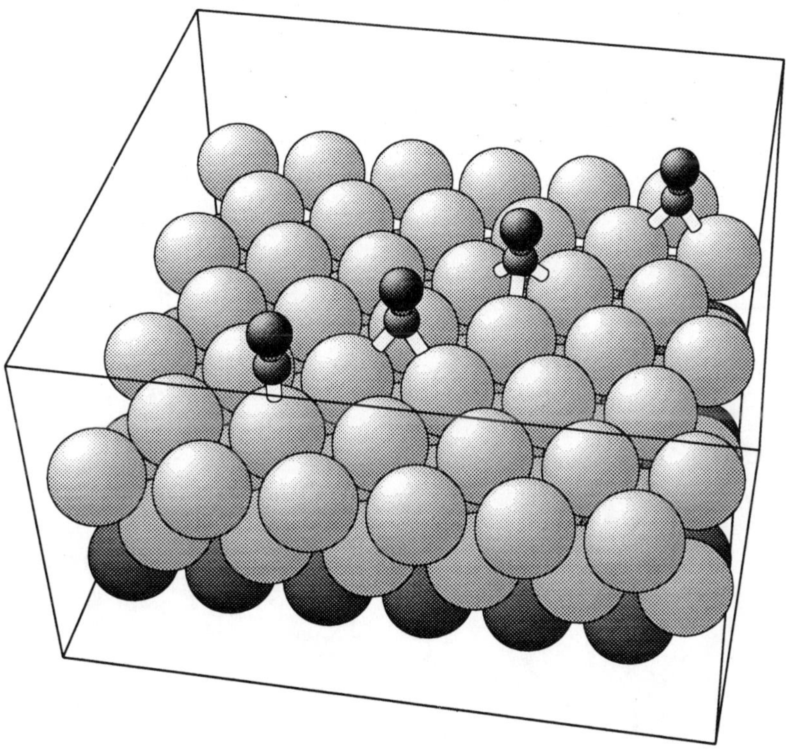

Fig. 61b : fcc(111)-CO high symmetry sites (perspective)

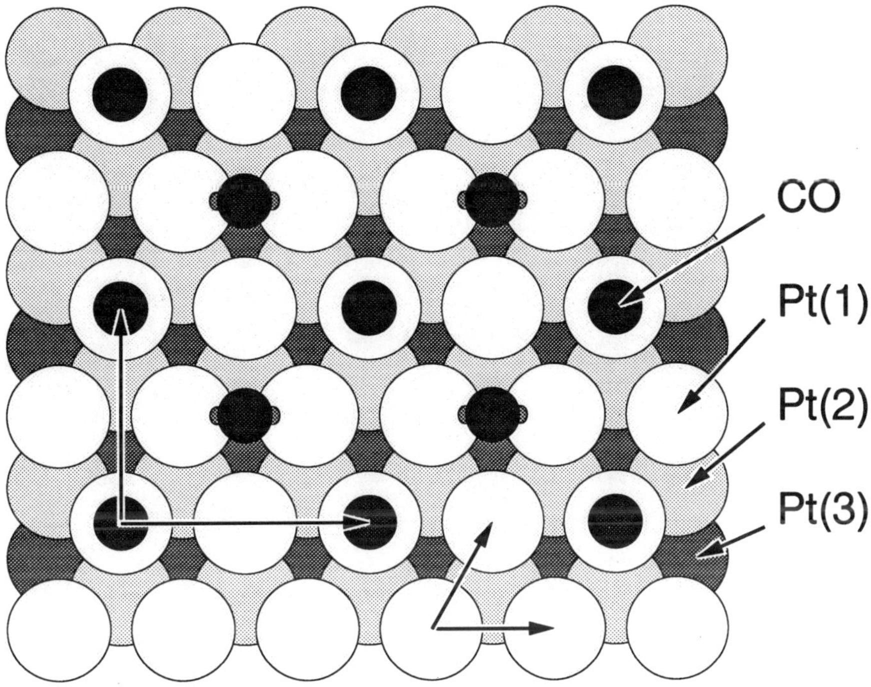

Fig. 62a : Pt(111)-c(4x2)-2CO (top view)

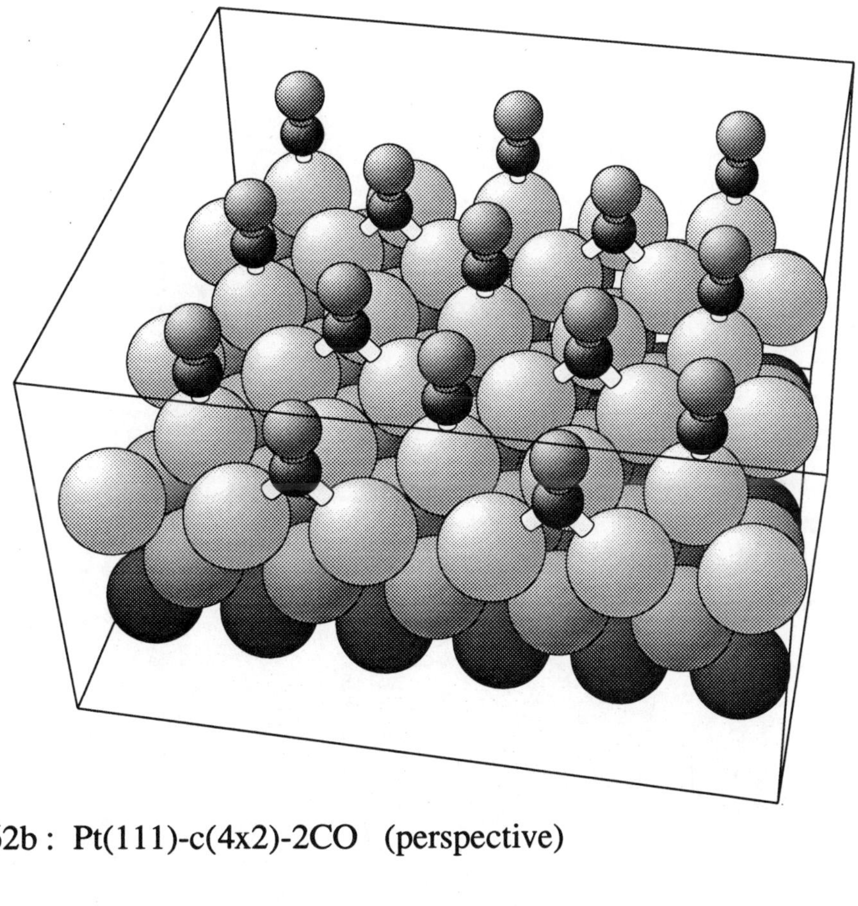

Fig. 62b : Pt(111)-c(4x2)-2CO (perspective)

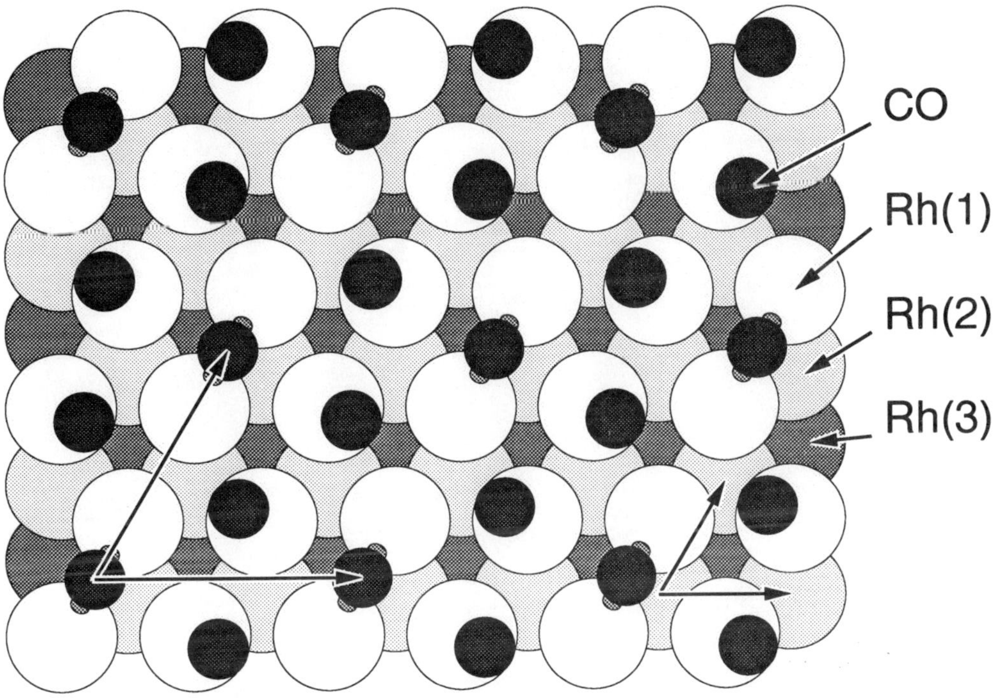

Fig. 63a : Rh(111)-(2x2)-3CO (top view)

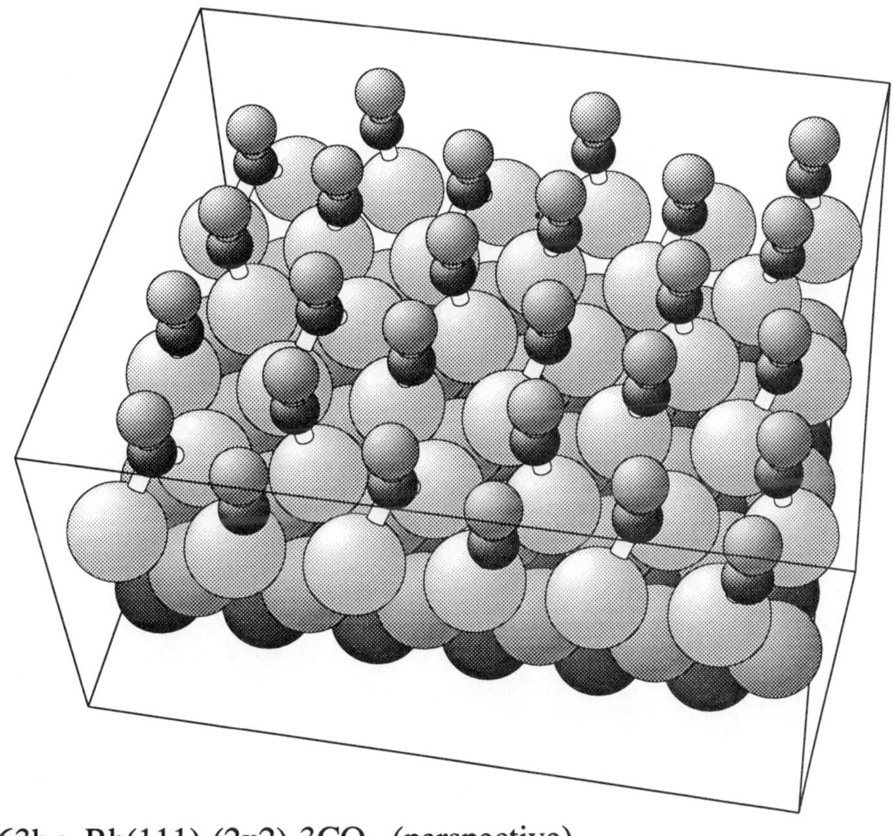

Fig. 63b : Rh(111)-(2x2)-3CO (perspective)

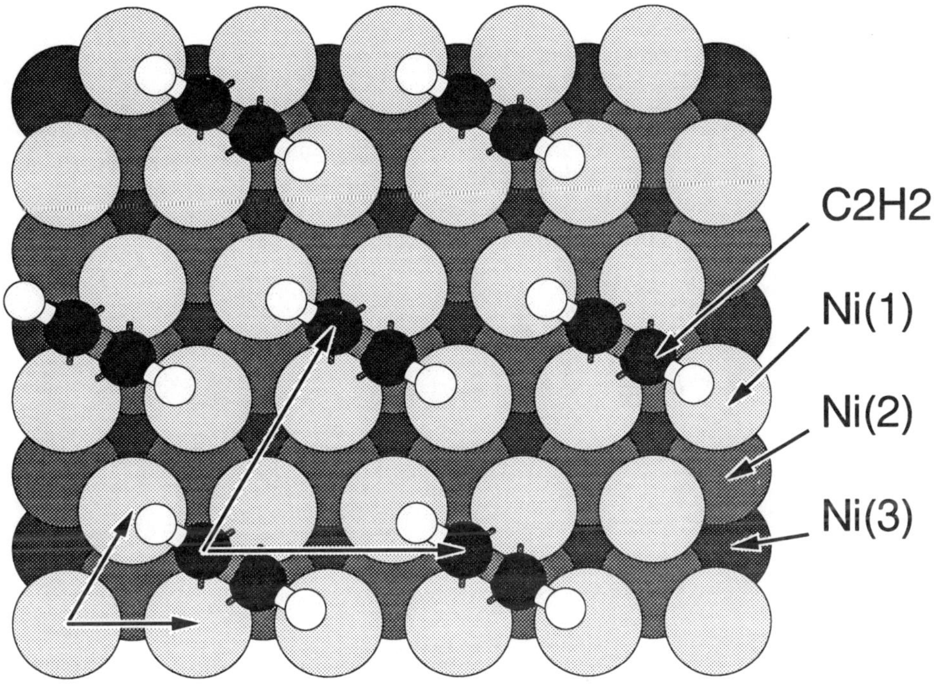

Fig. 64a : Ni(111)-(2x2)-C2H2 (top view)

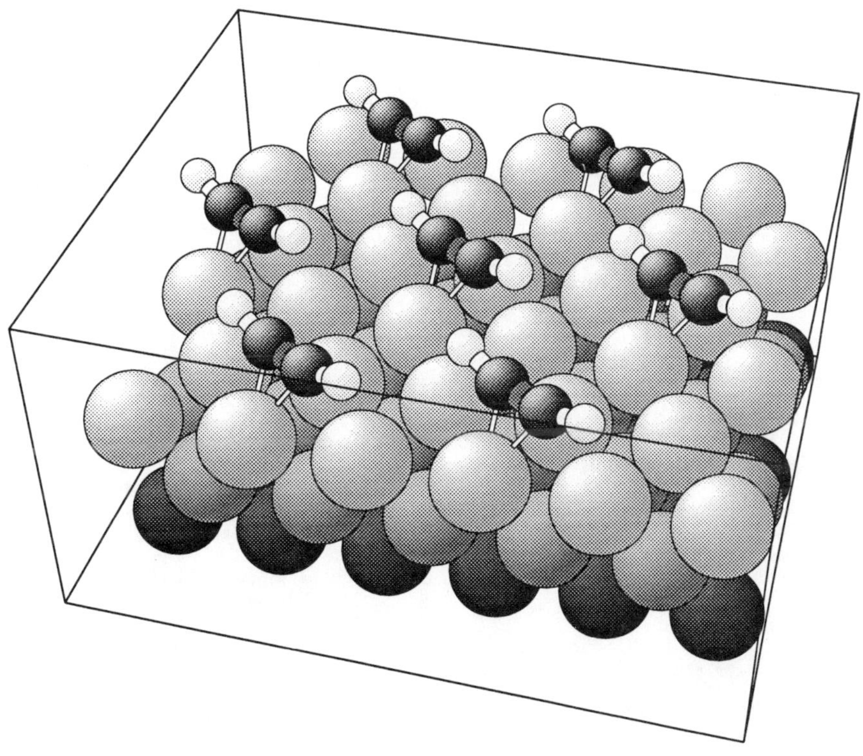

Fig. 64b : Ni(111)-(2x2)-C2H2 (perspective)

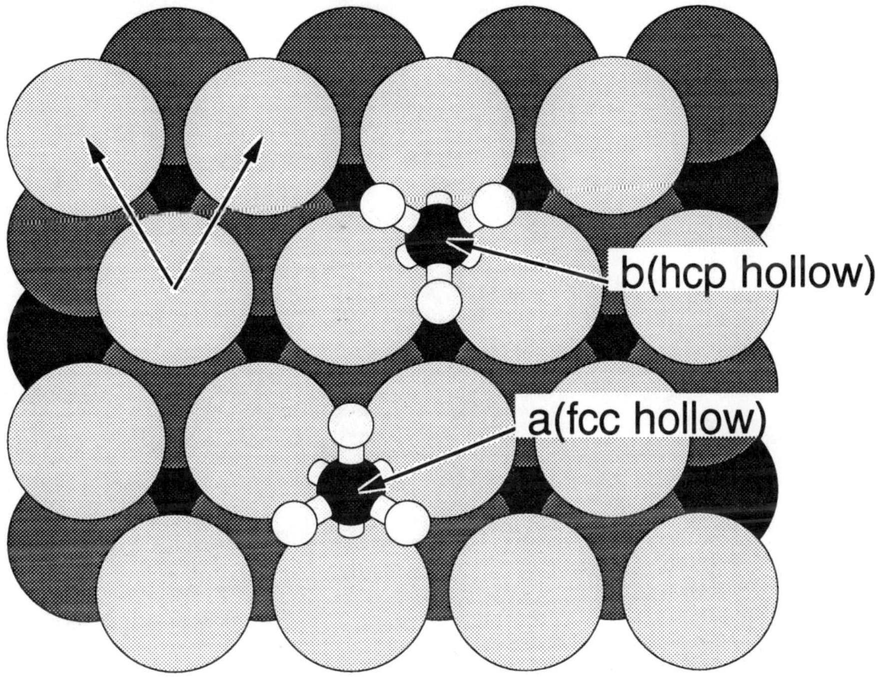

Fig. 65a : fcc(111)-C2H3 high symmetry sites (top view)

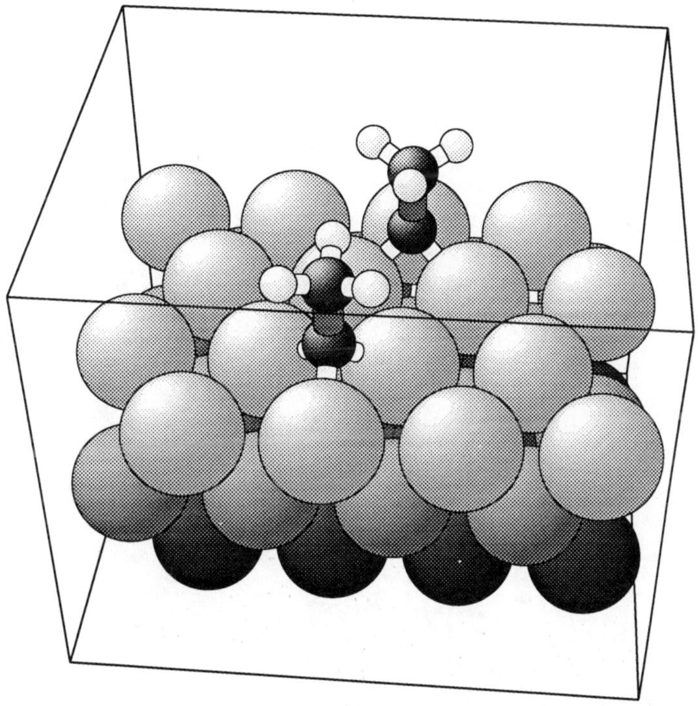

Fig. 65b : fcc(111)-C2H3 high symmetry sites (perspective)

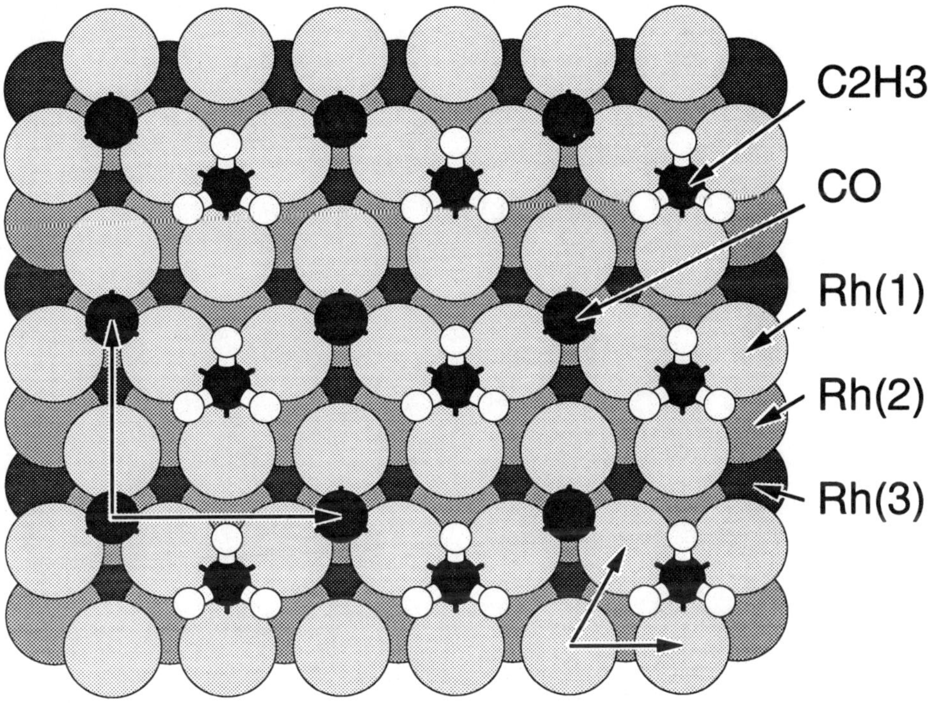

Fig. 66a : Rh(111)-c(4x2)-C2H3+CO (top view)

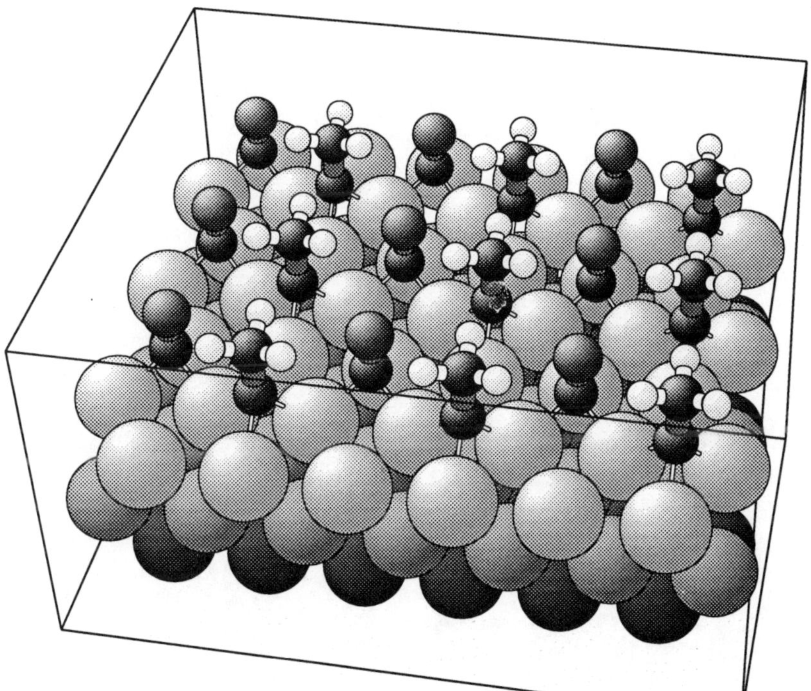

Fig. 66b : Rh(111)-c(4x2)-C2H3+CO (perspective)

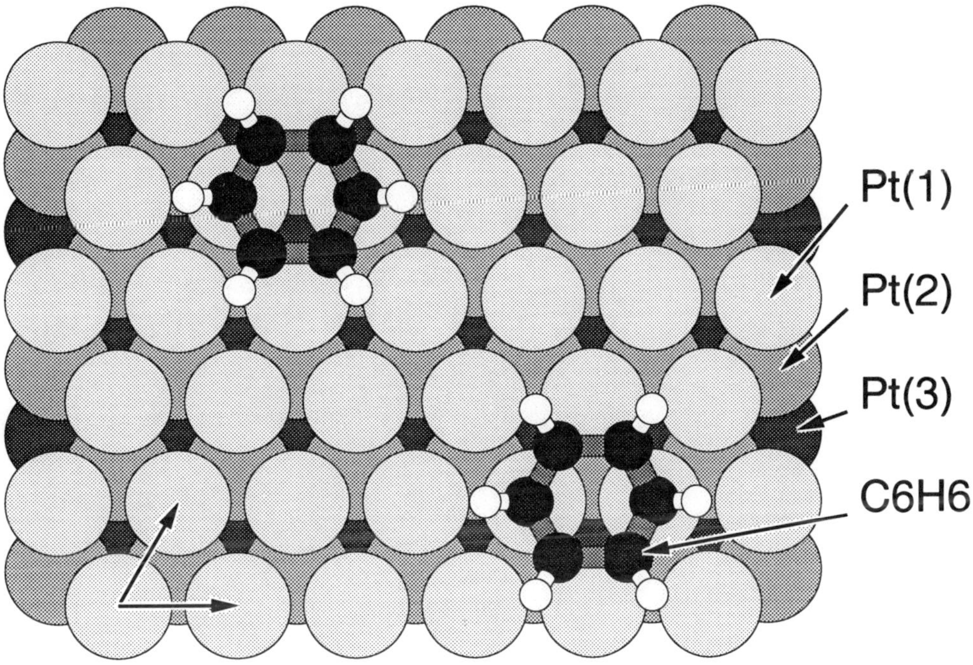

Fig. 67a : Pt(111)-C6H6 disordered (top view)

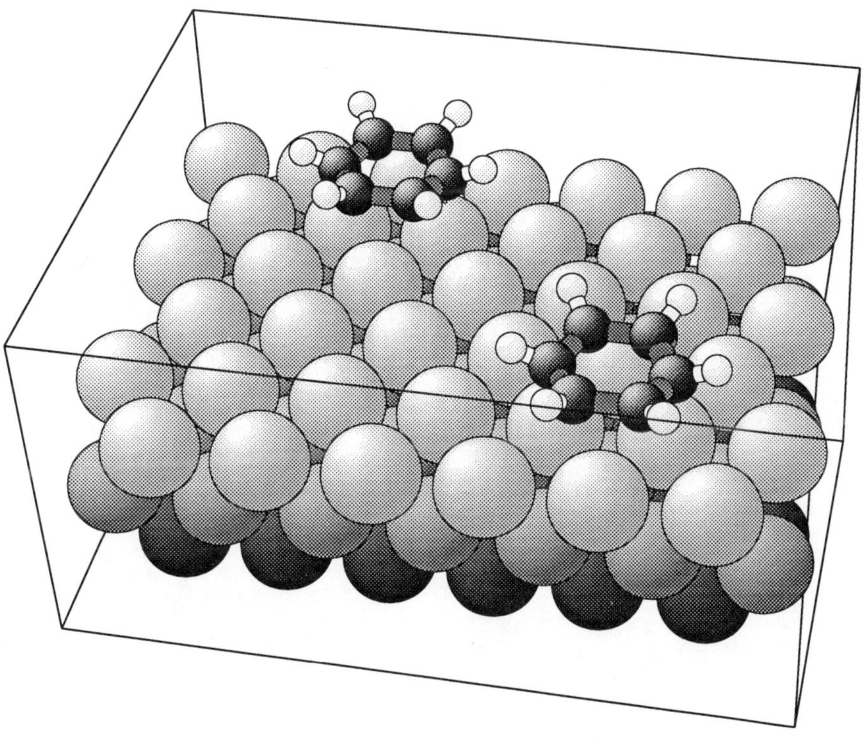

Fig. 67b : Pt(111)-C6H6 disordered (perspective)

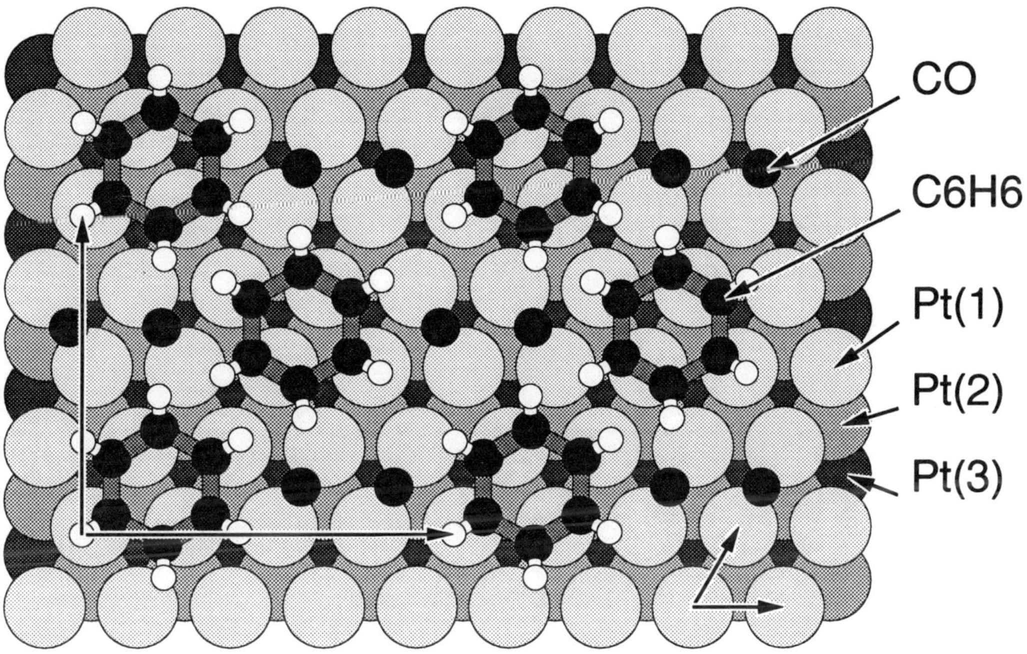

Fig. 68a : Pt(111)-(2√3x4)rect-2C6H6+4CO (top view)

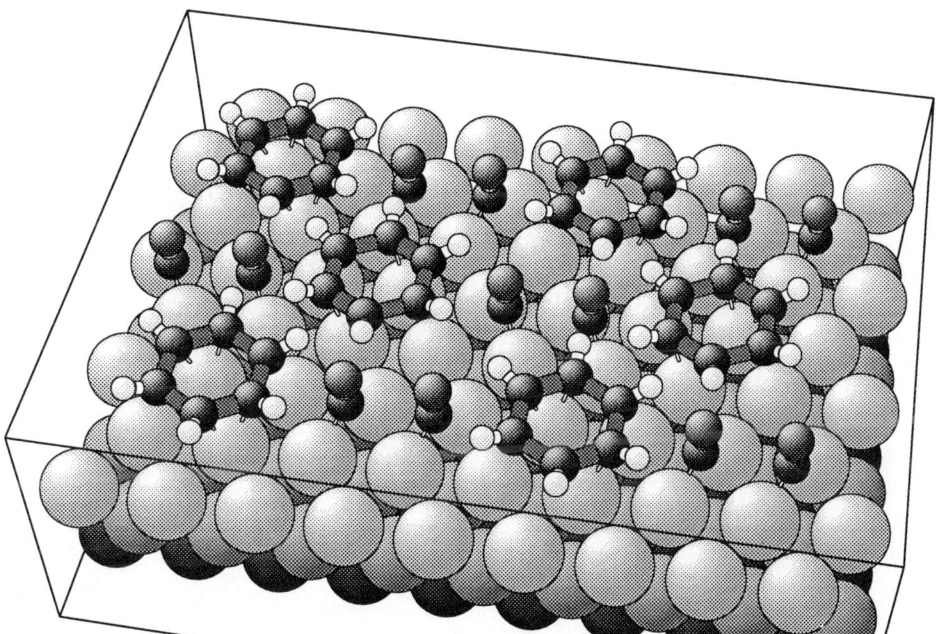

Fig. 68b : Pt(111)-(2√3x4)rect-2C6H6+4CO (perspective)

Fig. [illegible]

Fig. [illegible]

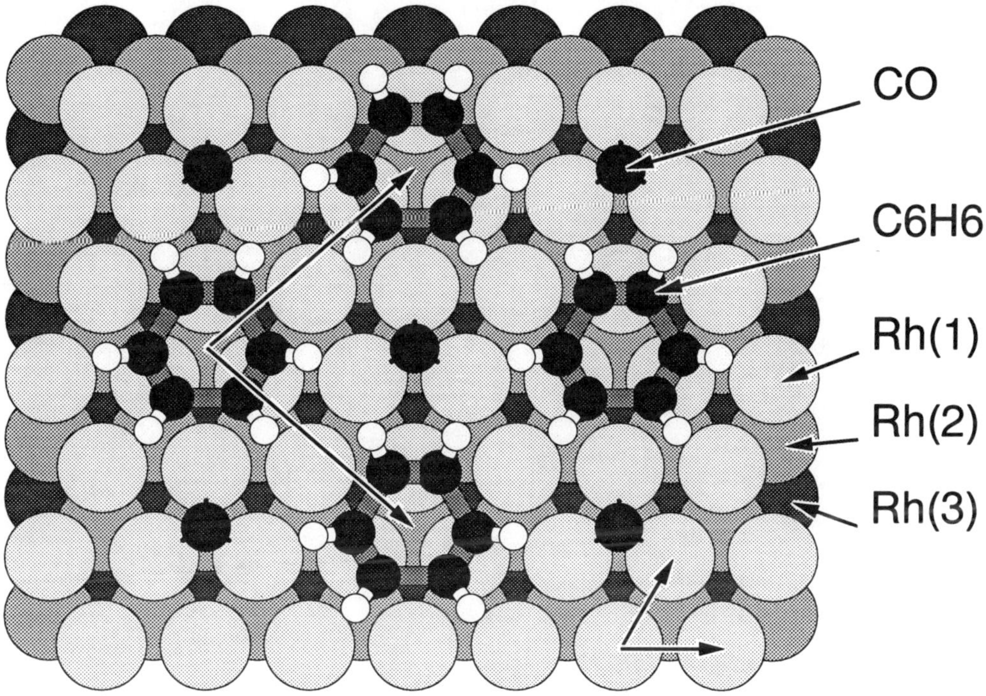

Fig. 69a : Rh(111)-c(2√3x4)rect-C6H6+CO (top view)

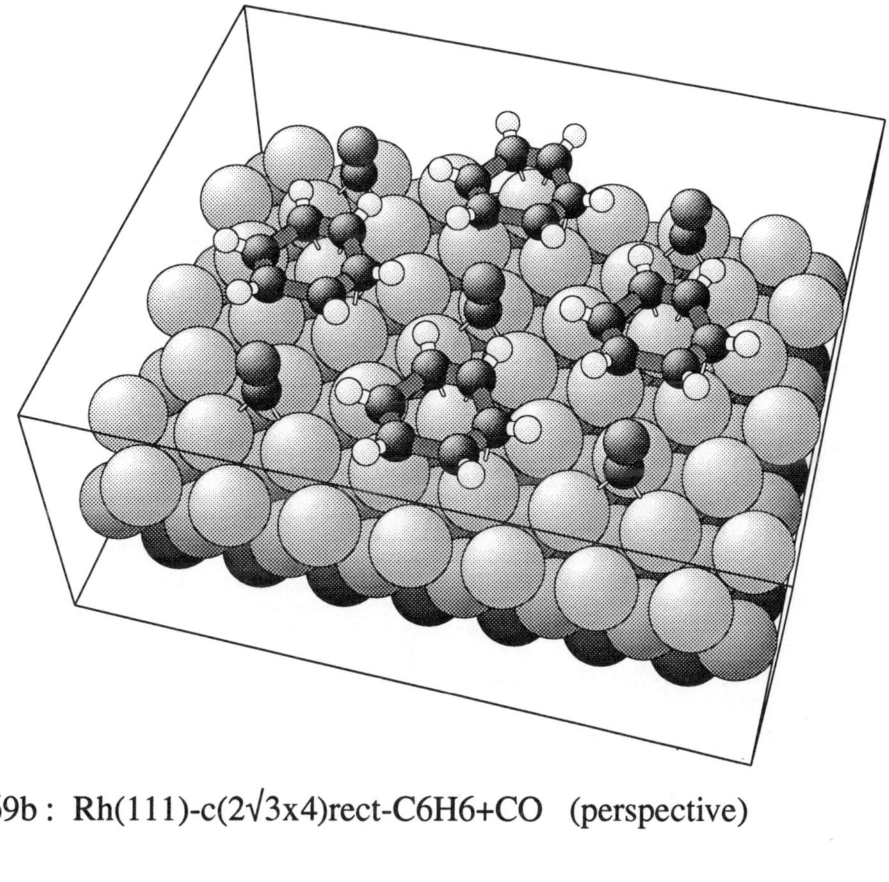

Fig. 69b : Rh(111)-c(2√3x4)rect-C6H6+CO (perspective)

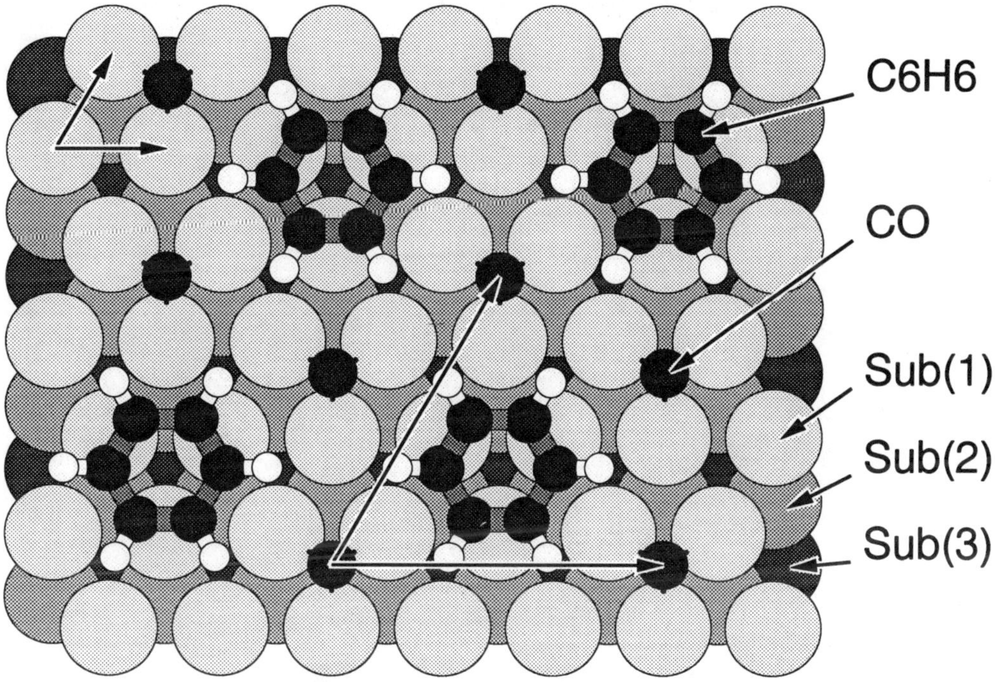

Fig. 70a : fcc(111)-(3x3)-C6H6+2CO (top view)

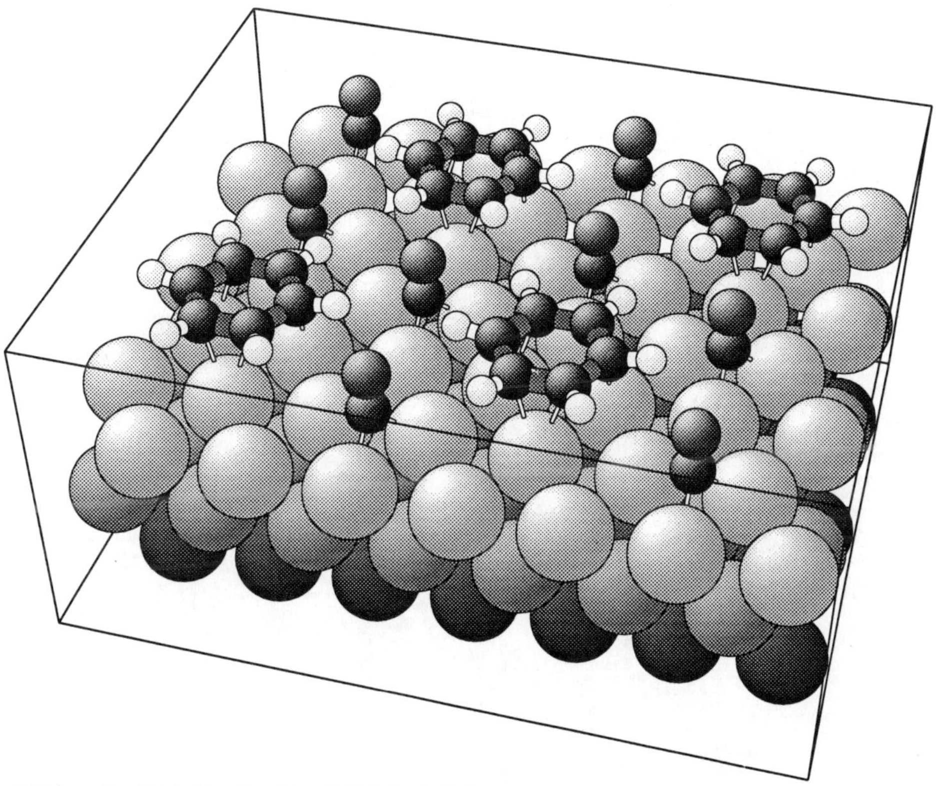

Fig. 70b : fcc(111)-(3x3)-C6H6+2CO (perspective)

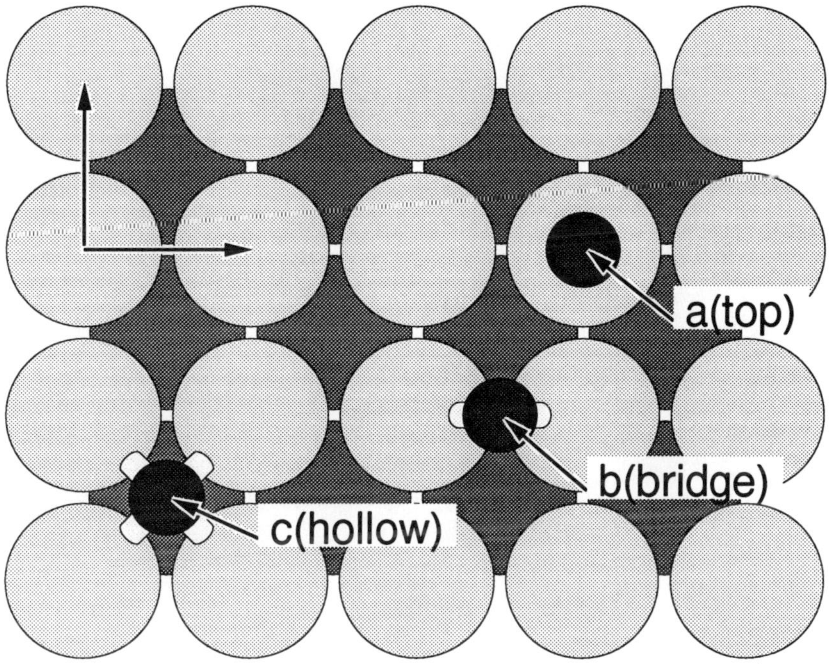

Fig. 71a : fcc(100)-CO high symmetry sites (top view)

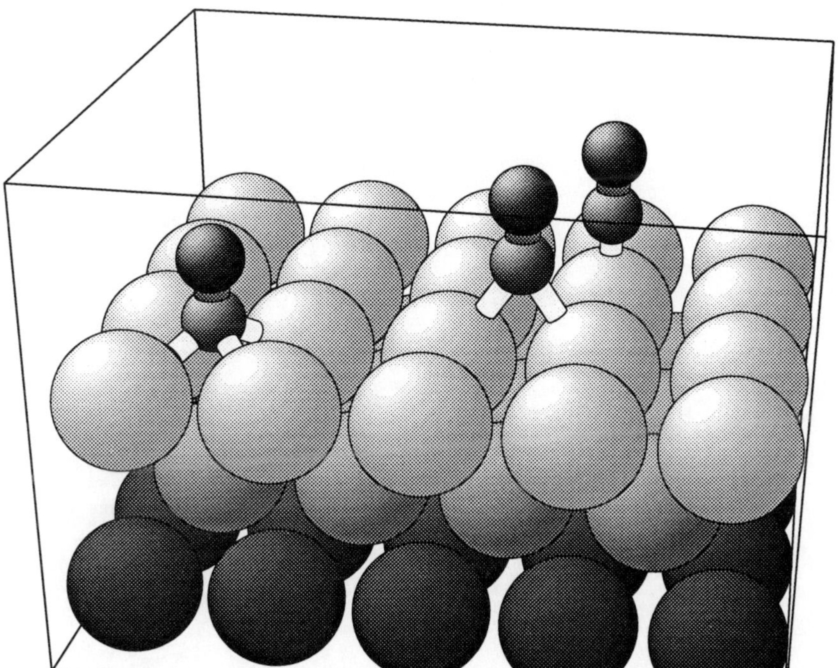

Fig. 71b : fcc(100)-CO high symmetry sites (perspective)

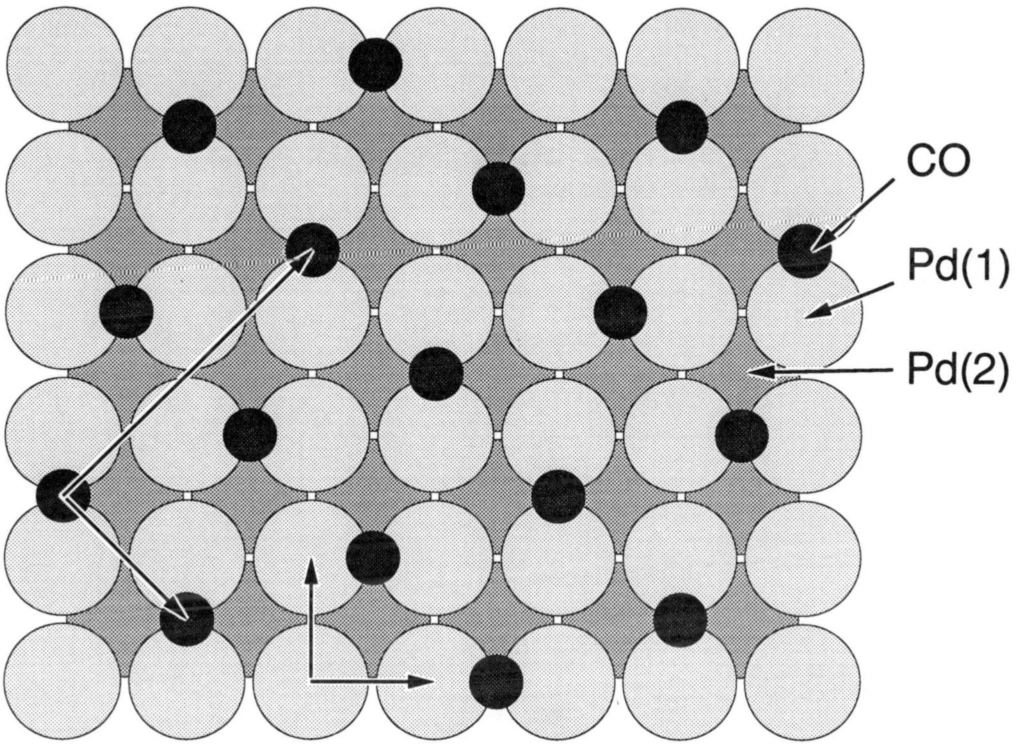

Fig. 72a : Pd(100)-(2√2x√2)R45°-2CO (top view)

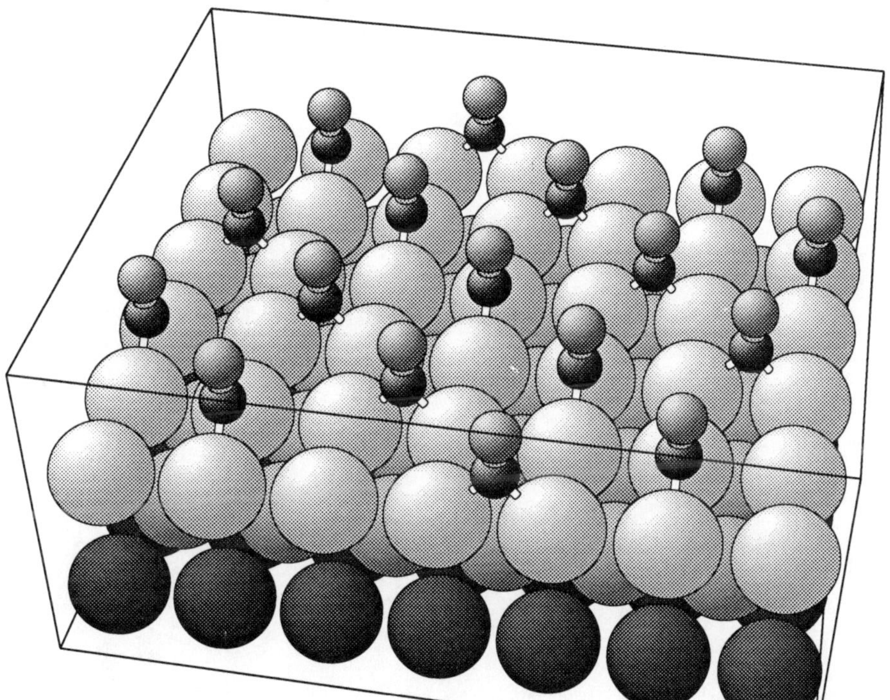

Fig. 72b : Pd(100)-(2√2x√2)R45°-2CO (perspective)

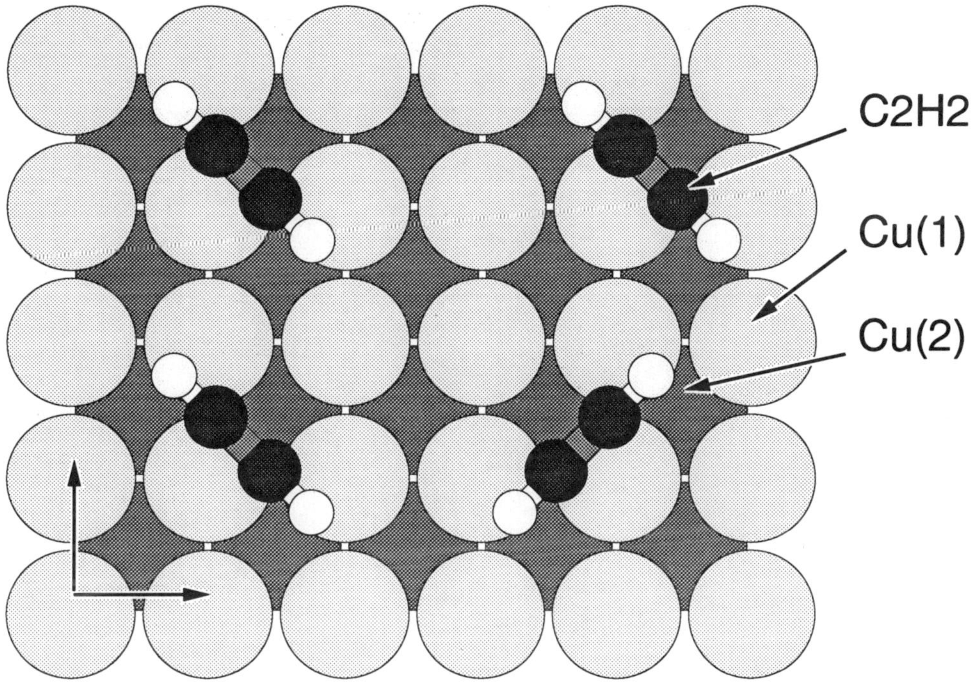

Fig. 73a : Cu(100)-C2H2 disordered (top view)

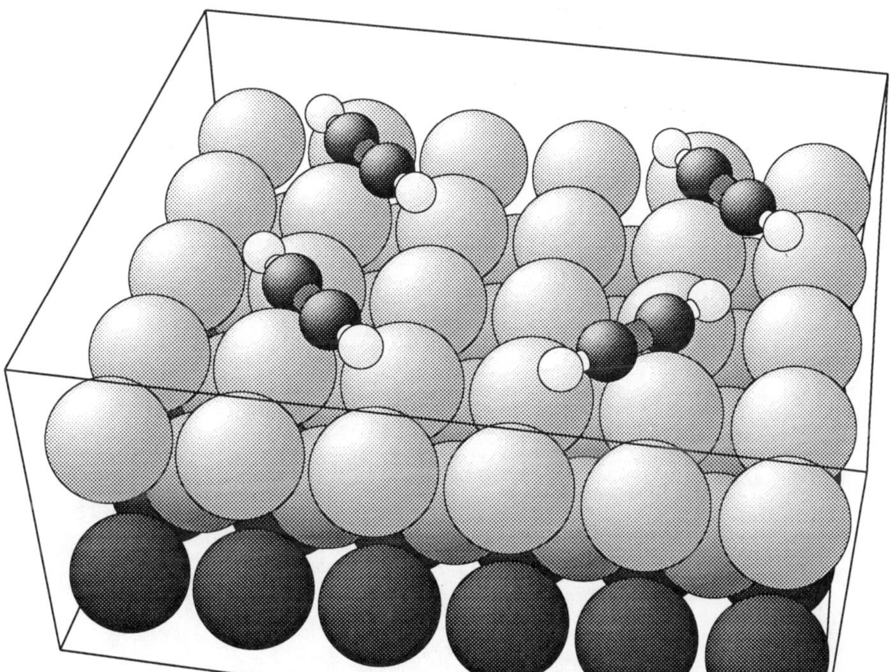

Fig. 73b : Cu(100)-C2H2 disordered (perspective)

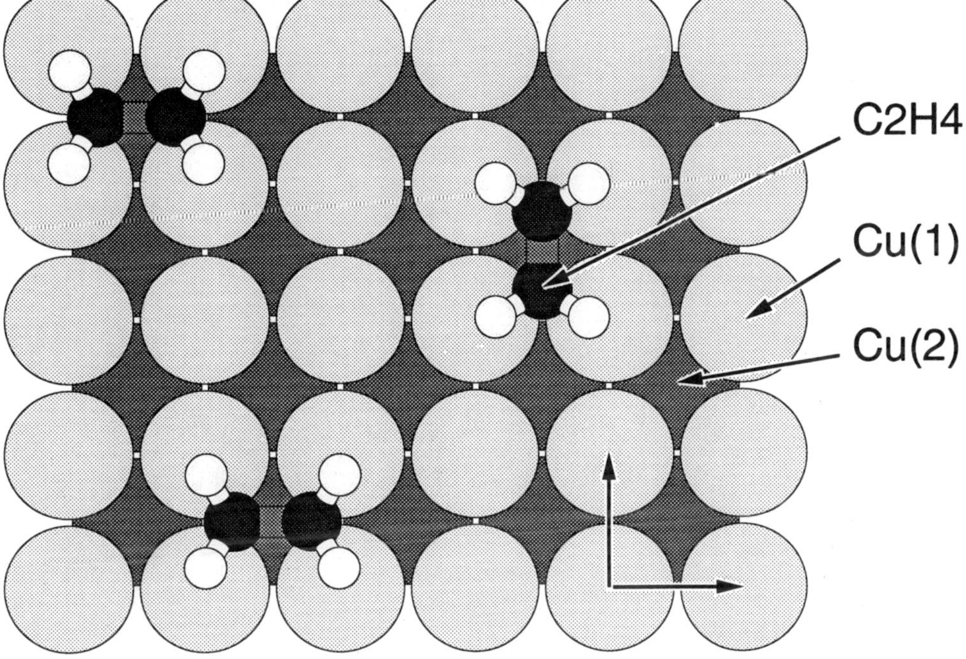

Fig. 74a : Cu(100)-C2H4 disordered (top view)

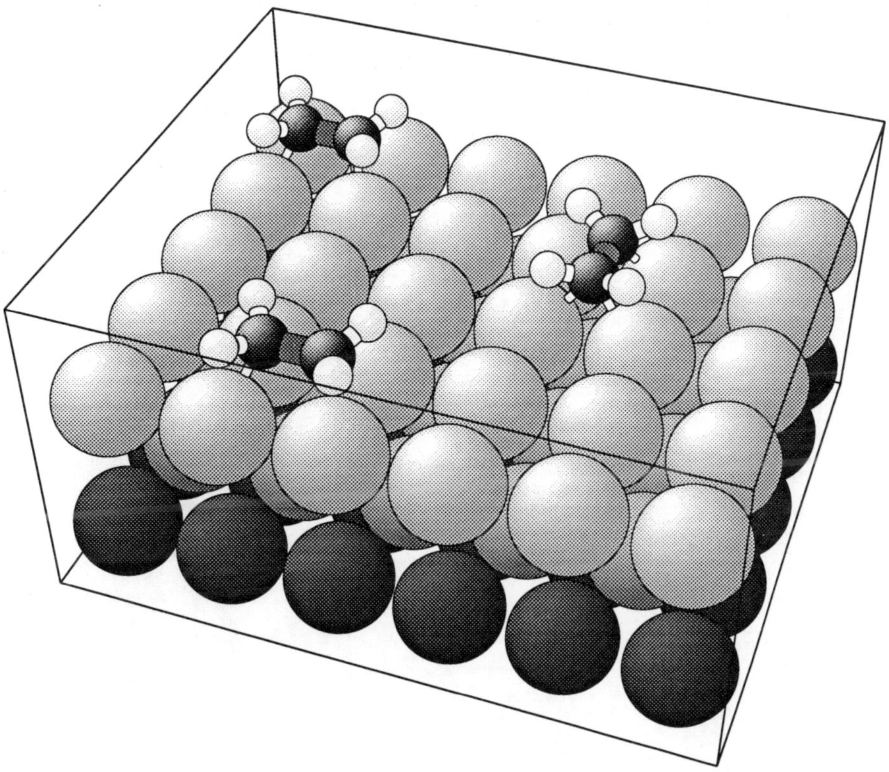

Fig. 74b : Cu(100)-C2H4 disordered (perspective)

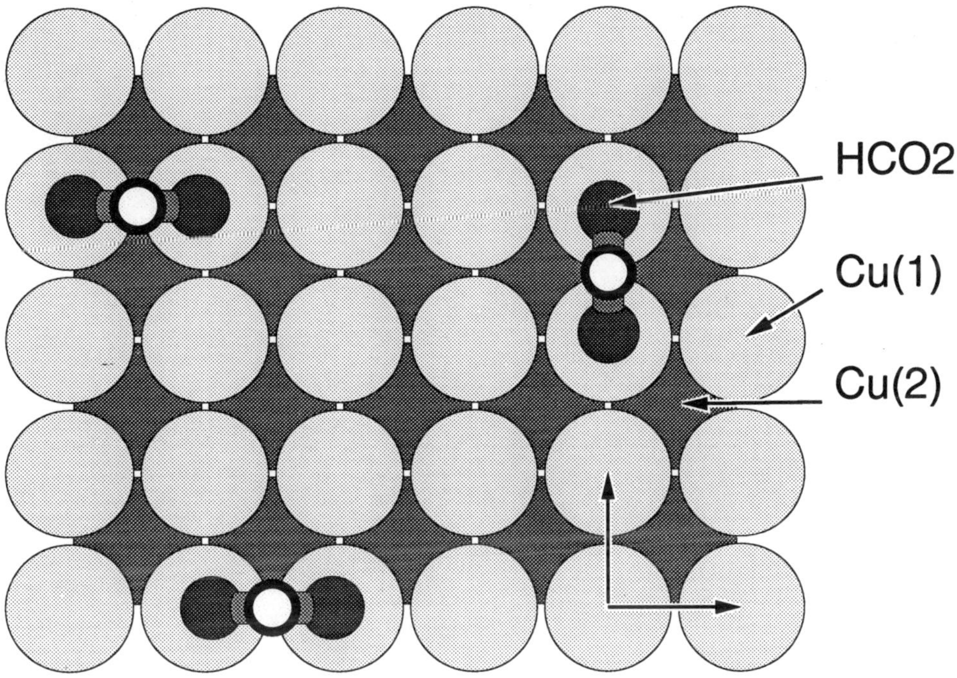

Fig. 75a : Cu(100)-HCO2 disordered (top view)

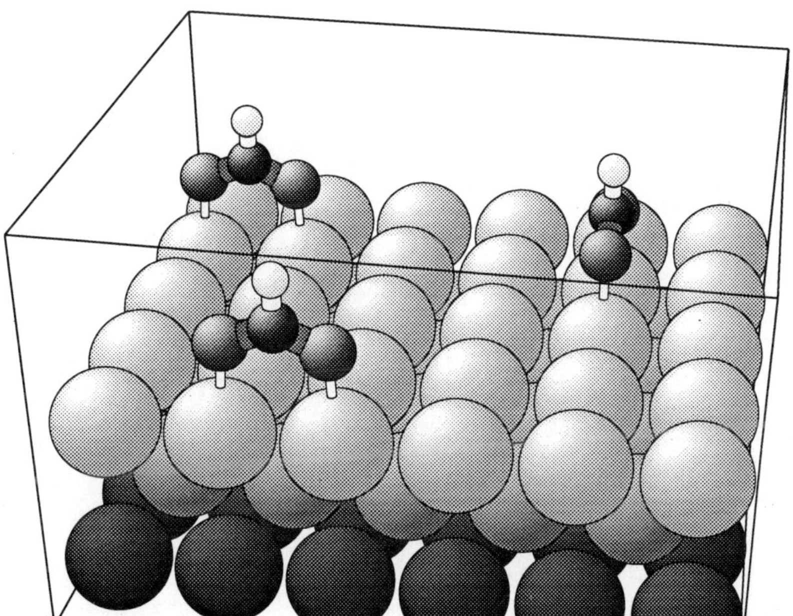

Fig. 75b : Cu(100)-HCO2 disordered (perspective)

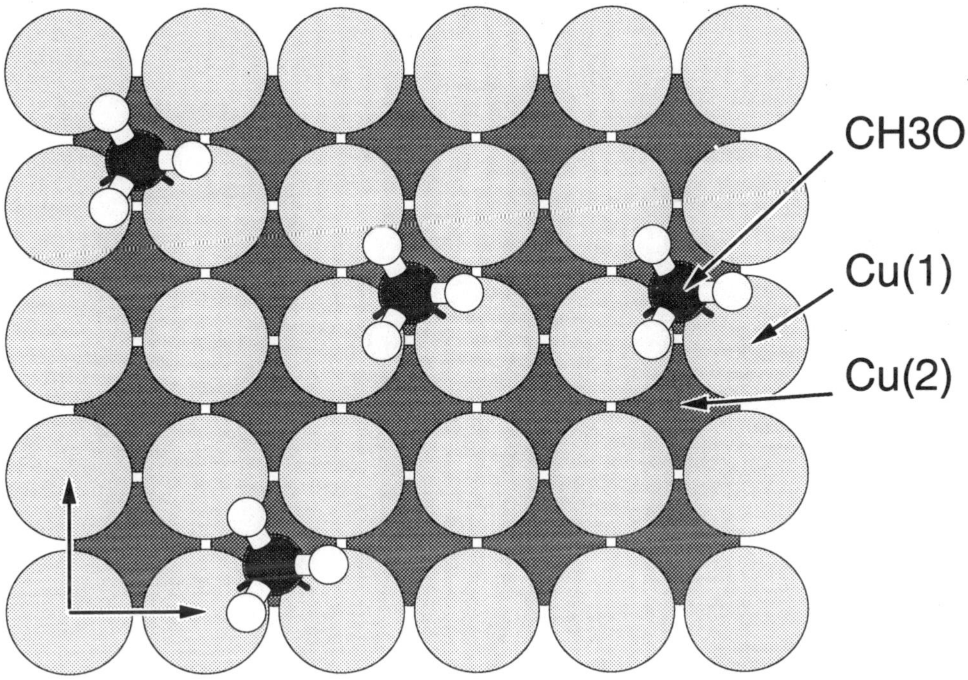

Fig. 76a : Cu(100)-CH3O disordered (top view)

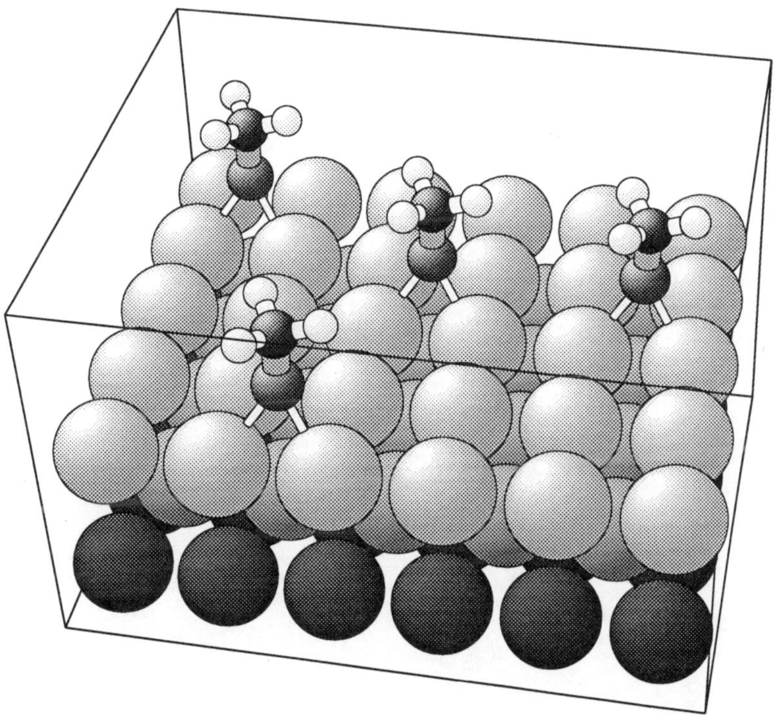

Fig. 76b : Cu(100)-CH3O disordered (perspective)

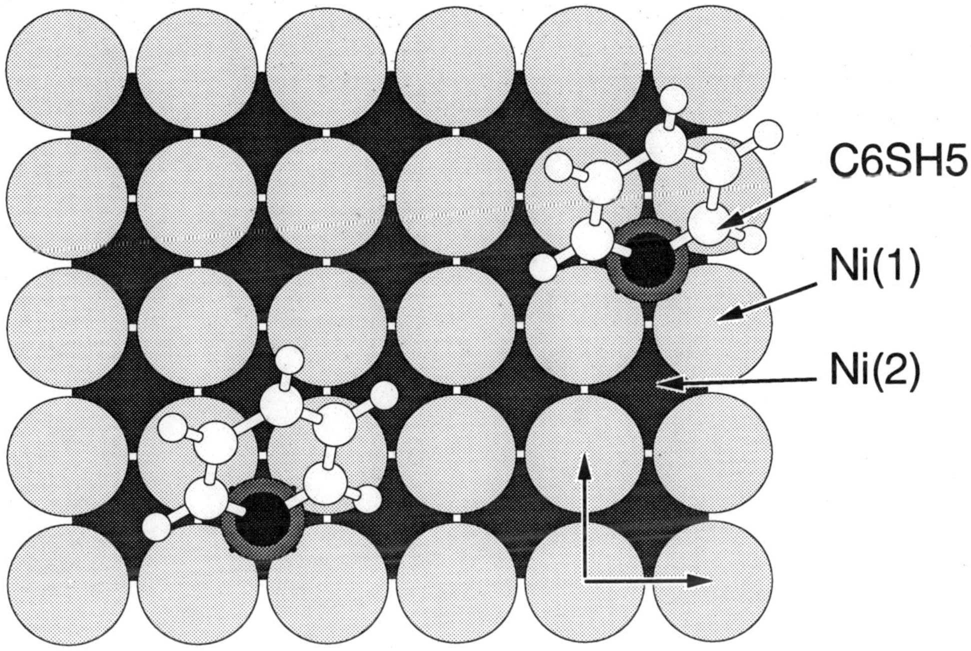

Fig. 77a : Ni(100)-C6SH5 disordered (top view)

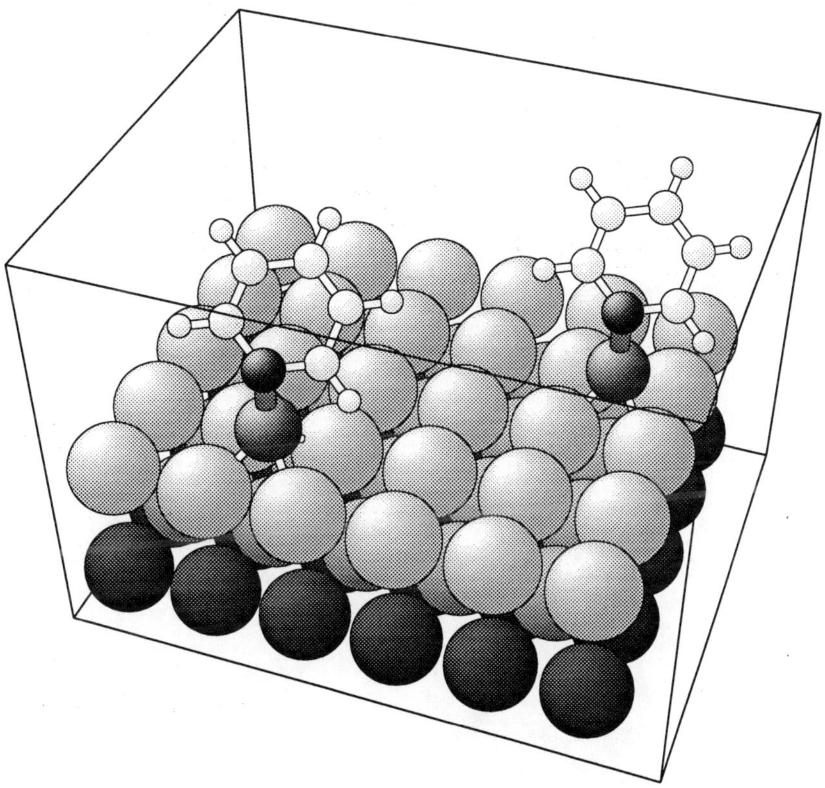

Fig. 77b : Ni(100)-C6SH5 disordered (perspective)

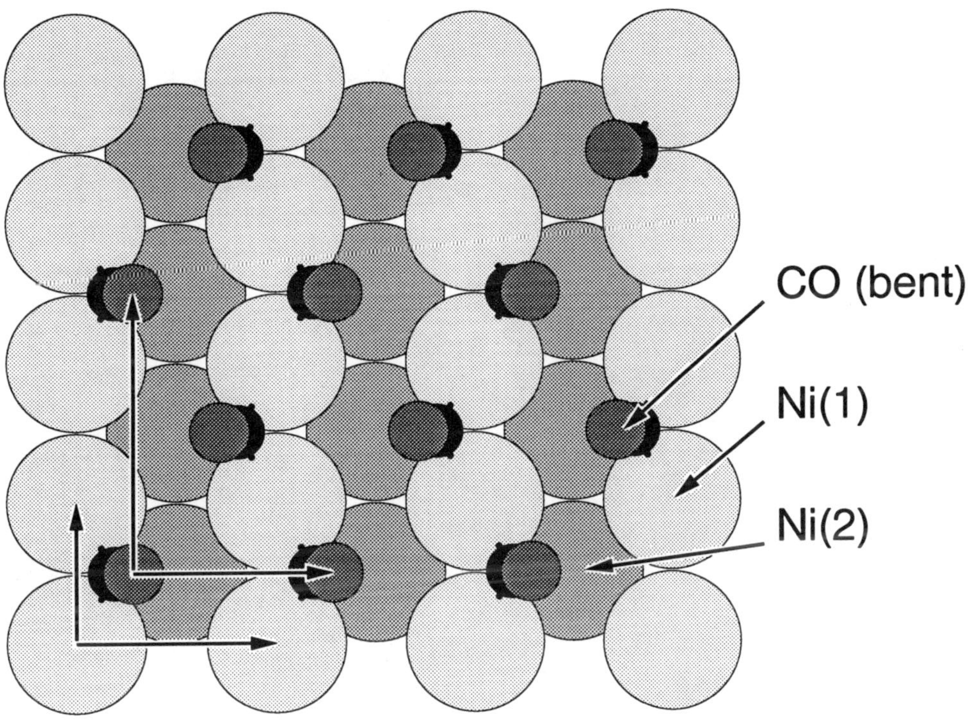

Fig. 78a : Ni(110)-p(2x1)-2CO (top view)

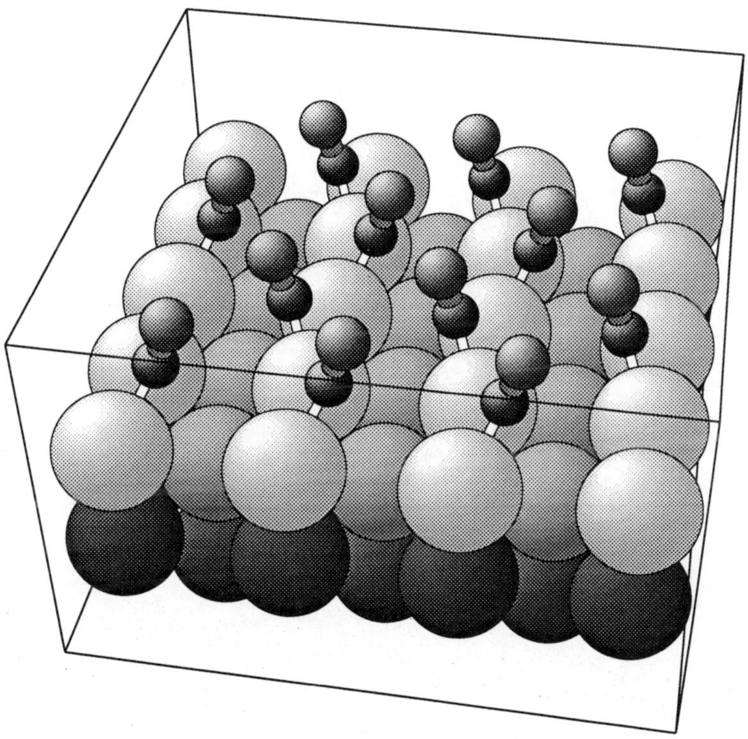

Fig. 78b : Ni(110)-p(2x1)-2CO (perspective)

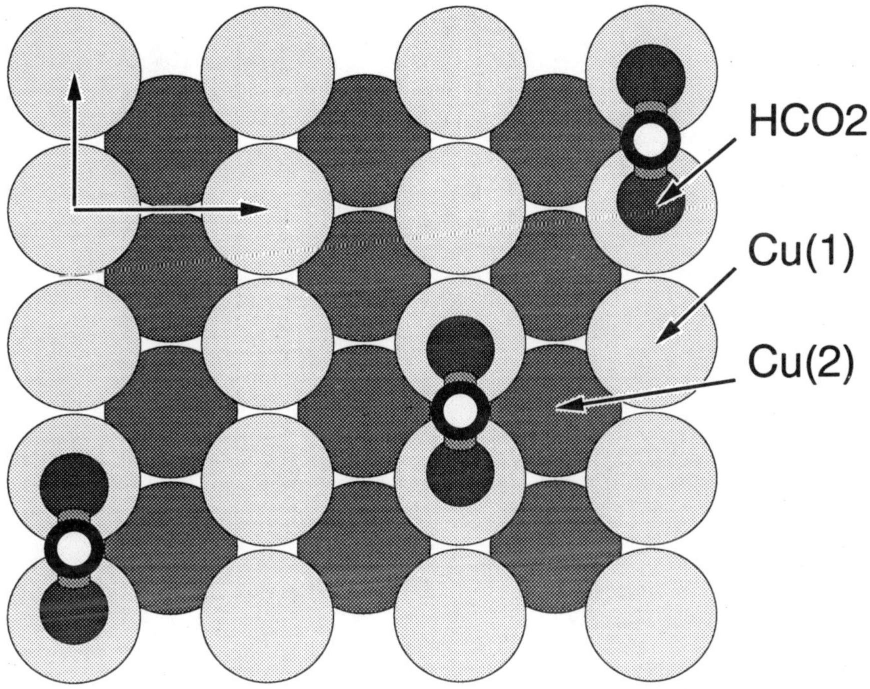

Fig. 79a : Cu(110)-HCO2 disordered (top view)

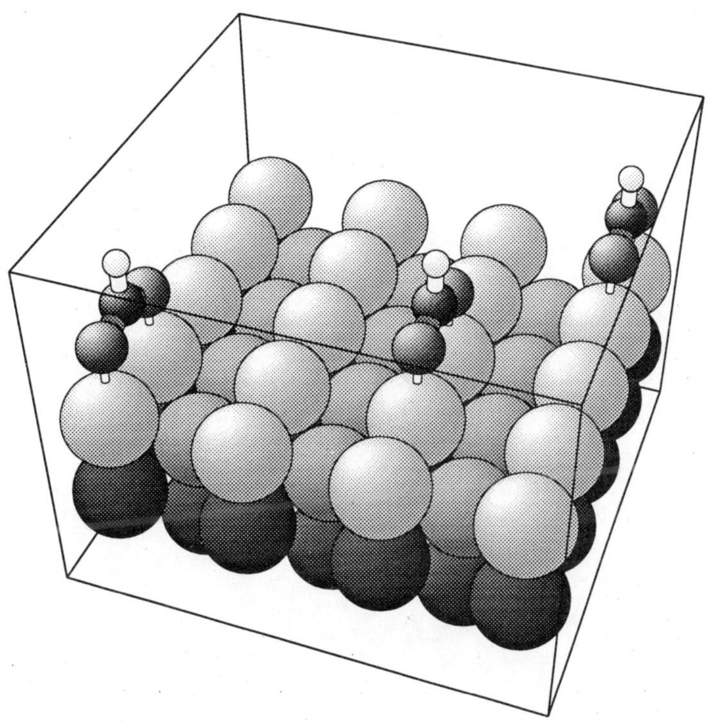

Fig. 79b : Cu(110)-HCO2 disordered (perspective)

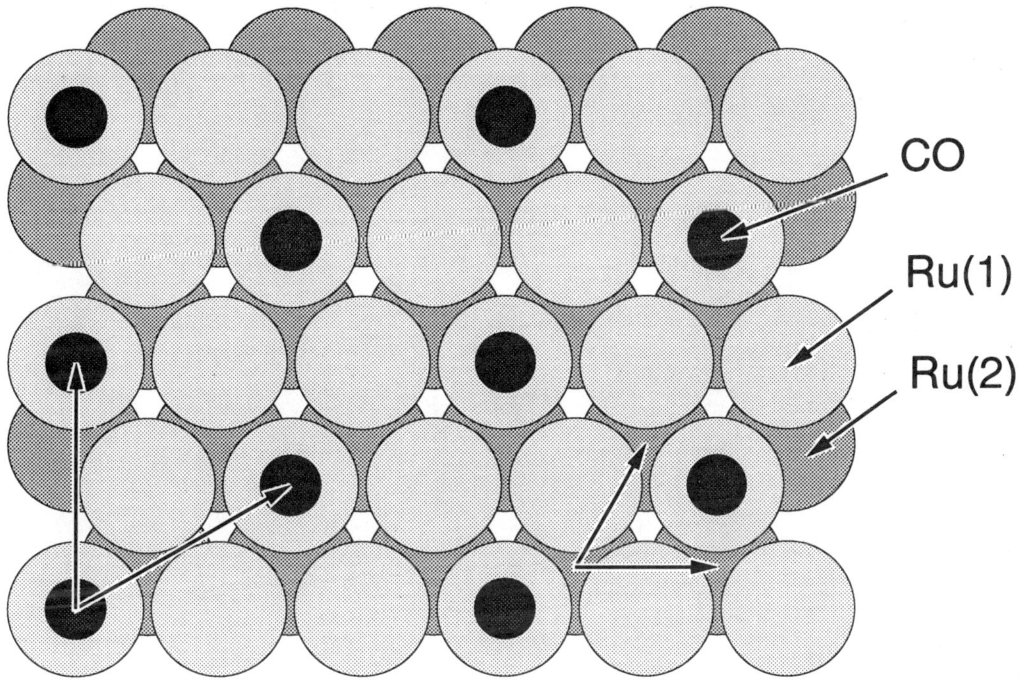

Fig. 80a : Ru(0001)-(√3x√3)R30°-CO (top view)

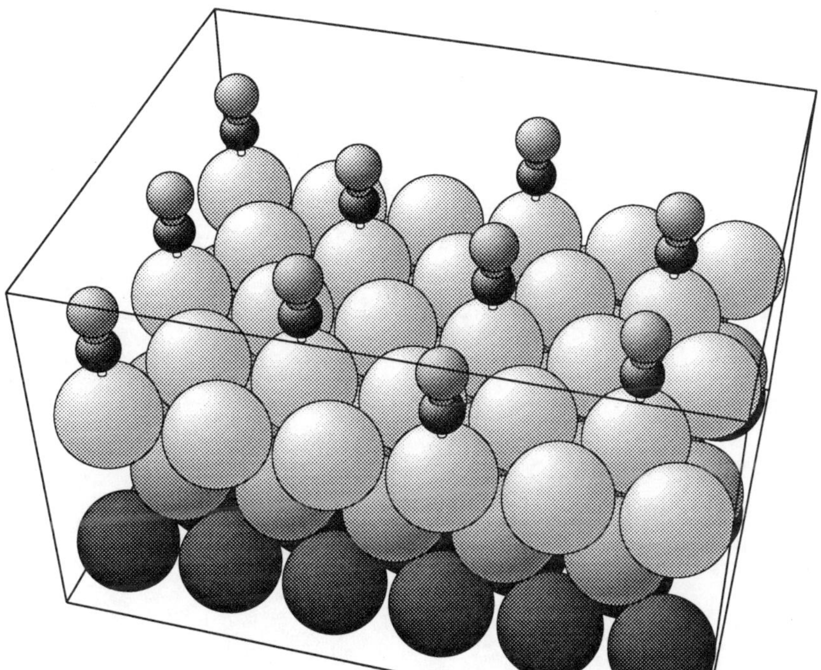

Fig. 80b : Ru(0001)-(√3x√3)R30°-CO (perspective)

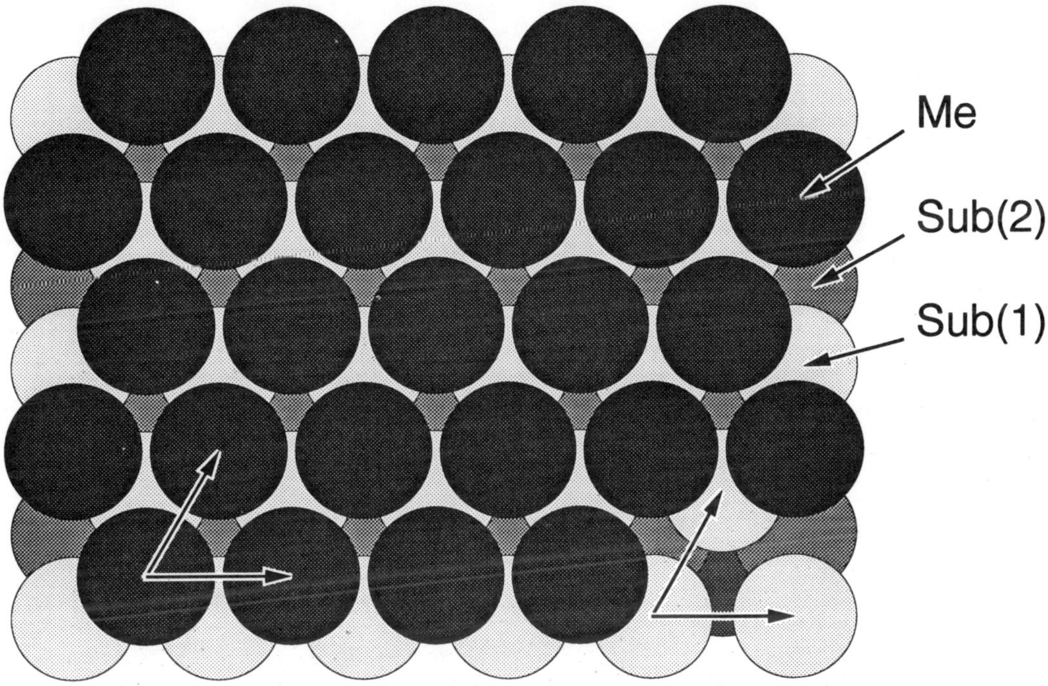

Fig. 81a : fcc(111)-(1x1)-1Me (top view)

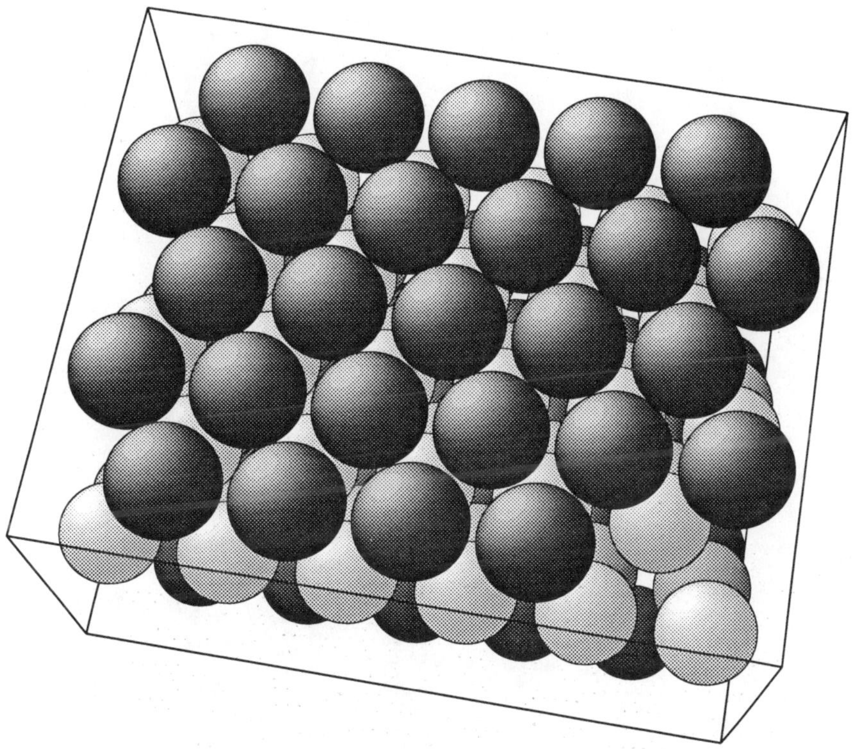

Fig. 81b : fcc(111)-(1x1)-1Me (perspective)

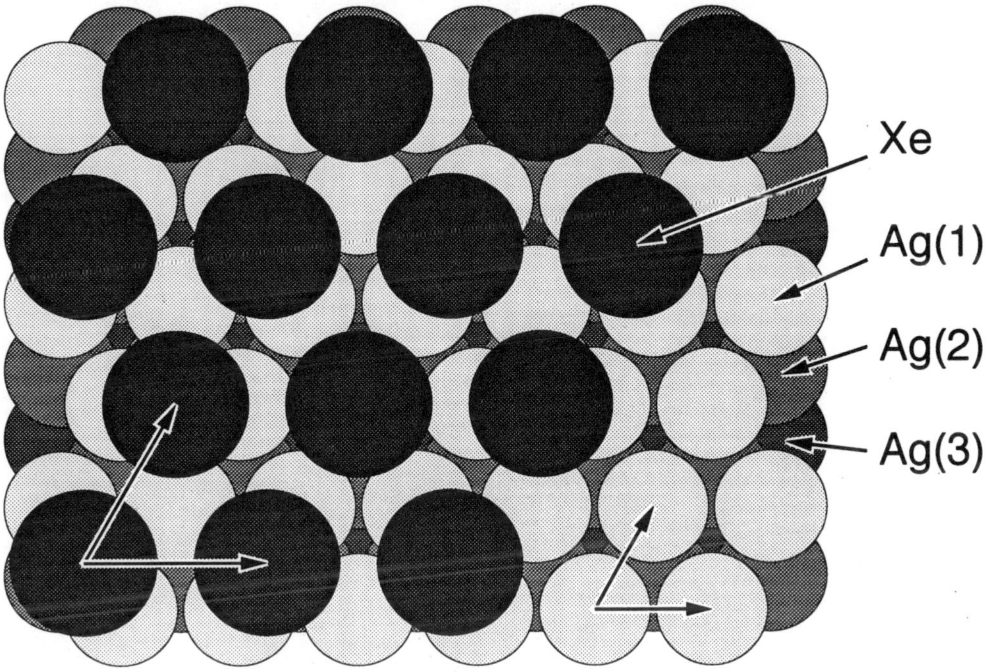

Fig. 82a : Ag(111)-Xe incommensurate (top view)

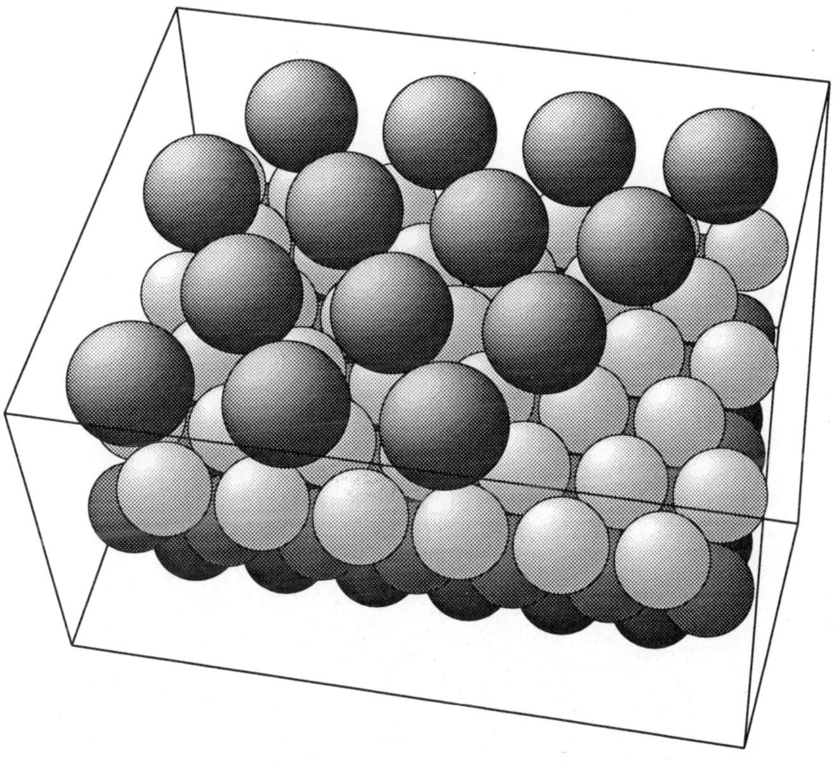

Fig. 82b : Ag(111)-Xe incommensurate (perspective)

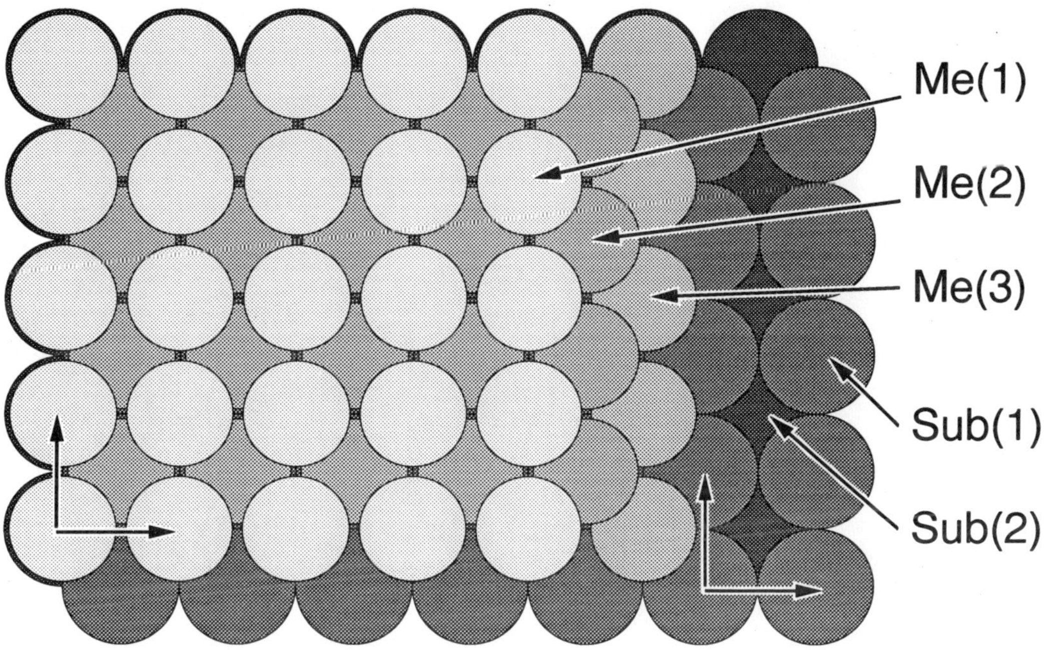

Fig. 83a : fcc(100)-(1x1)-nMe (e.g. n = 3) (top view)

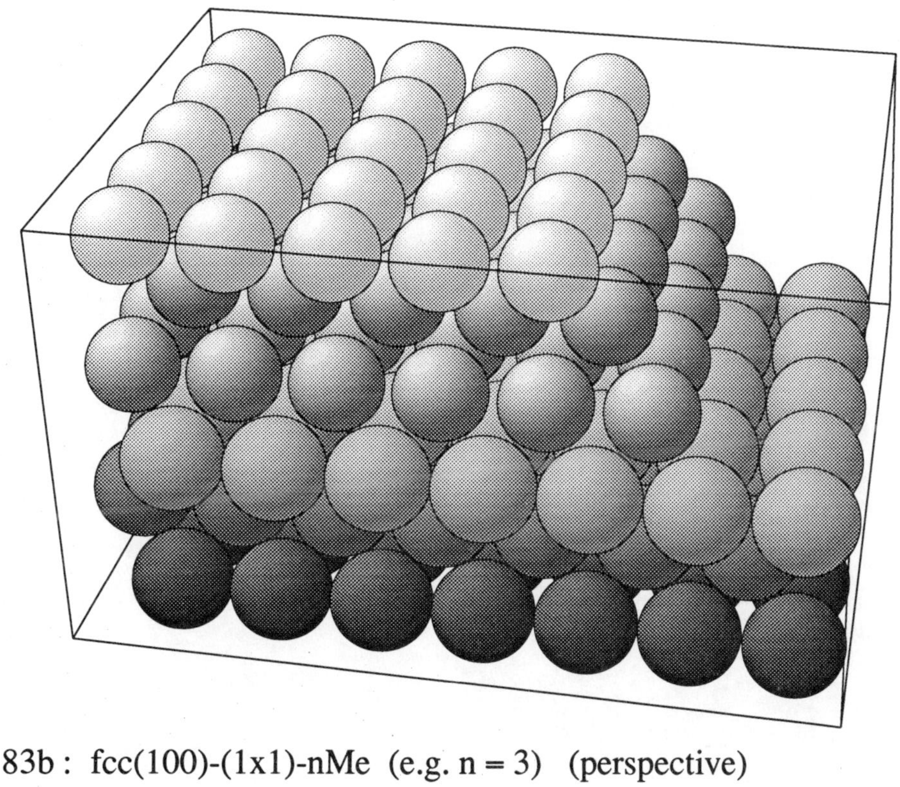

Fig. 83b : fcc(100)-(1x1)-nMe (e.g. n = 3) (perspective)

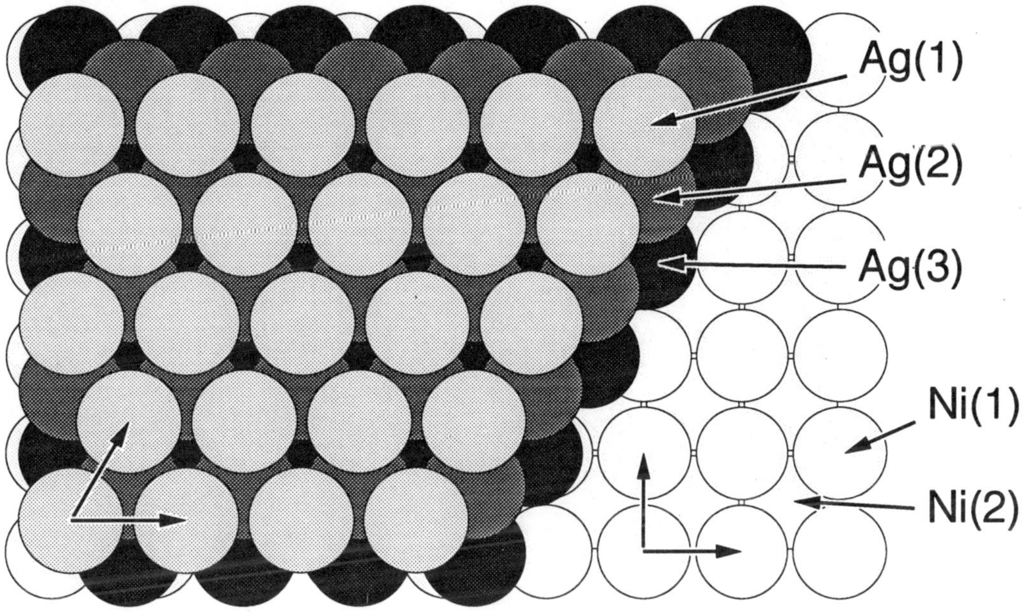

Fig. 84a : Ni(100)-Ag(111) multilayers (top view)

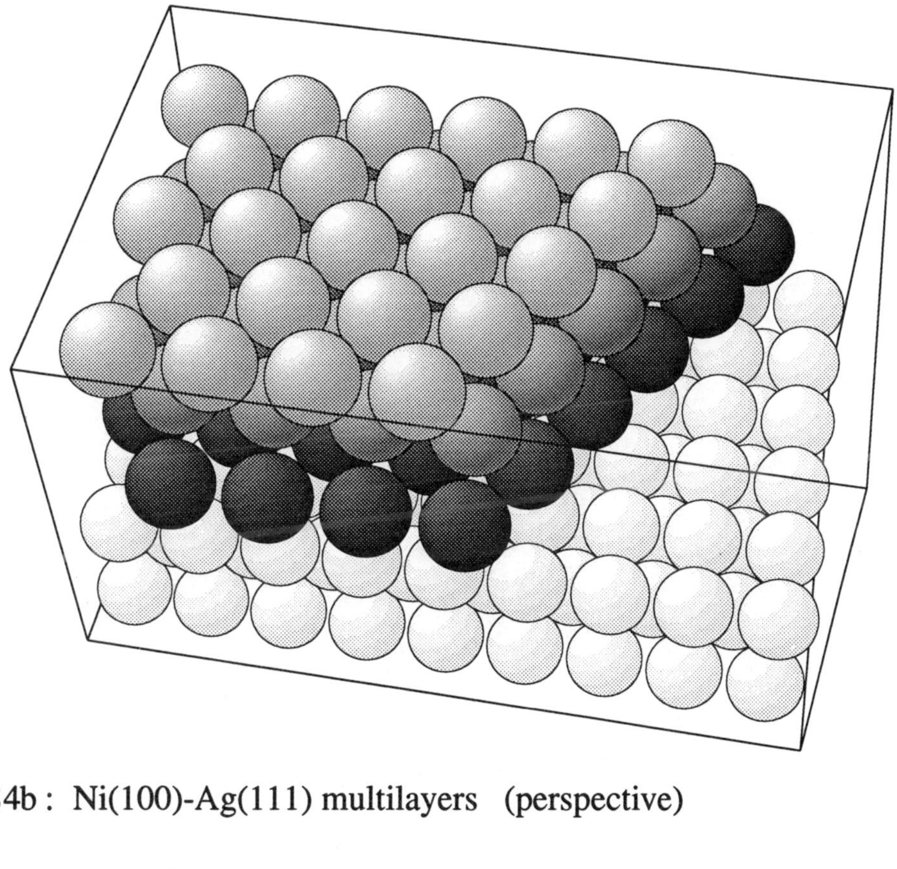

Fig. 84b : Ni(100)-Ag(111) multilayers (perspective)

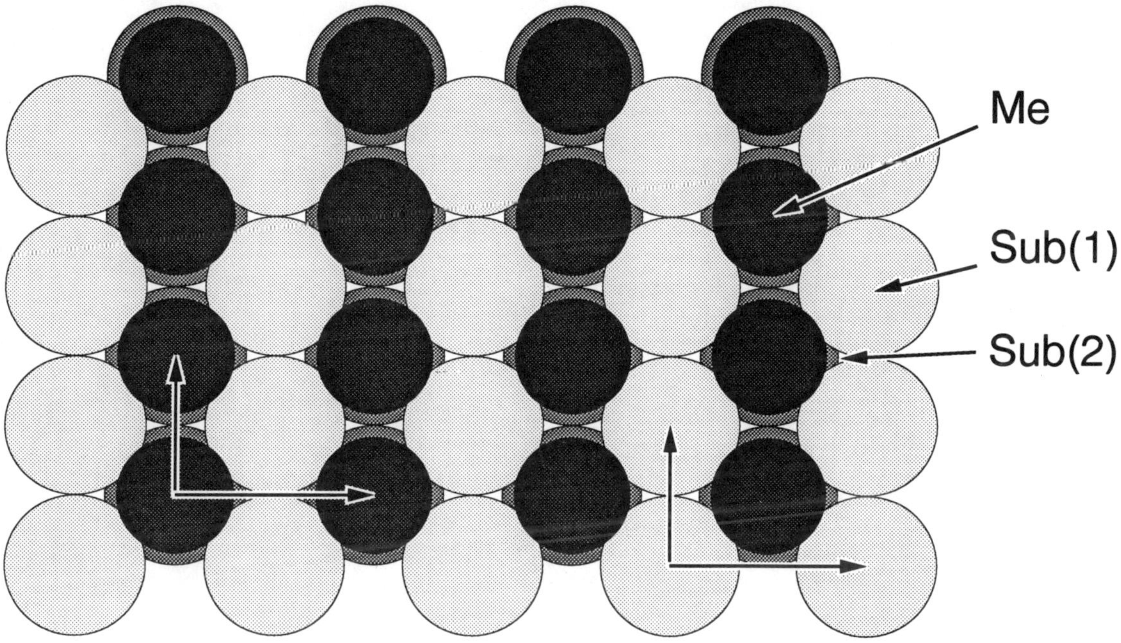

Fig. 85a : fcc(110)-(1x1)-1Me (top view)

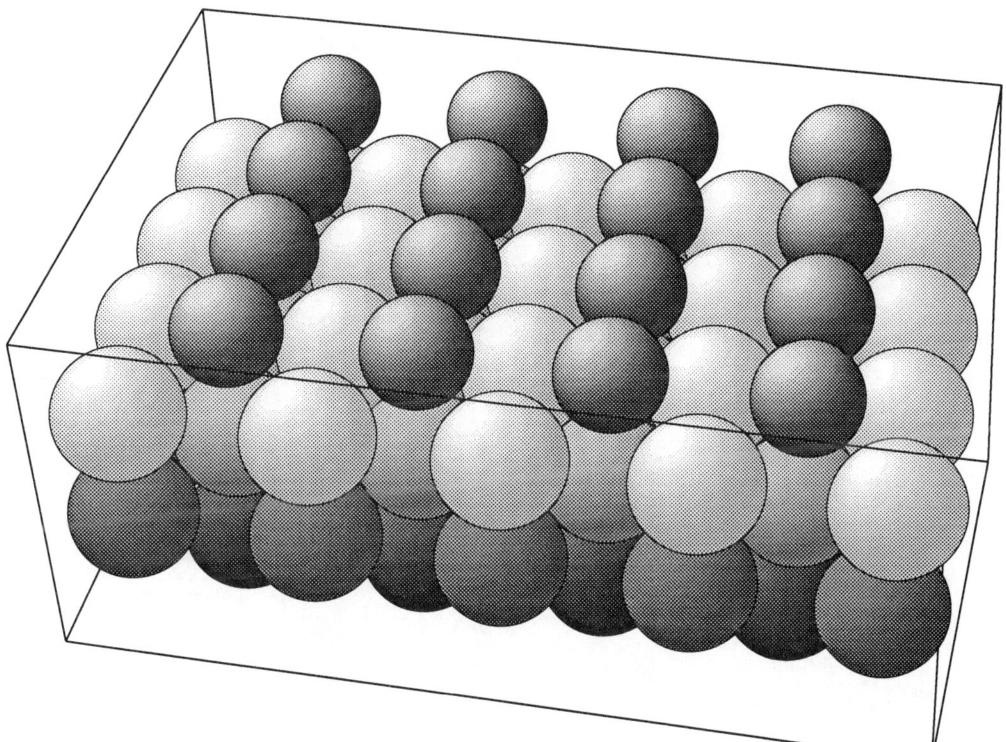

Fig. 85b : fcc(110)-(1x1)-1Me (perspective)

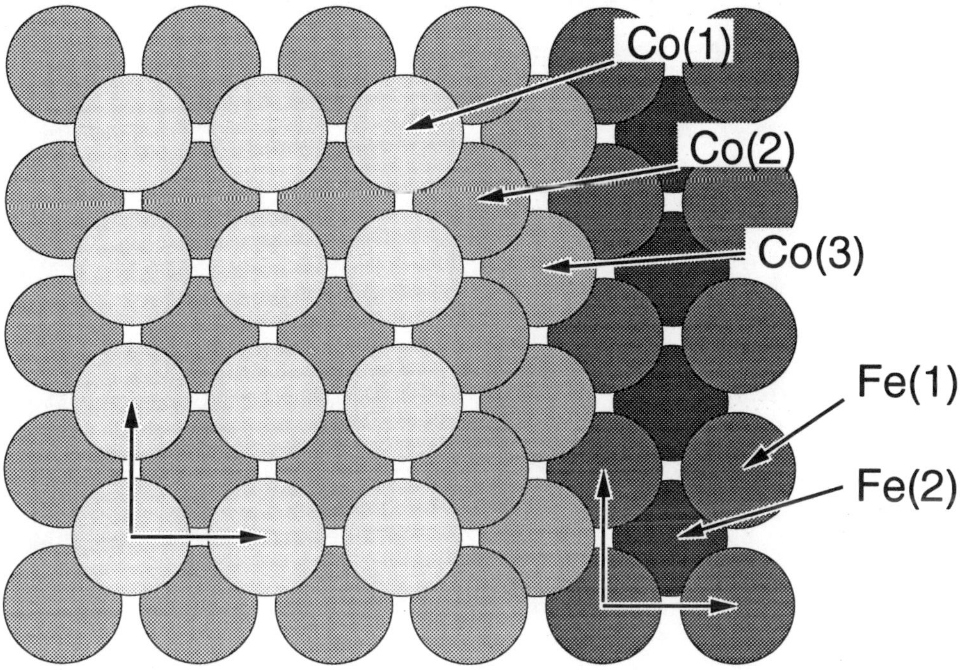

Fig. 86a : bcc(100)-(1x1)-nMe (e.g. n = 3) (top view)

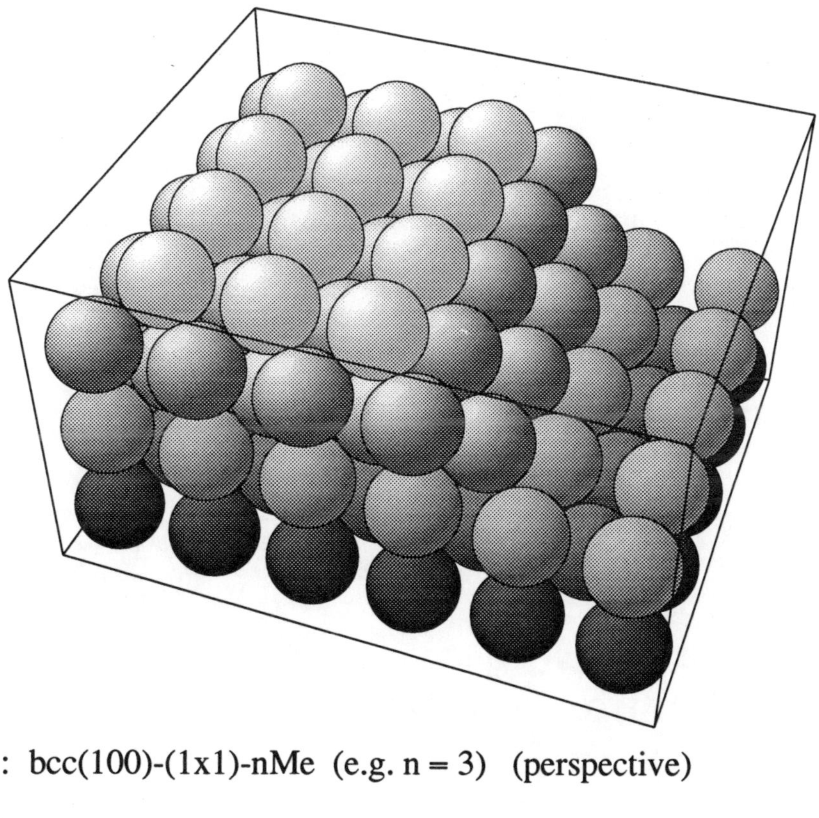

Fig. 86b : bcc(100)-(1x1)-nMe (e.g. n = 3) (perspective)

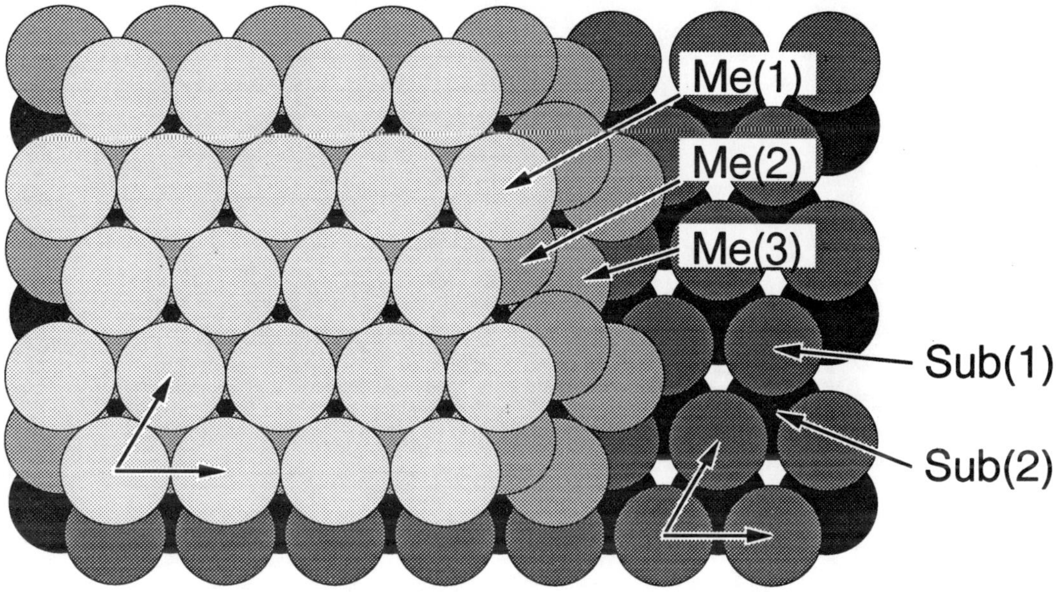

Fig. 87a : hcp(0001)-(1x1)-nMe (e.g. n = 3) (top view)

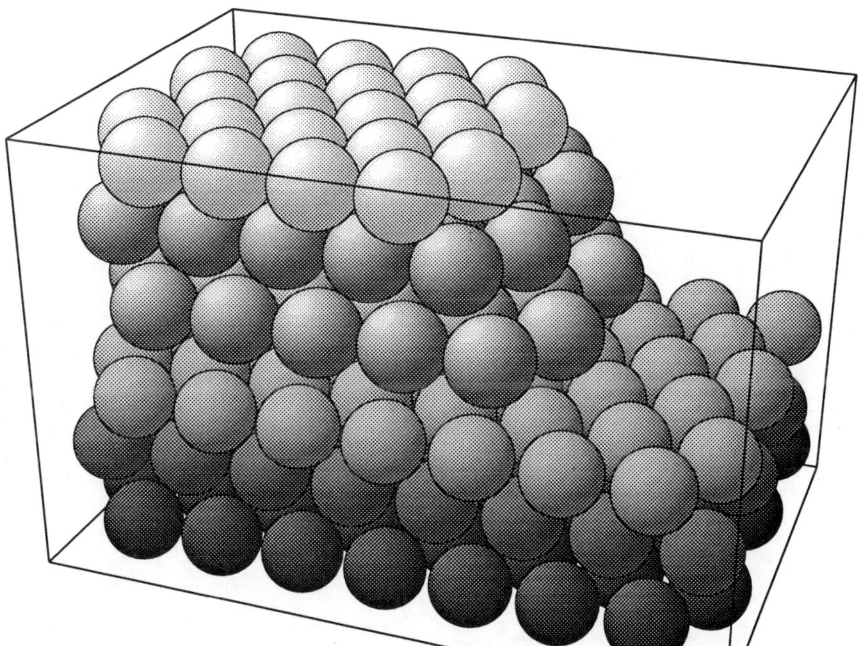

Fig. 87b : hcp(0001)-(1x1)-nMe (e.g. n = 3) (perspective)

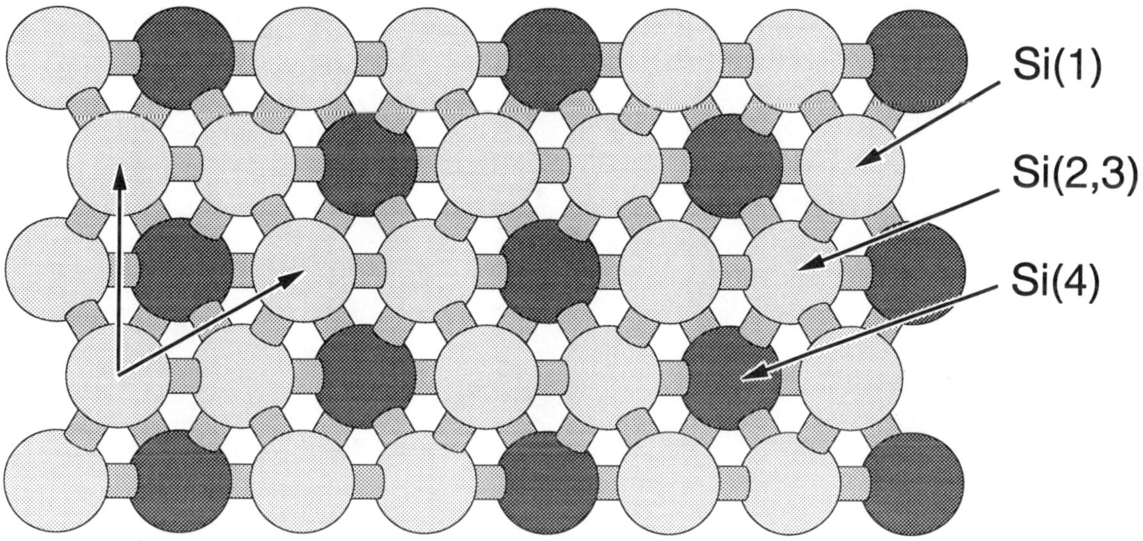

Fig. 88a : Si(111)-(1x1) (top view)

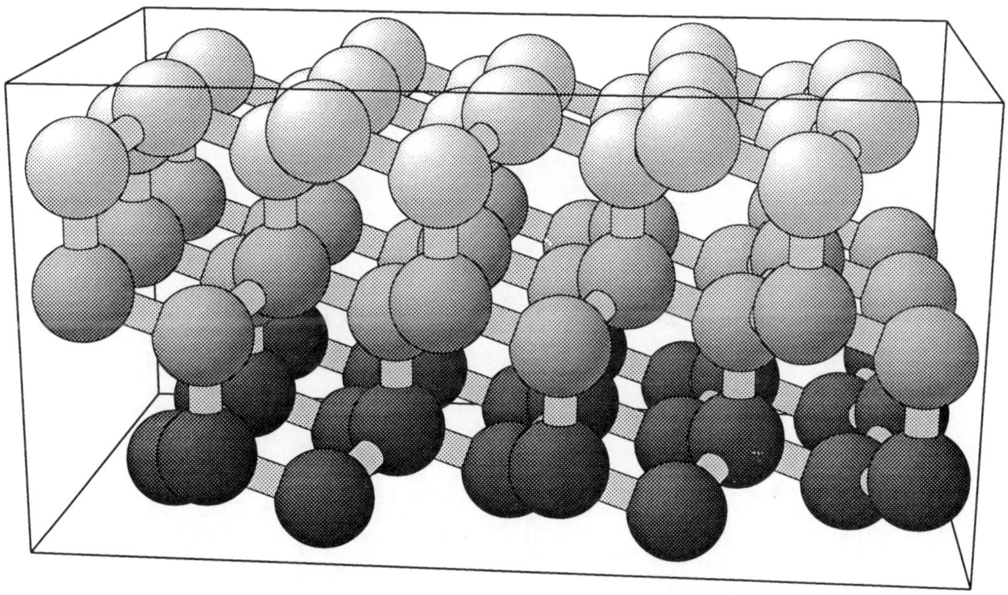

Fig. 88b : Si(111)-(1x1) (perspective)

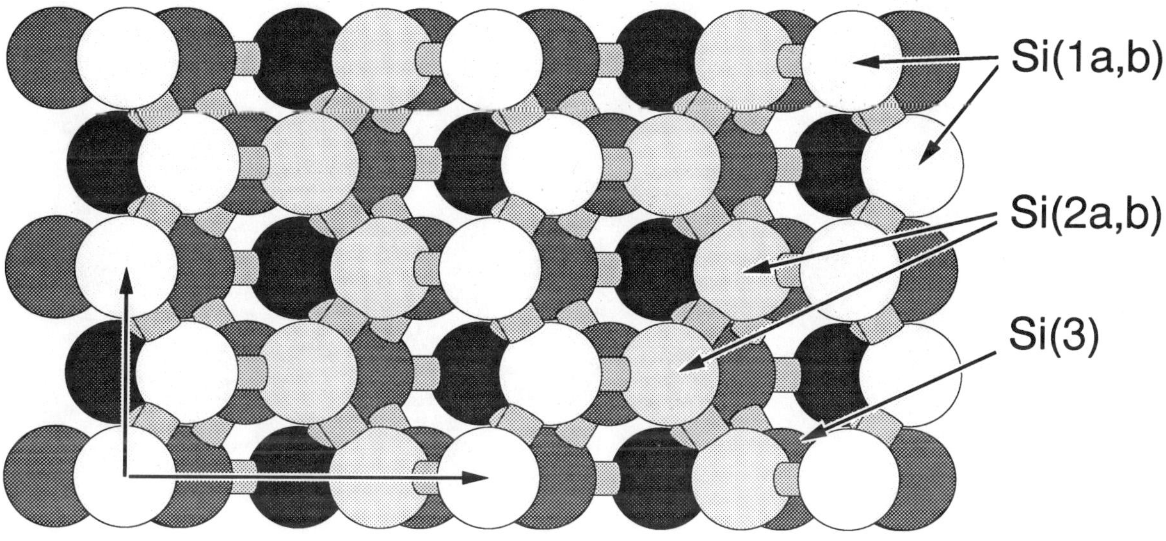

Fig. 89a : Si(111)-(2x1) (top view)

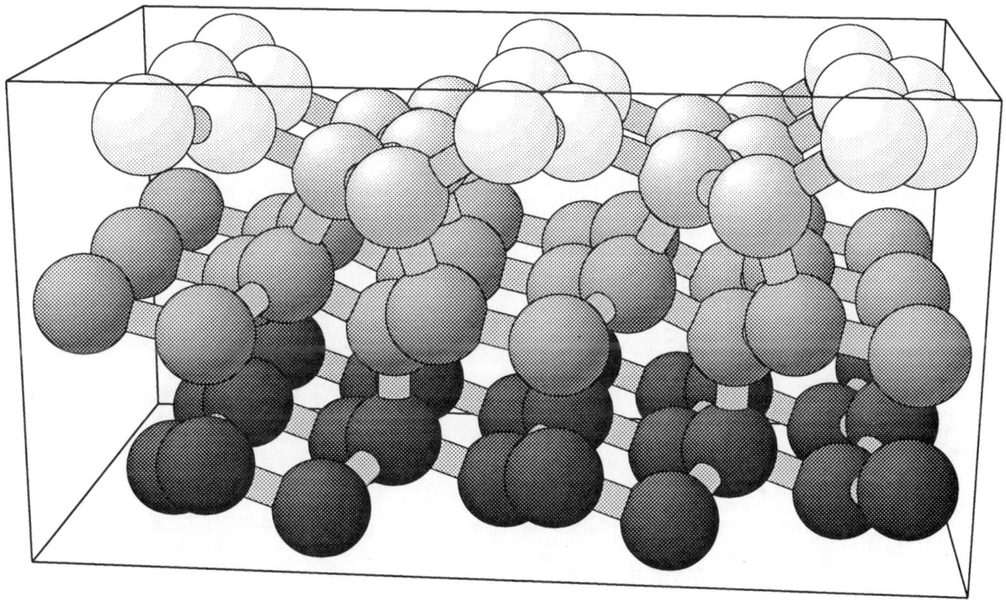

Fig. 89b : Si(111)-(2x1) (perspective)

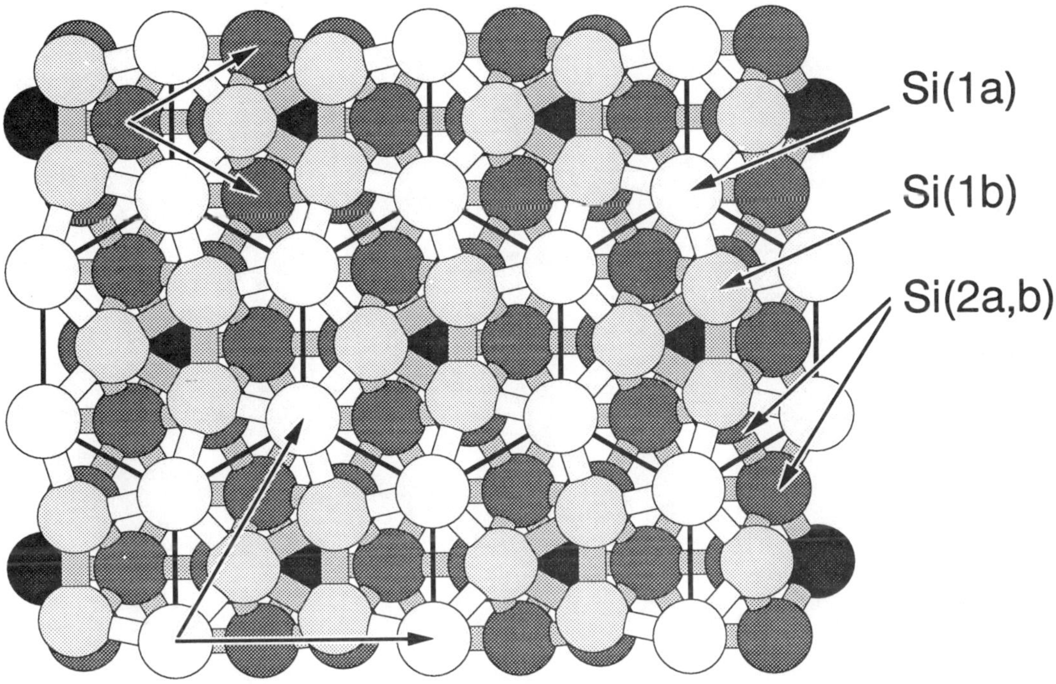

Fig. 90a : Si(111)-(√3x√3)R30° (top view)

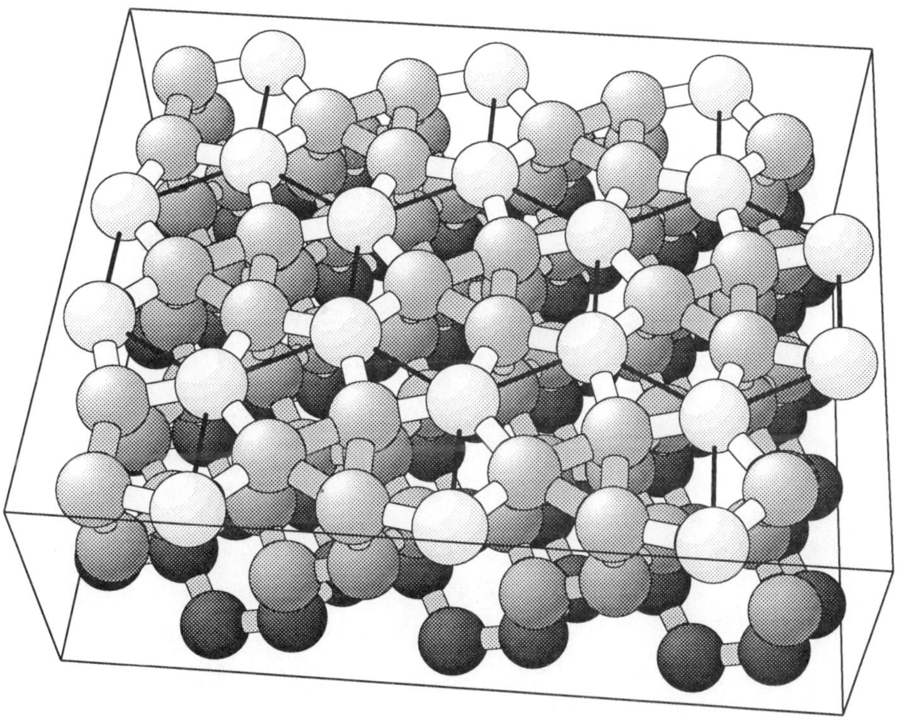

Fig. 90b : Si(111)-(√3x√3)R30° (perspective)

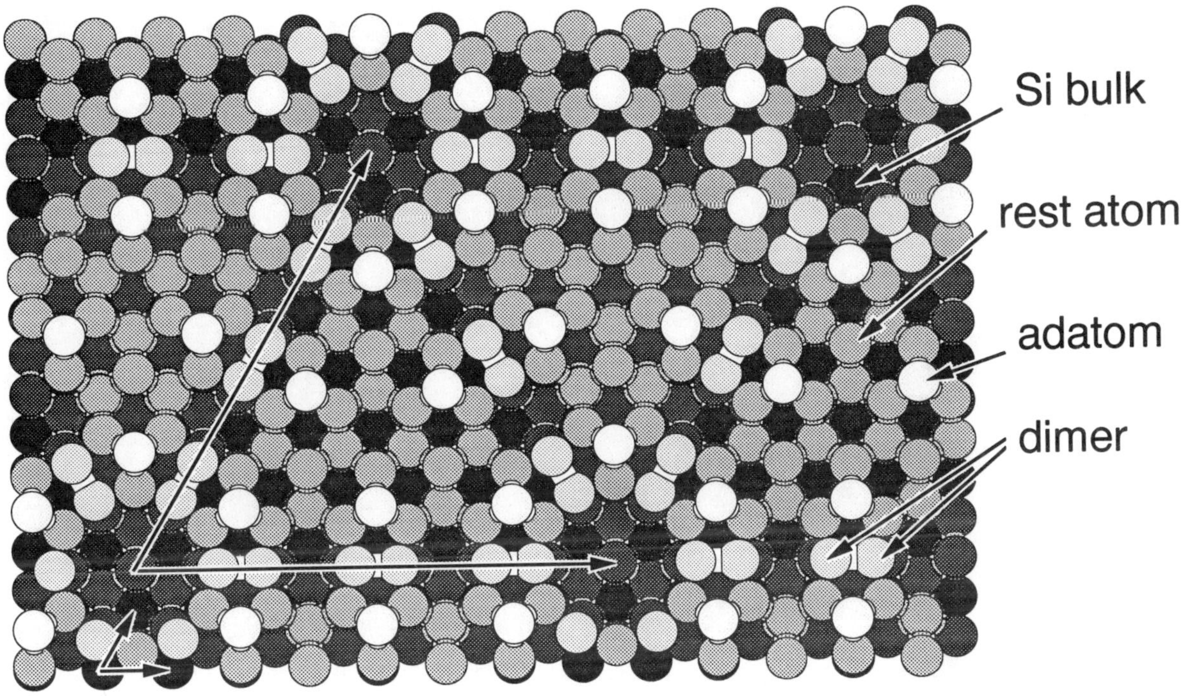

Fig. 91a : Si(111)-(7x7) (top view)

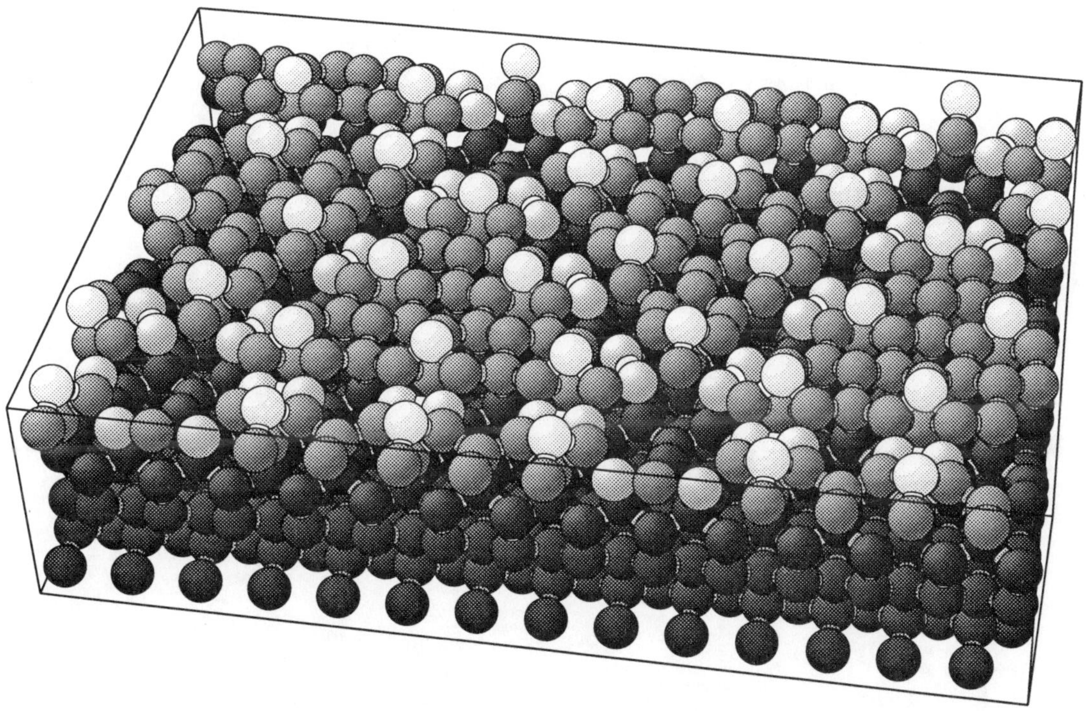

Fig. 91b : Si(111)-(7x7) (perspective)

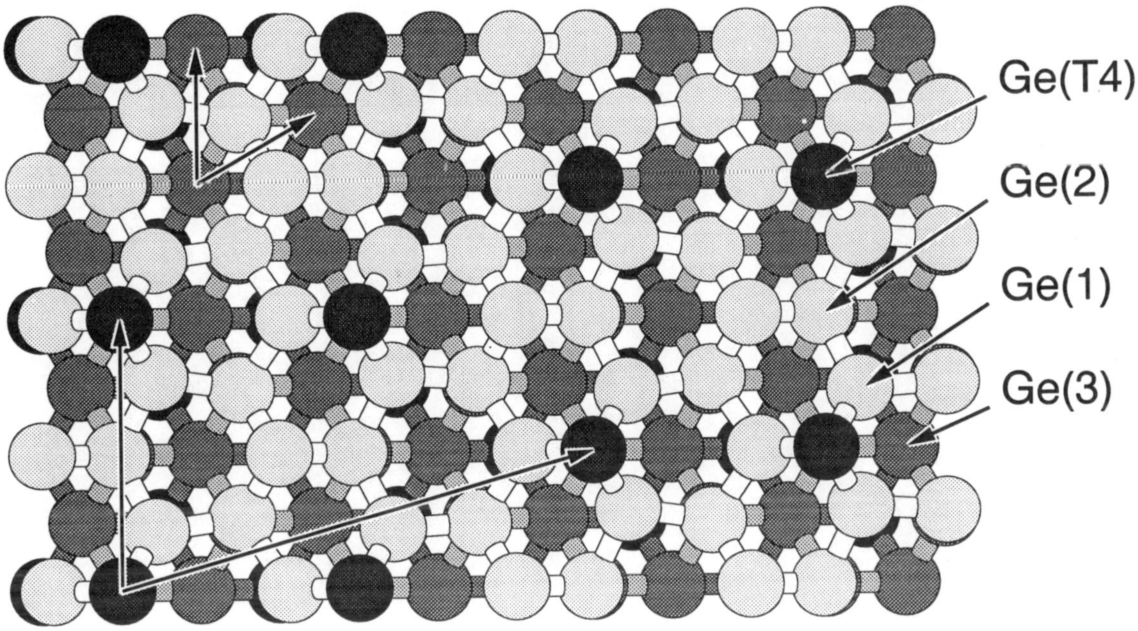

Fig. 92a : Ge(111)-c(2x8) (top view)

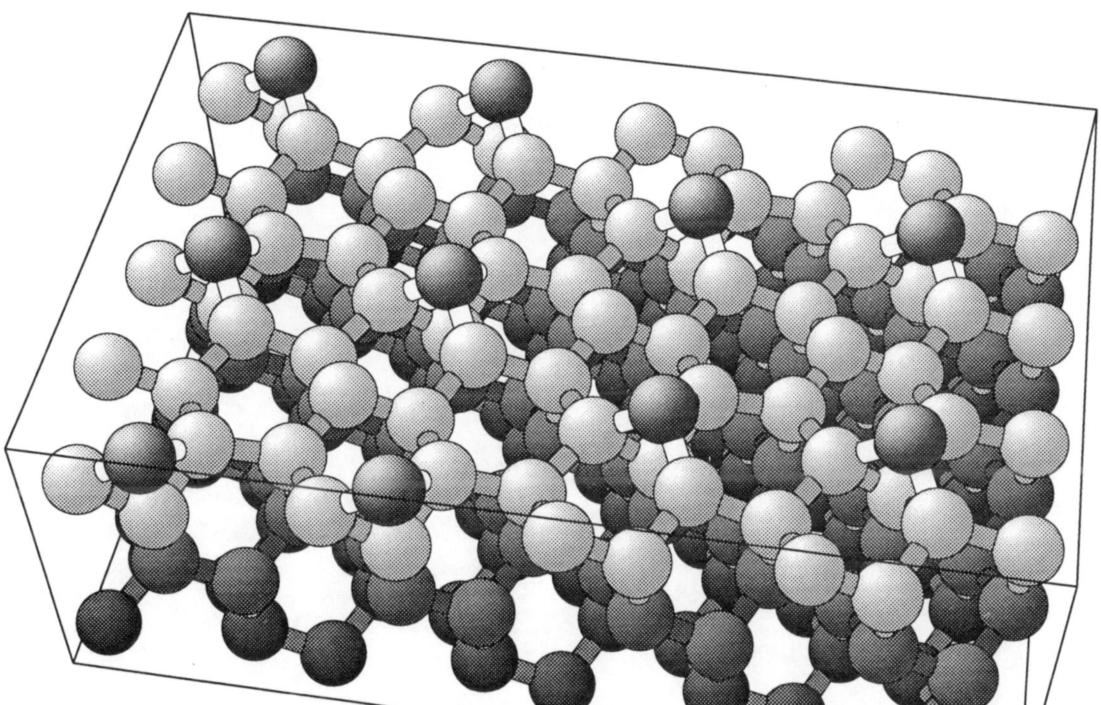

Fig. 92b : Ge(111)-c(2x8) (perspective)

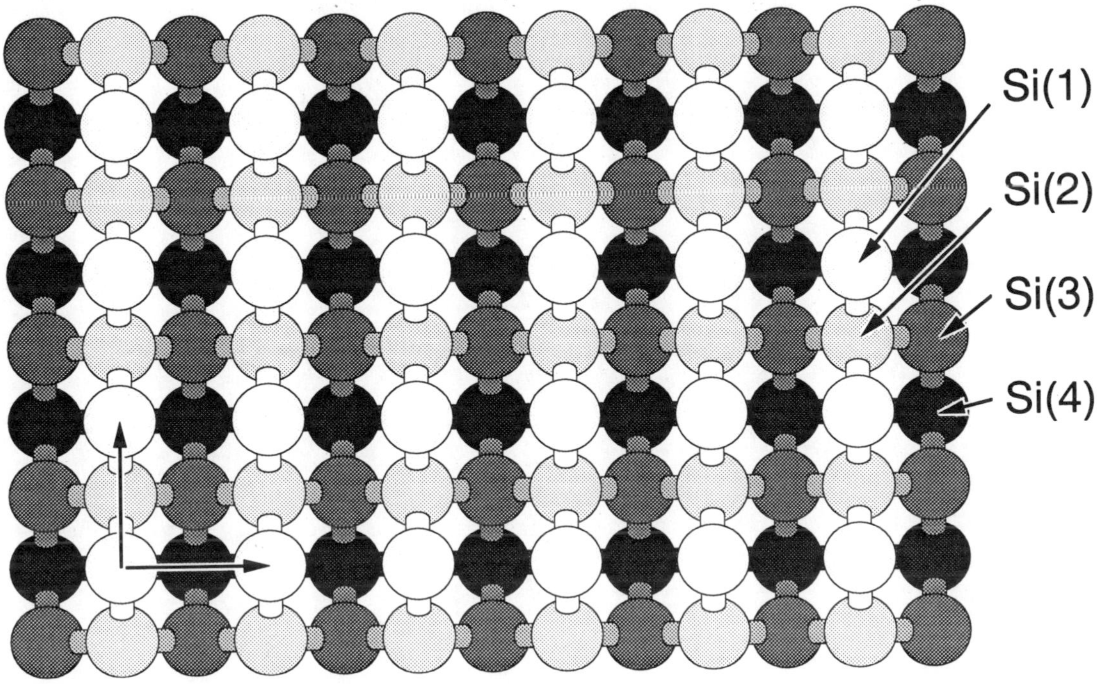

Fig. 93a : Si(100)-(1x1) (top view)

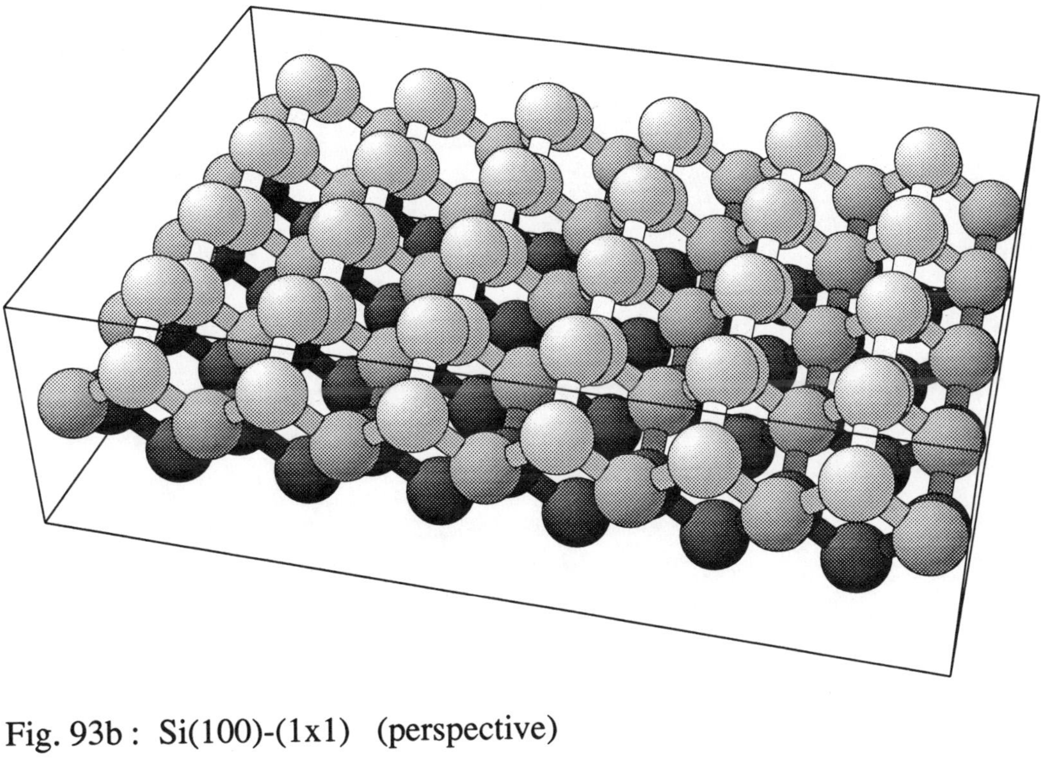

Fig. 93b : Si(100)-(1x1) (perspective)

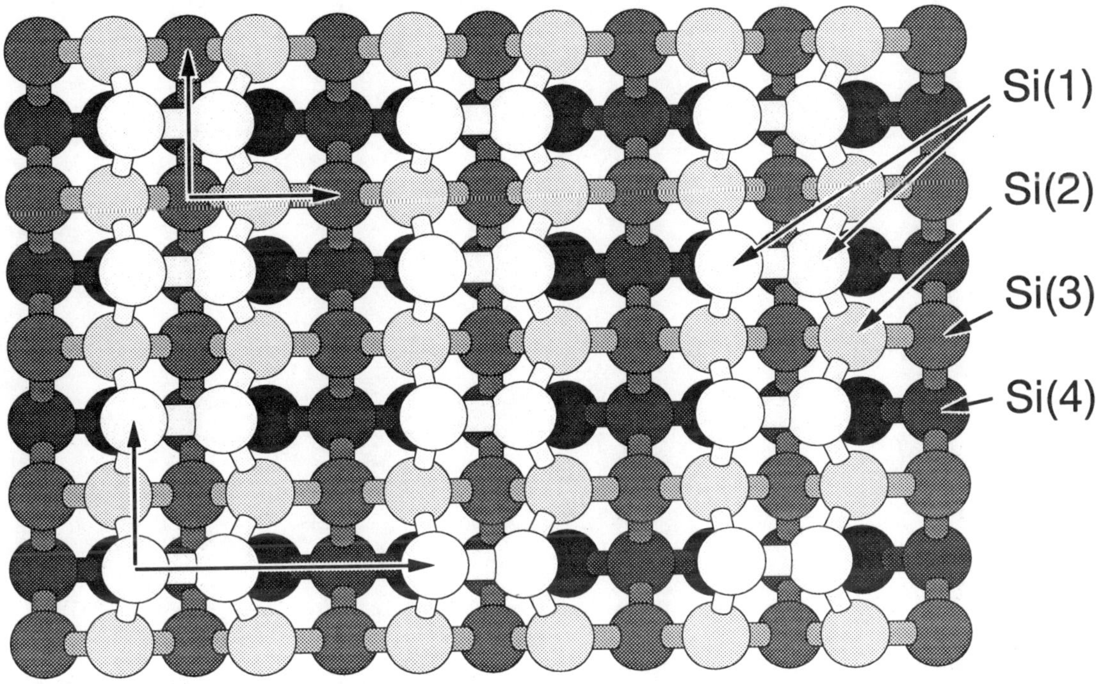

Fig. 94a : Si(100)-(2x1) (top view)

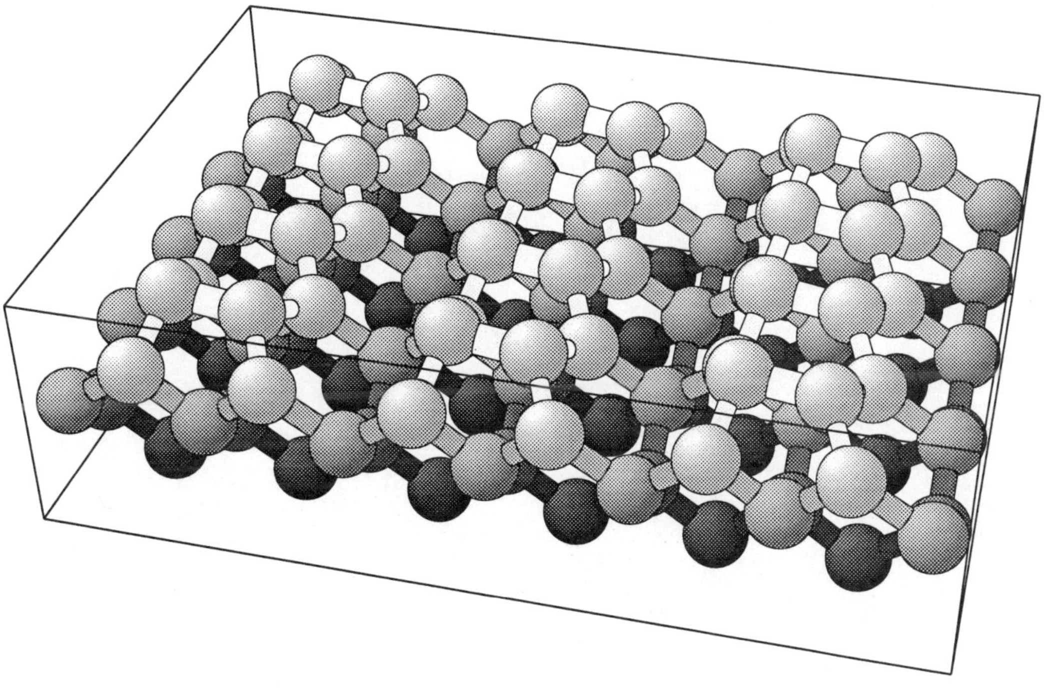

Fig. 94b : Si(100)-(2x1) (perspective)

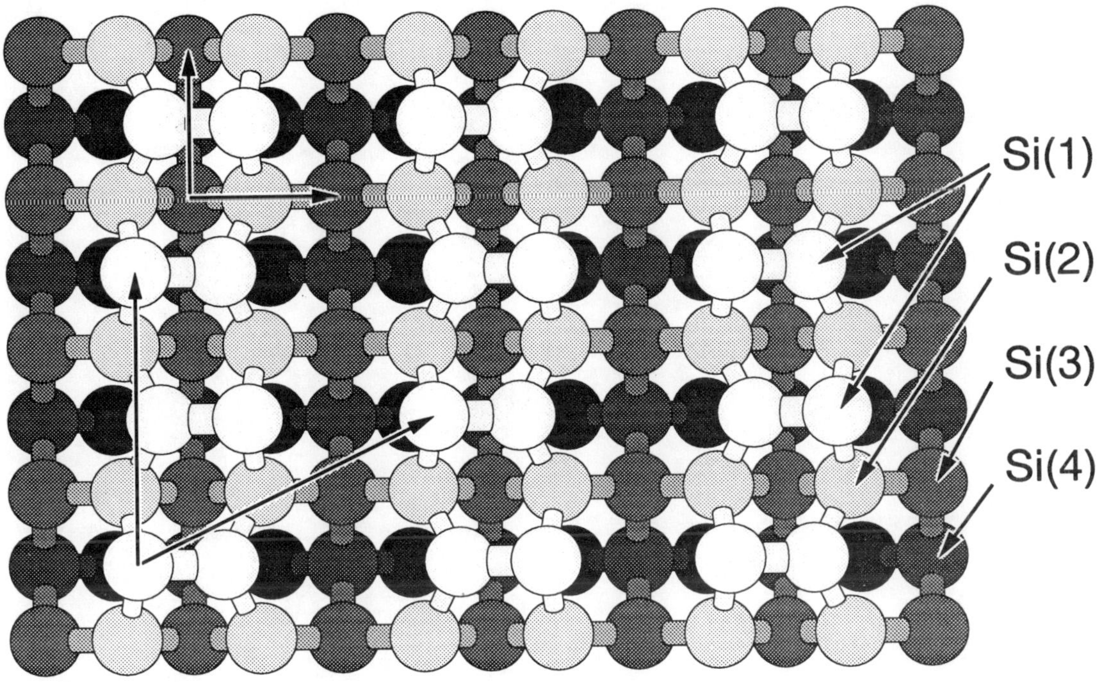

Fig. 95a : Si(100)-c(4x2) (top view)

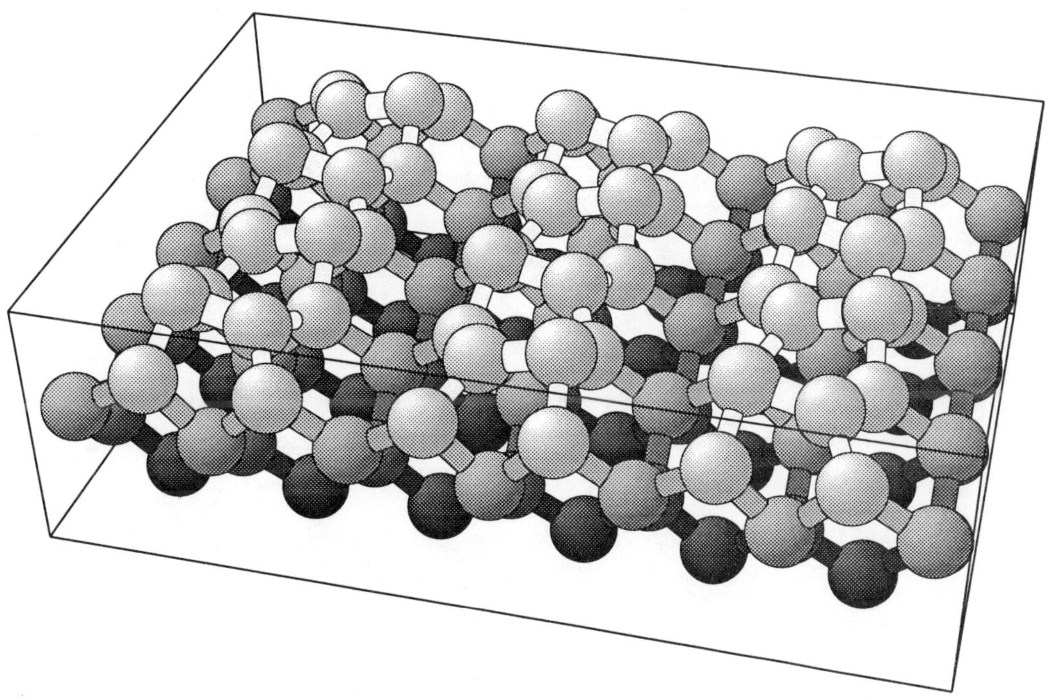

Fig. 95b : Si(100)-c(4x2) (perspective)

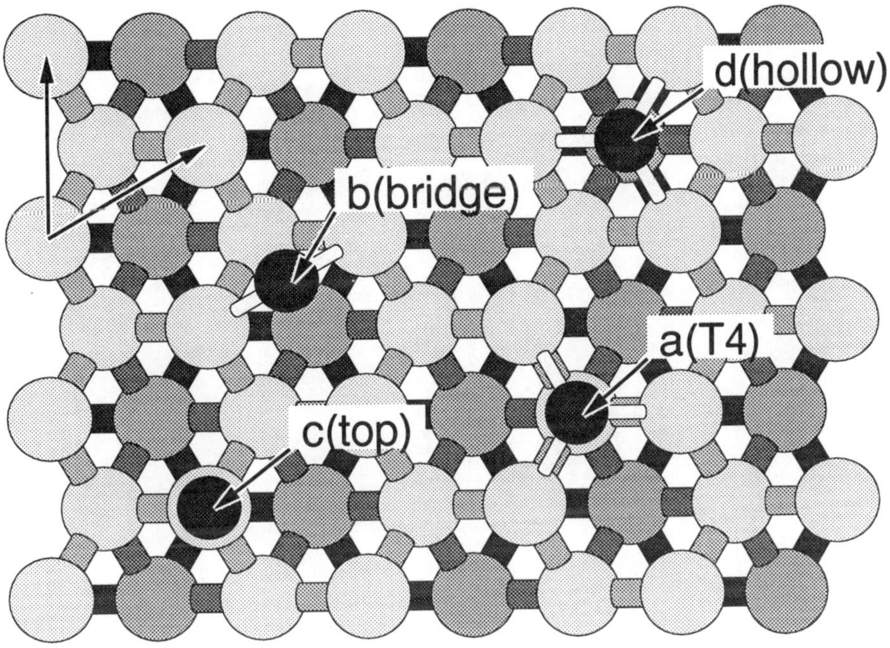

Fig. 96a : diamond(111) high symmetry adsorbate sites (top view)

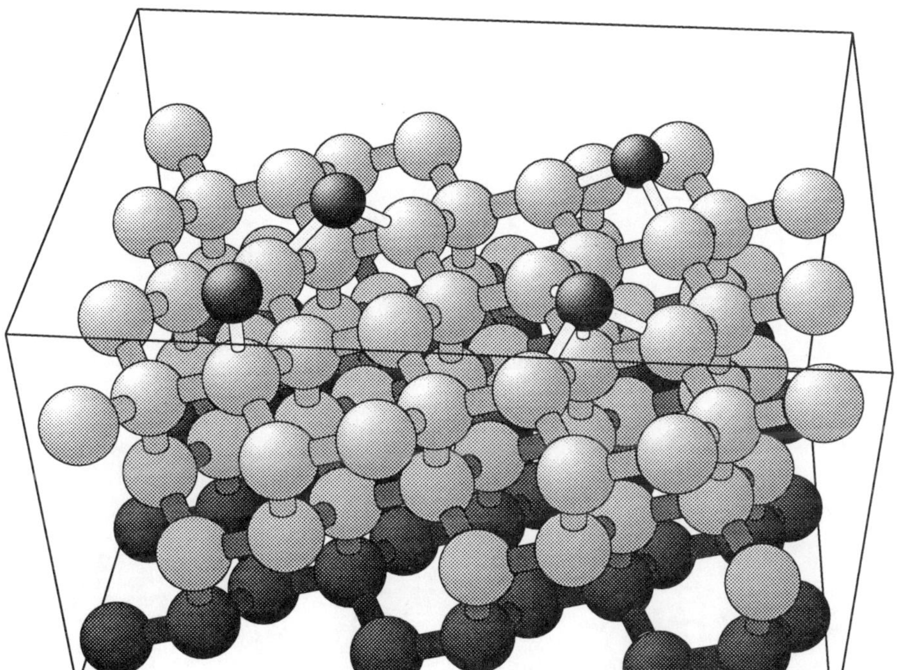

Fig. 96b : diamond(111) high symmetry adsorbate sites (perspective)

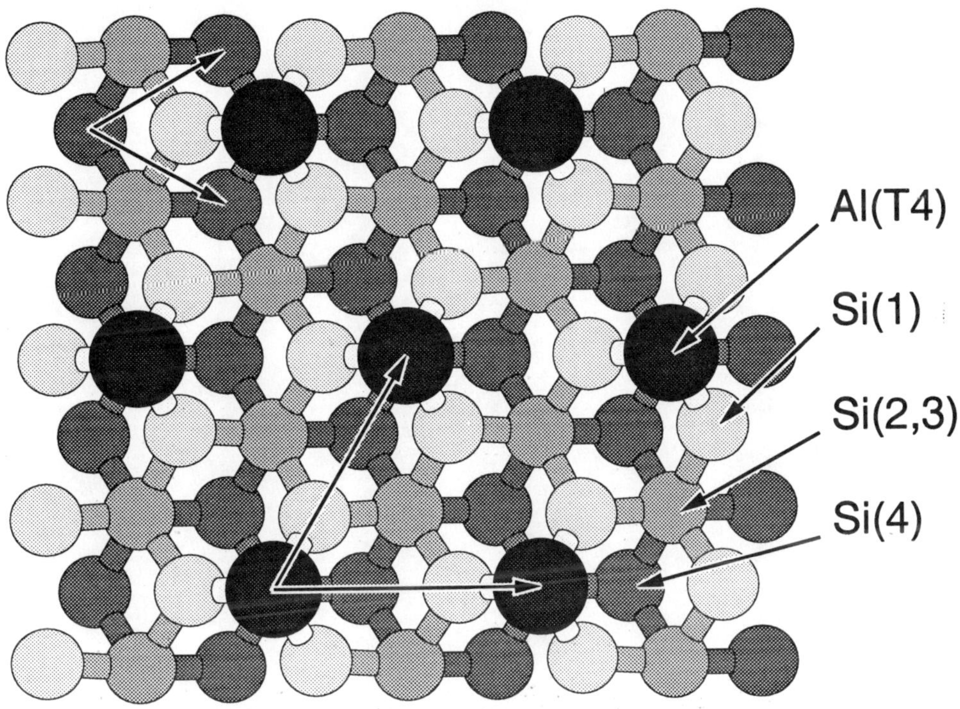

Fig. 97a : Si(111)-(√3x√3)R30°-Al (top view)

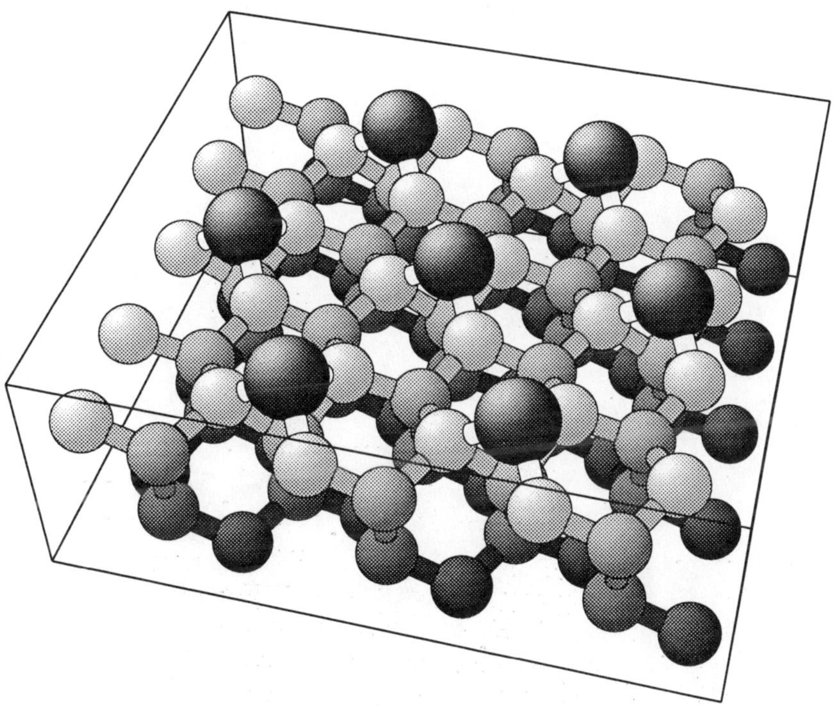

Fig. 97b : Si(111)-(√3x√3)R30°-Al (perspective)

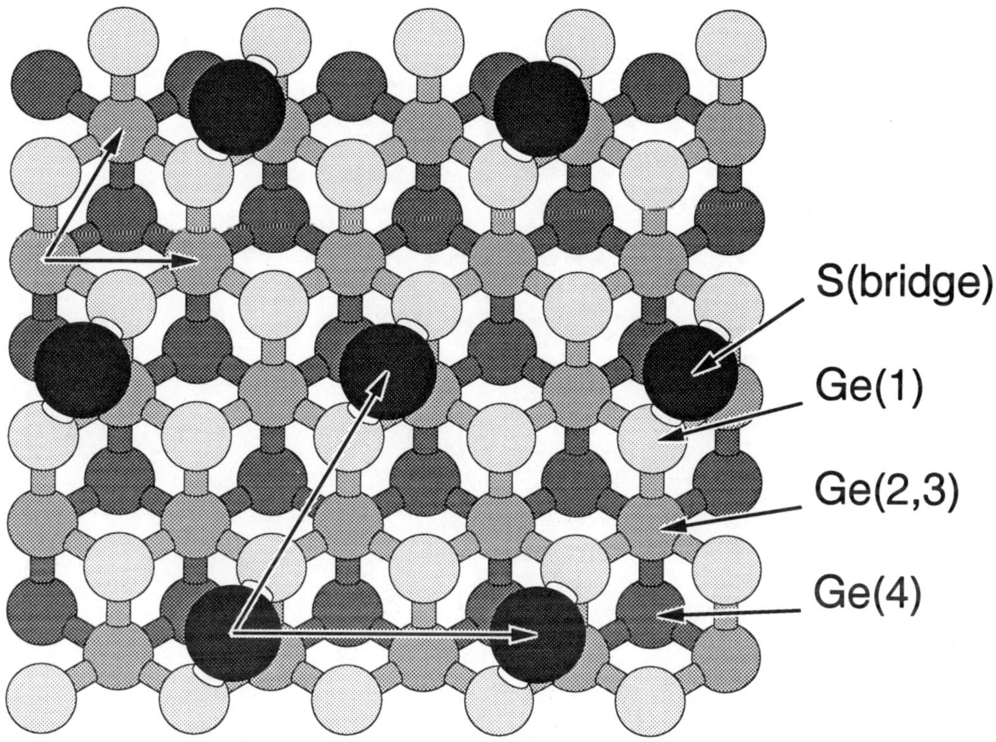

Fig. 98a : Ge(111)-(2x2)-S (top view)

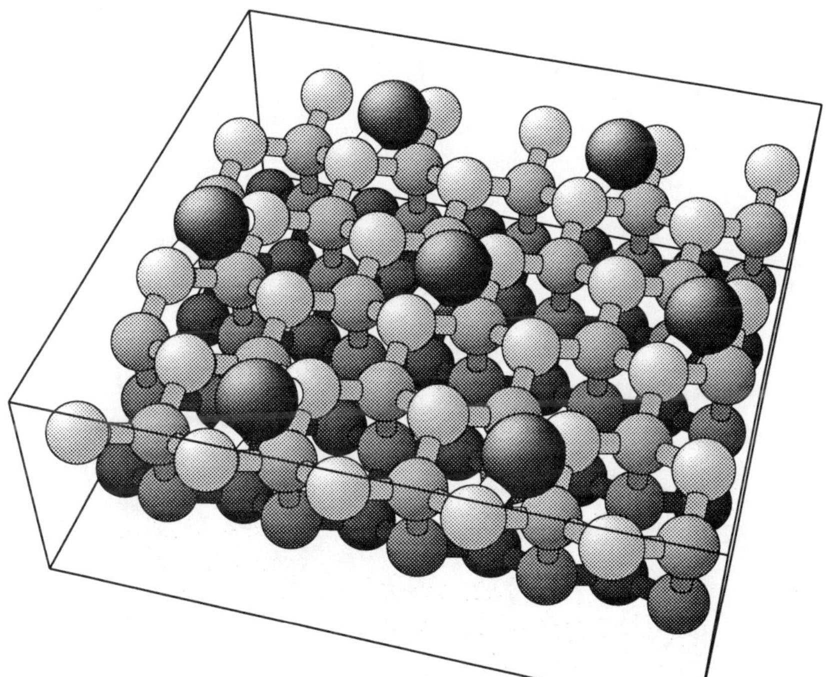

Fig. 98b : Ge(111)-(2x2)-S (perspective)

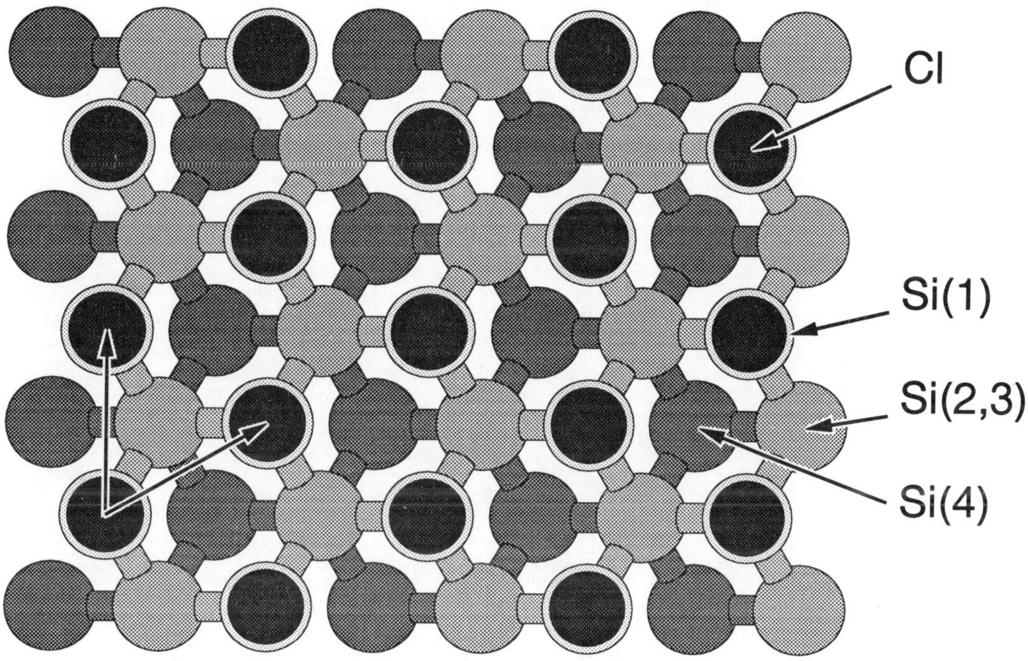

Fig. 99a : Si(111)-(1x1)-Cl (top view)

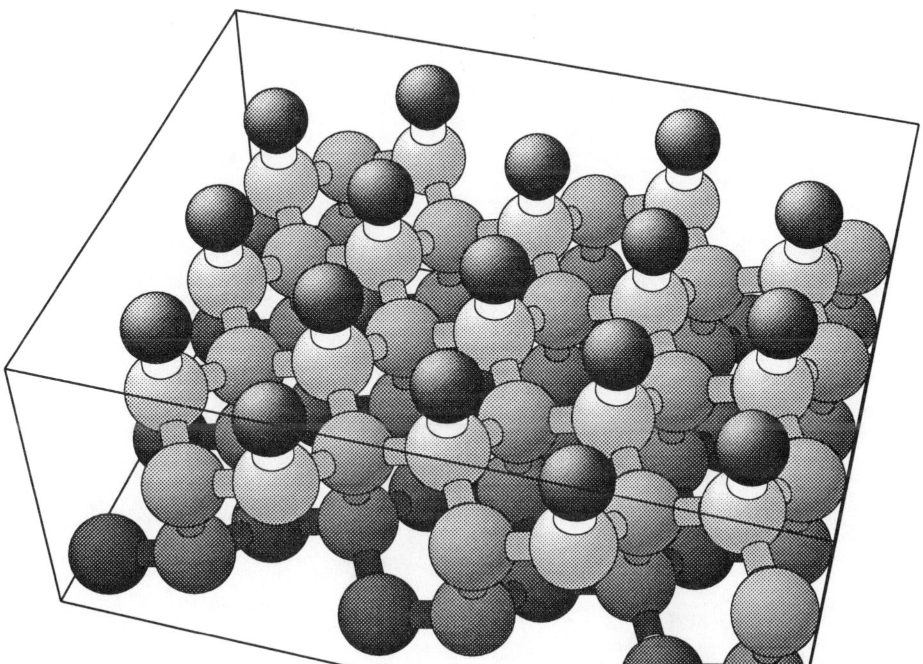

Fig. 99b : Si(111)-(1x1)-Cl (perspective)

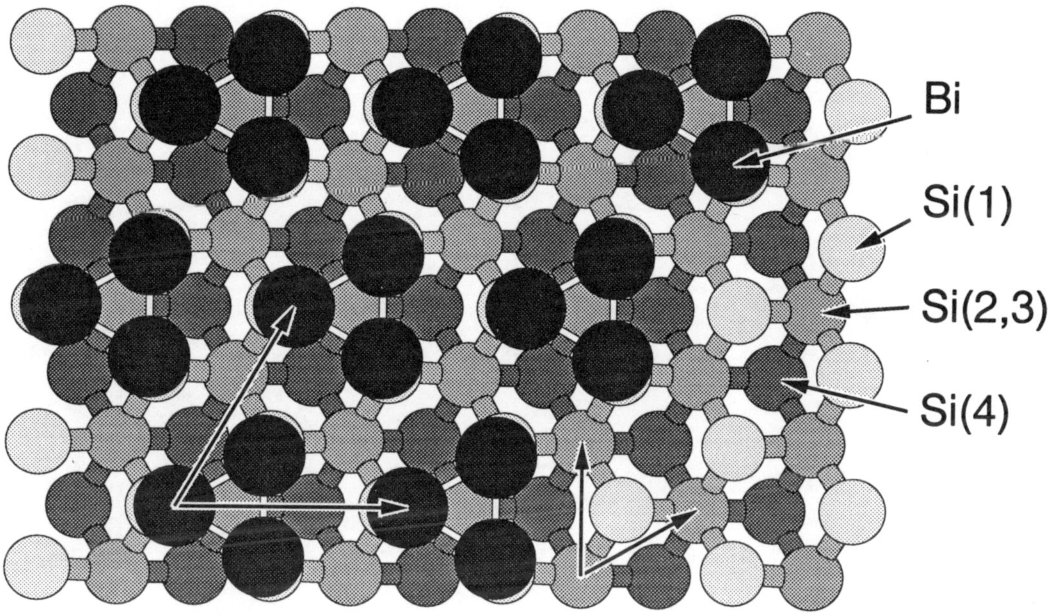

Fig. 100a : Si(111)-($\sqrt{3}$x$\sqrt{3}$)R30°-Bi (1 ML) (top view)

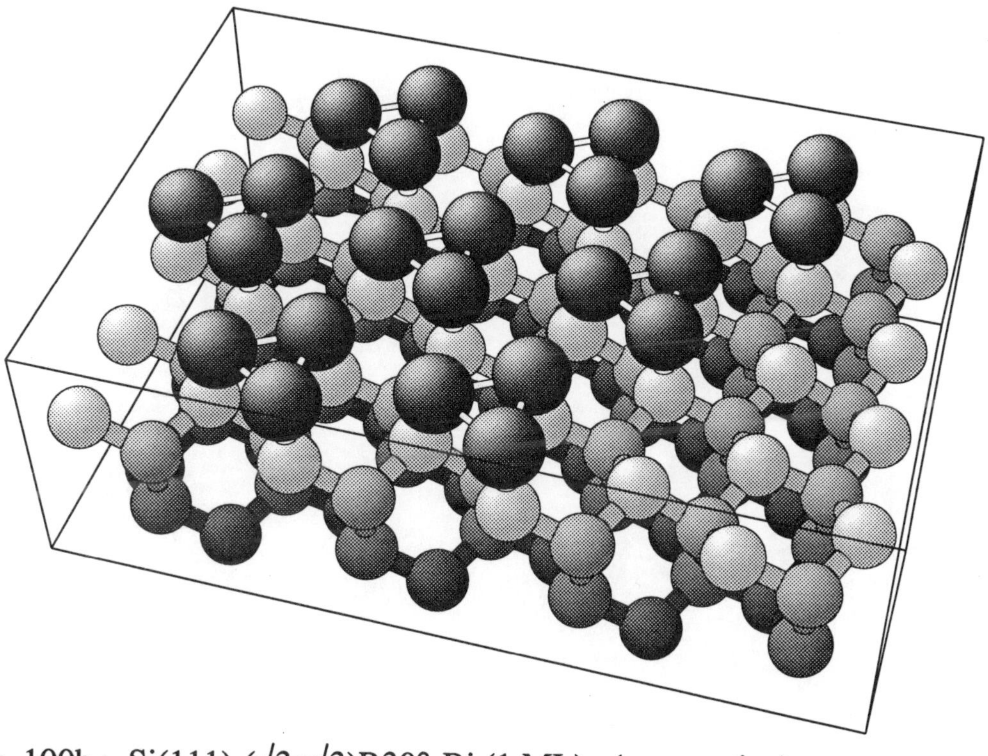

Fig. 100b : Si(111)-($\sqrt{3}$x$\sqrt{3}$)R30°-Bi (1 ML) (perspective)

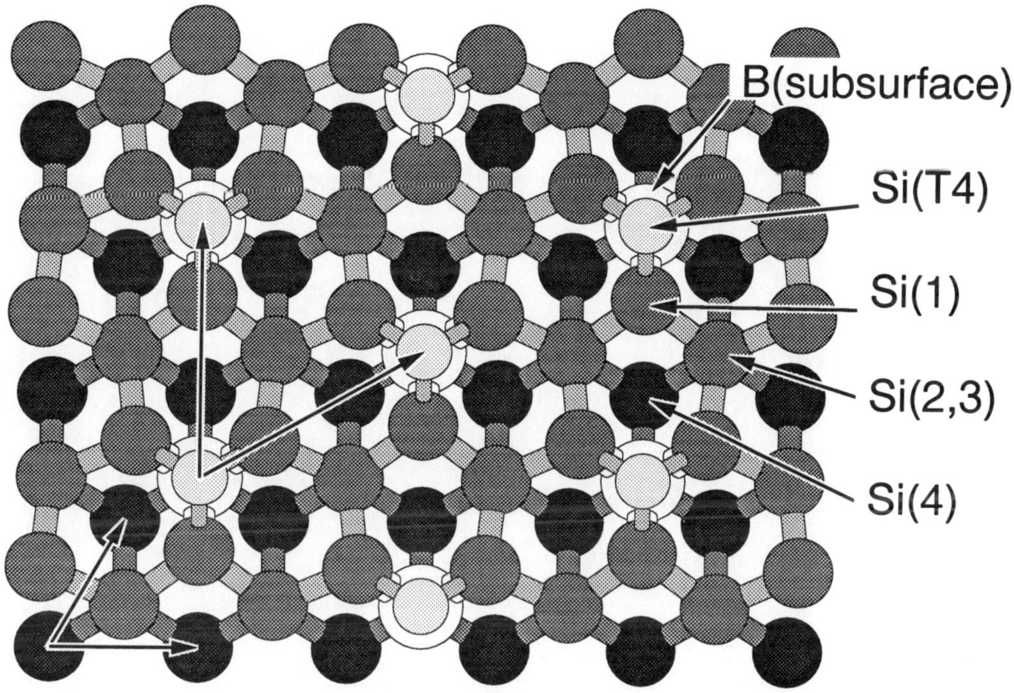

Fig. 101a : Si(111)-(√3x√3)R30°-B (top view)

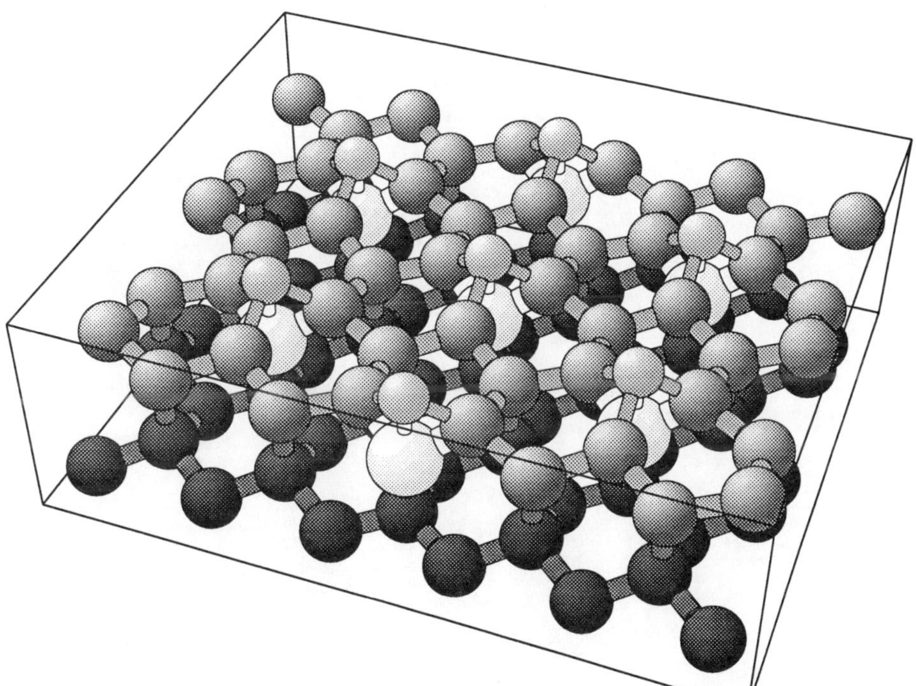

Fig. 101b : Si(111)-(√3x√3)R30°-B (perspective)

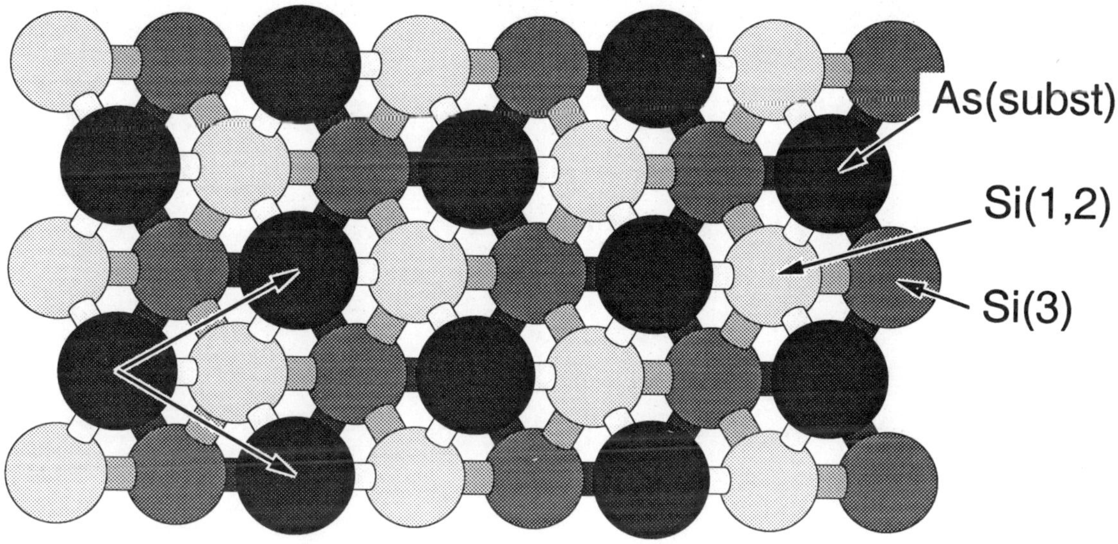

Fig. 102a : Si(111)-(1x1)-As (top view)

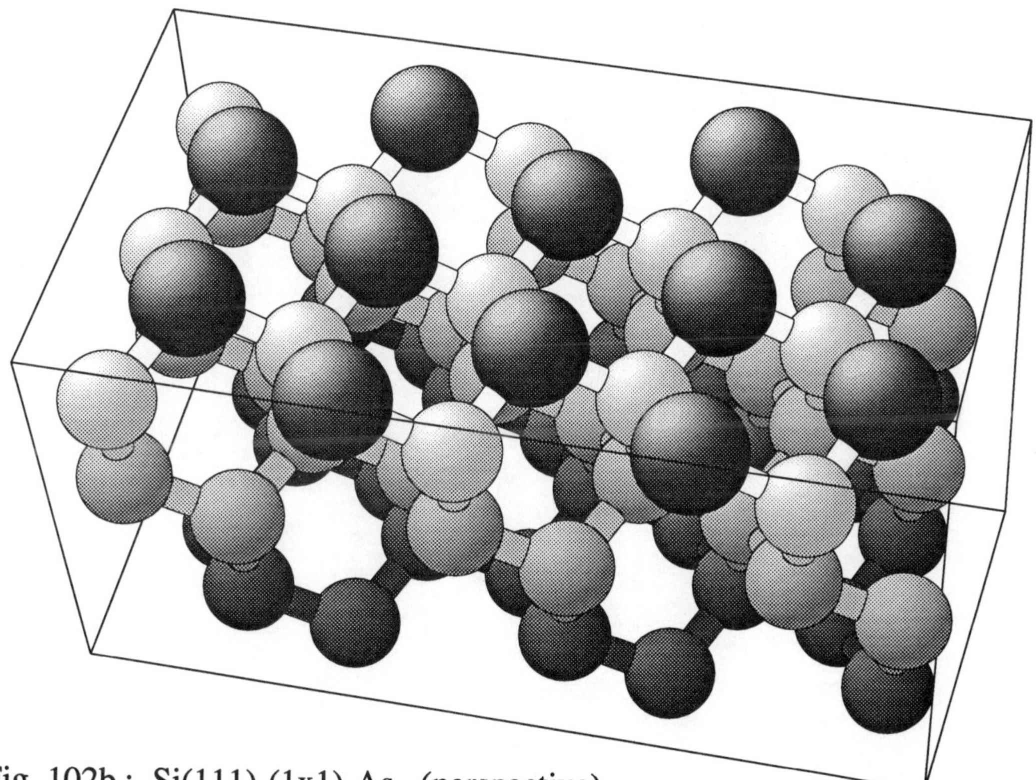

Fig. 102b : Si(111)-(1x1)-As (perspective)

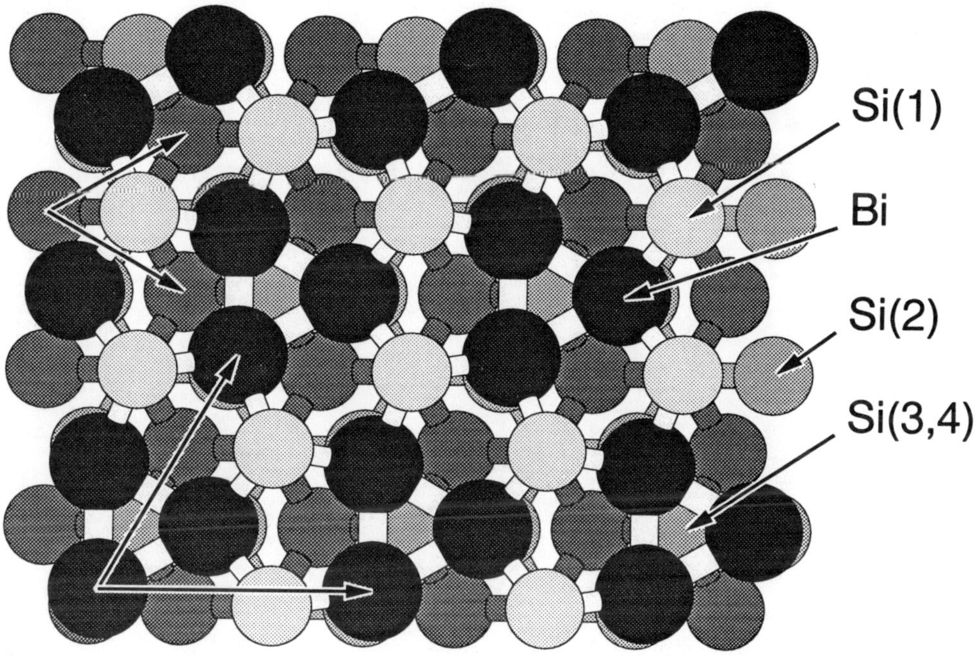

Fig. 103a : Si(111)-(√3x√3)R30°-Bi (top view)

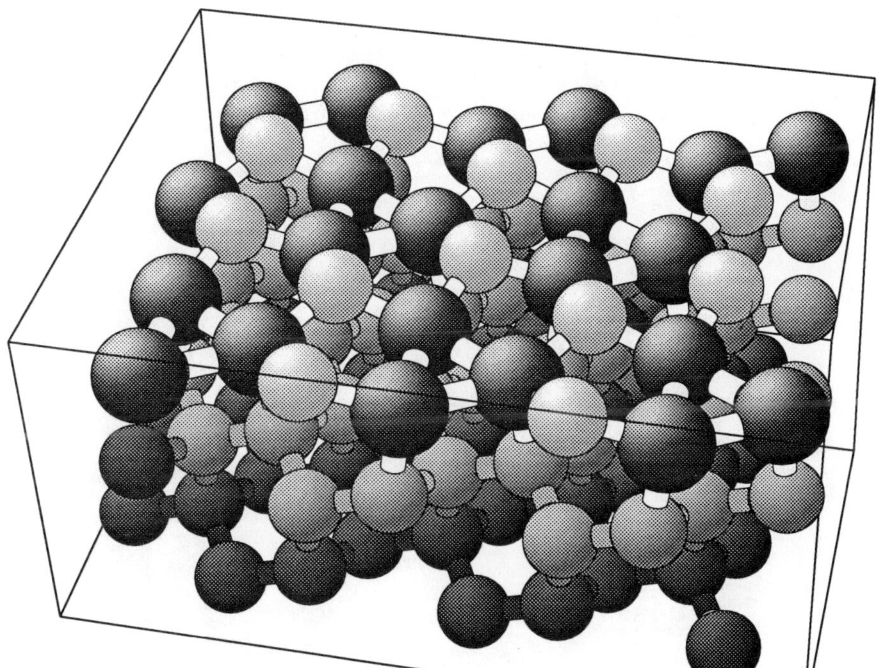

Fig. 103b : Si(111)-(√3x√3)R30°-Bi (perspective)

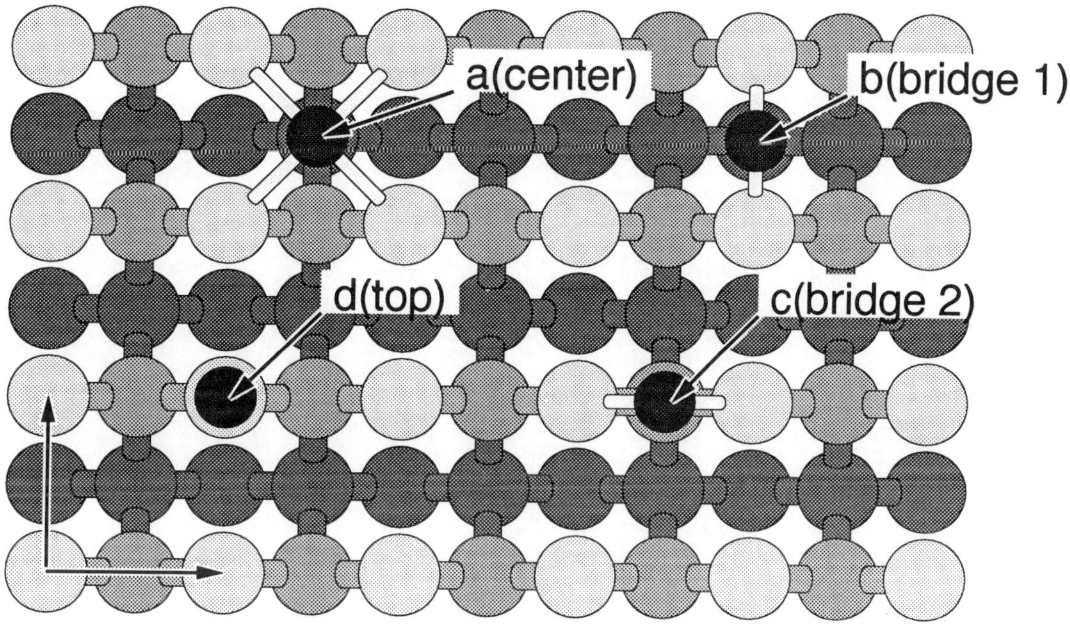

Fig. 104a : diamond(100) high symmetry adsorbate sites (top view)

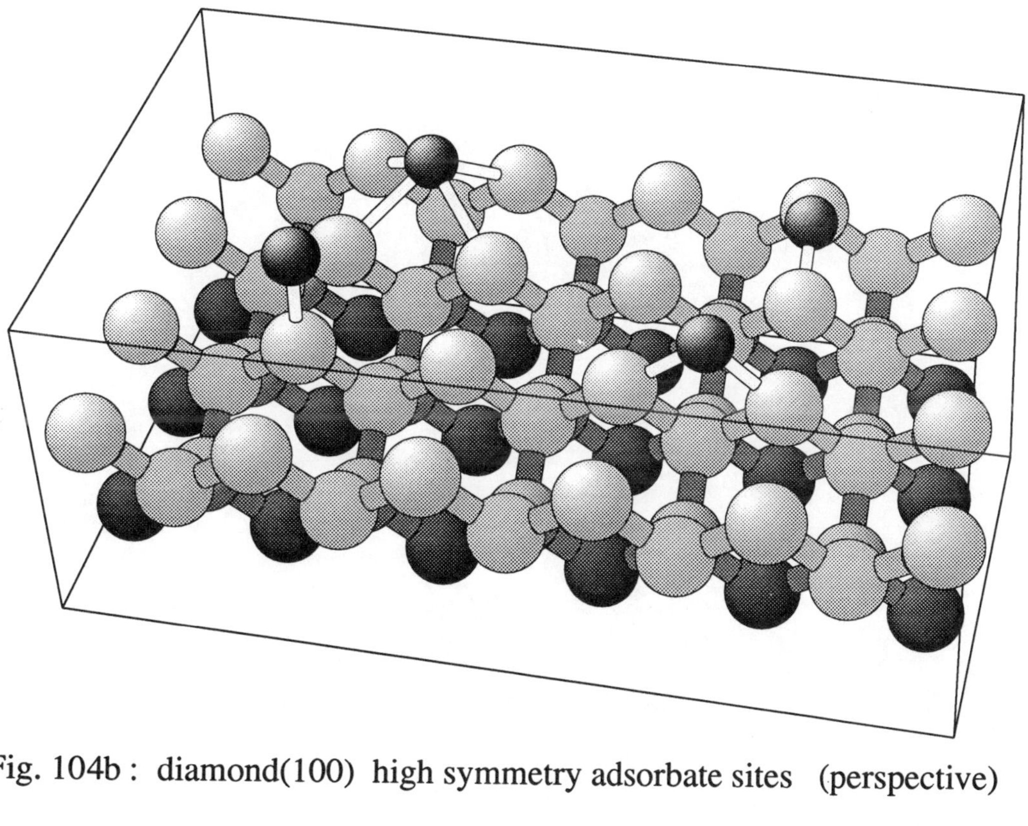

Fig. 104b : diamond(100) high symmetry adsorbate sites (perspective)

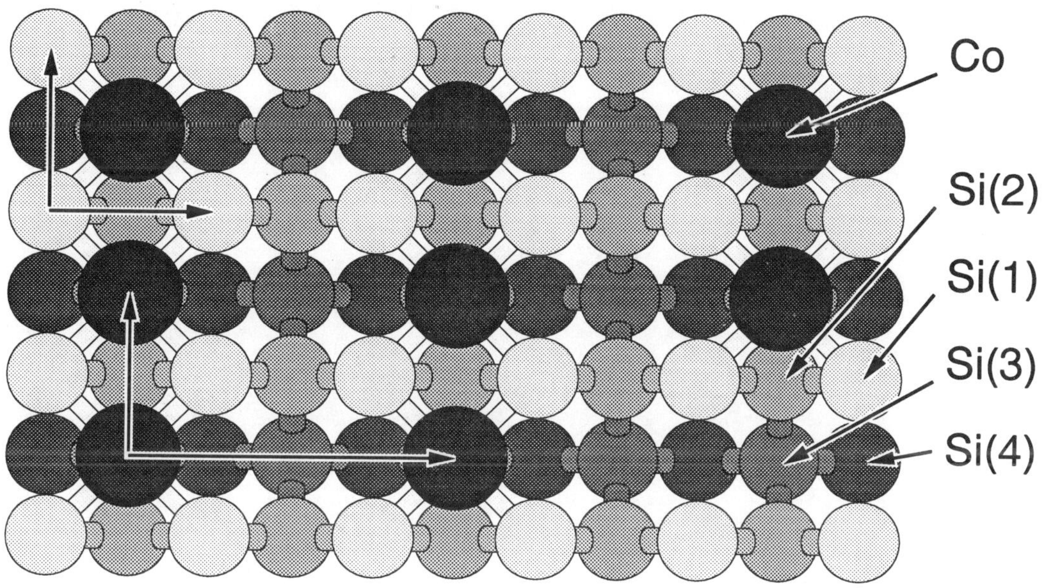

Fig. 105a : Si(100)-Co (0.4ML) (top view)

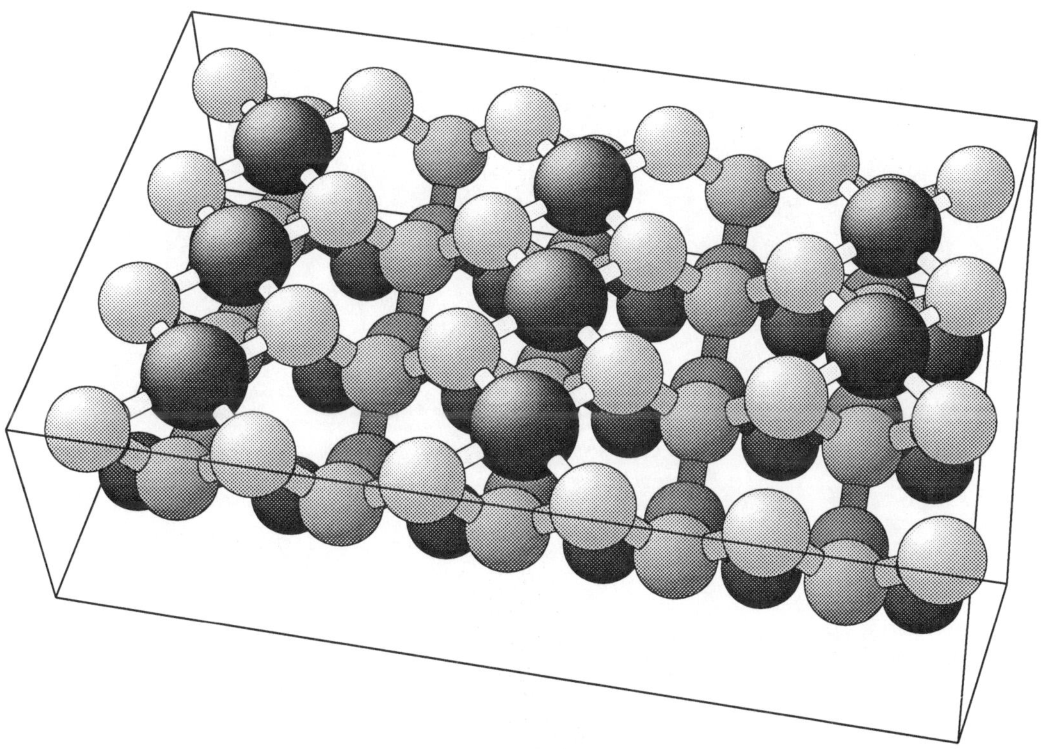

Fig. 105b : Si(100)-Co (0.4ML) (perspective)

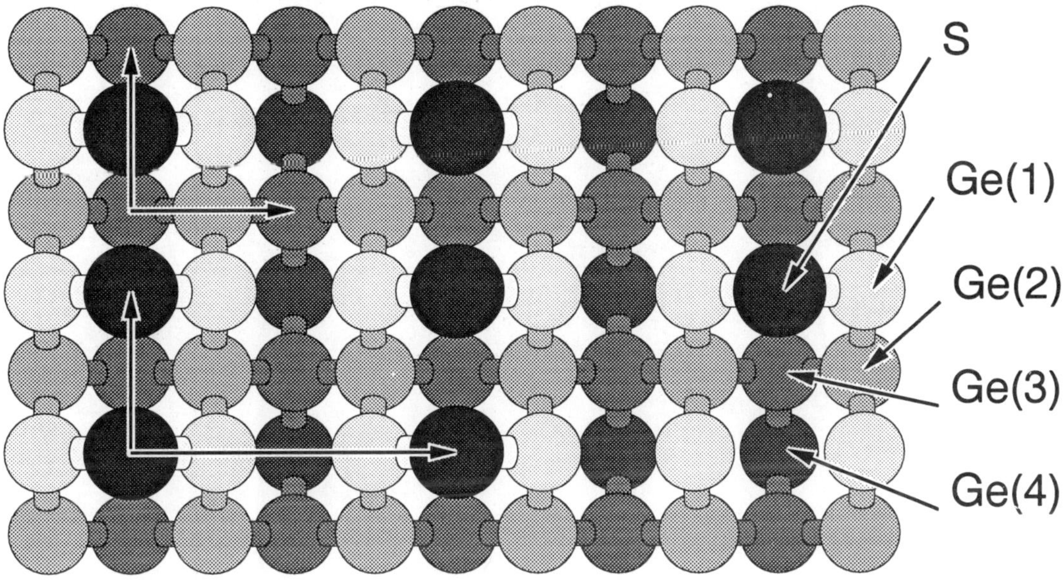

Fig. 106a : Ge(100)-(2x1)-S (top view)

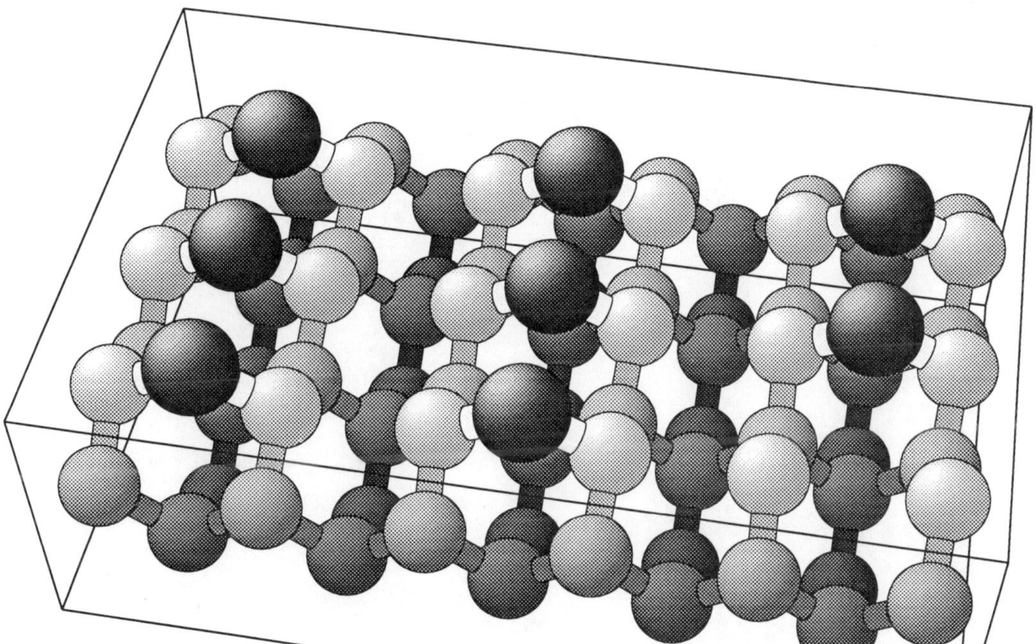

Fig. 106b : Ge(100)-(2x1)-S (perspective)

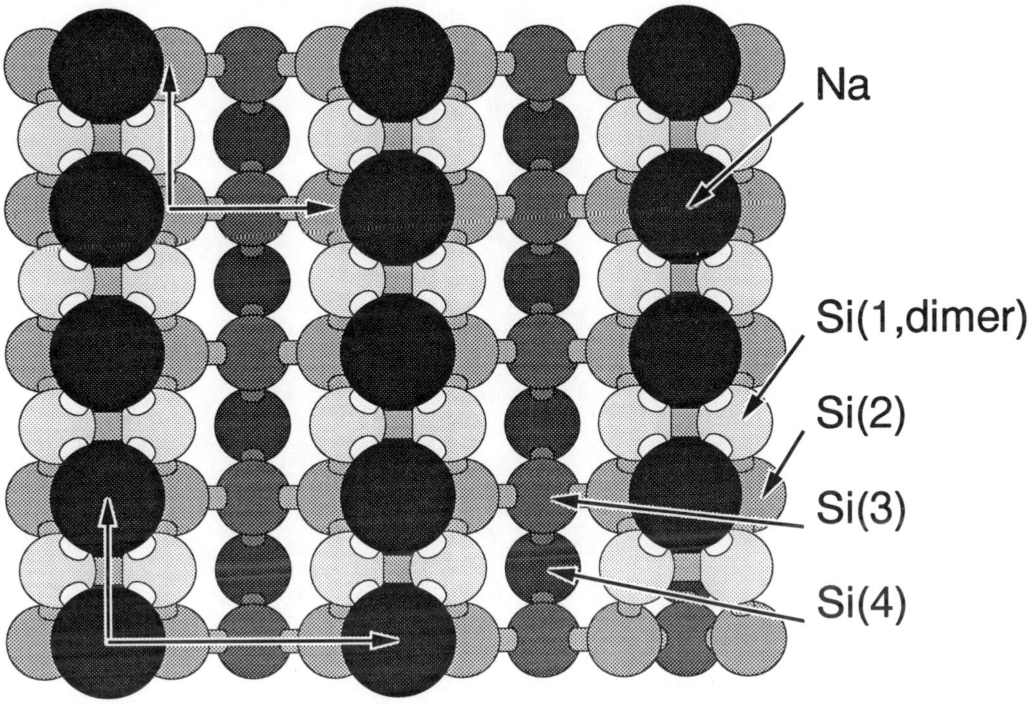

Fig. 107a : Si(100)-(2x1)-Na (top view)

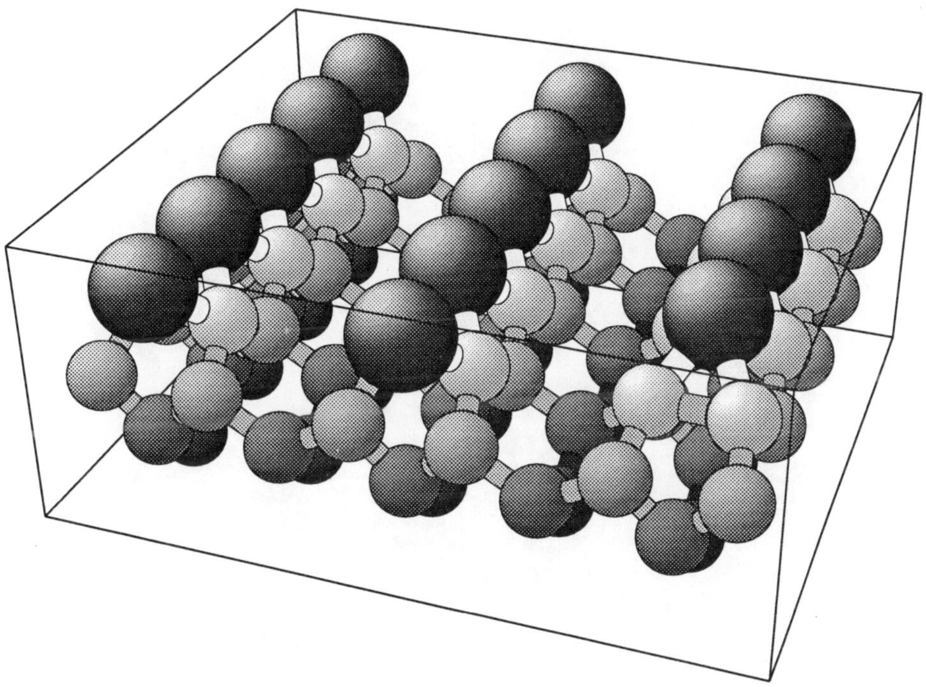

Fig. 107b : Si(100)-(2x1)-Na (perspective)

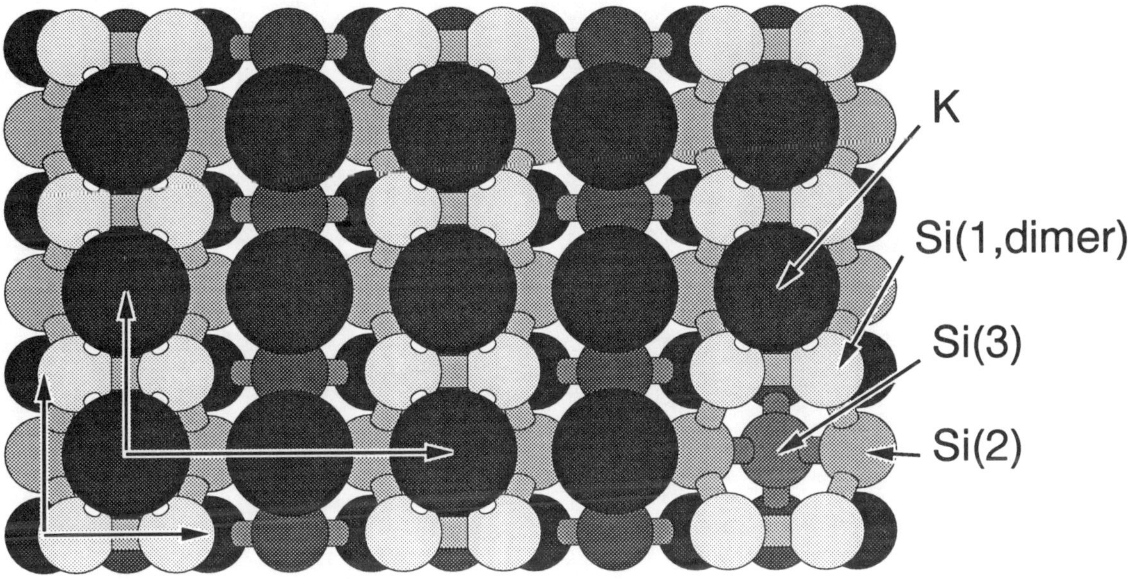

Fig. 108a : Si(100)-(2x1)-2K (top view)

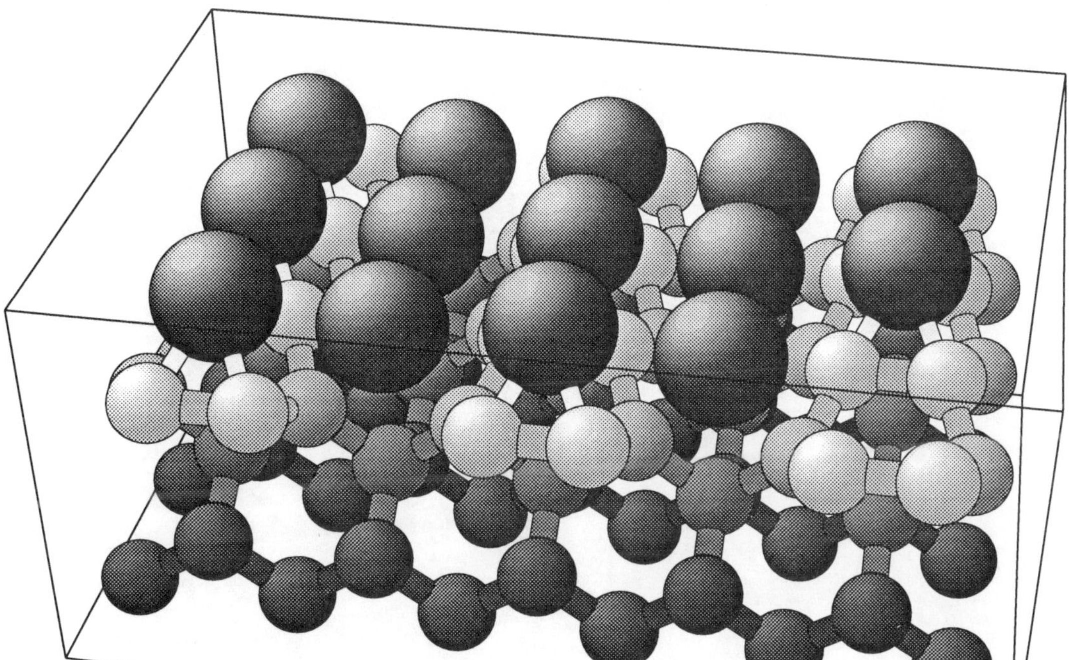

Fig. 108b : Si(100)-(2x1)-2K (perspective)

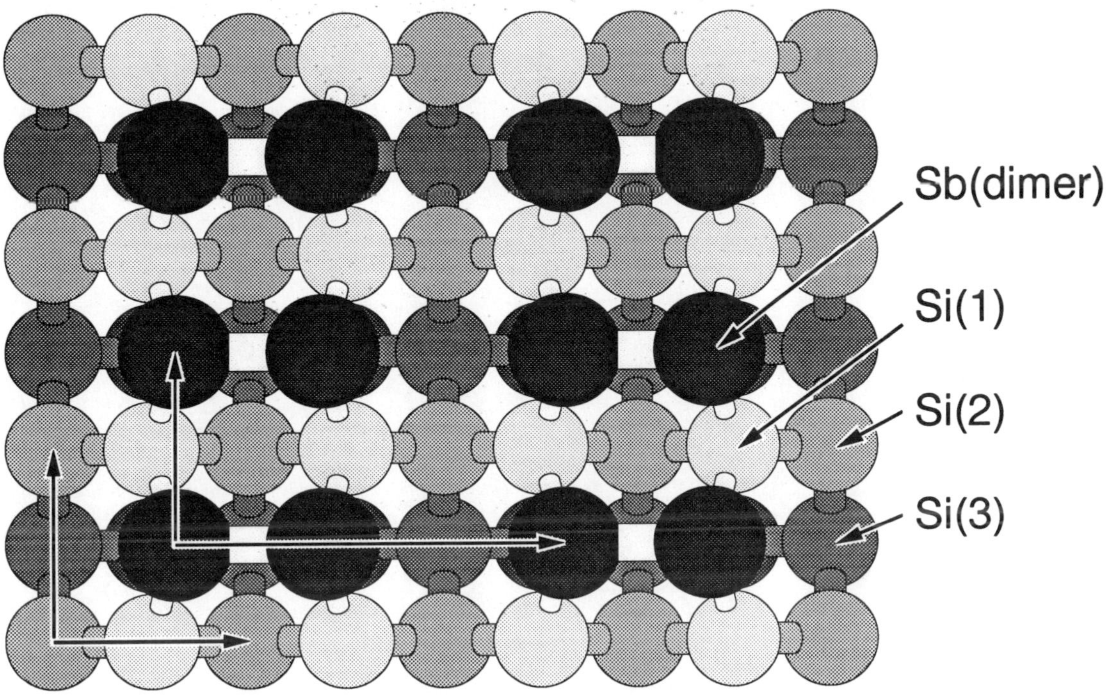

Fig. 109a : Si(100)-(2x1)-2Sb (top view)

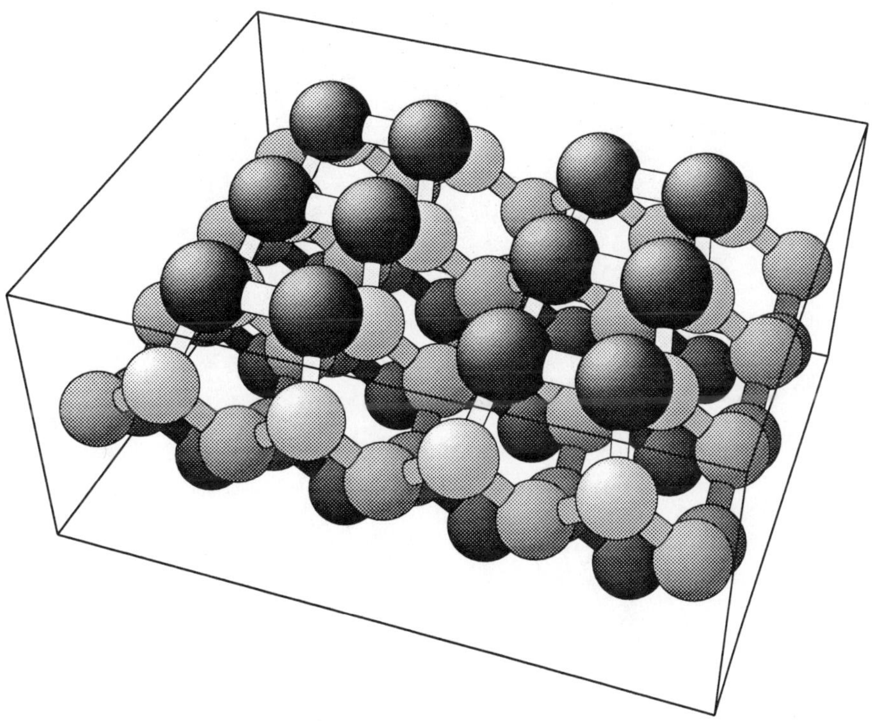

Fig. 109b : Si(100)-(2x1)-2Sb (perspective)

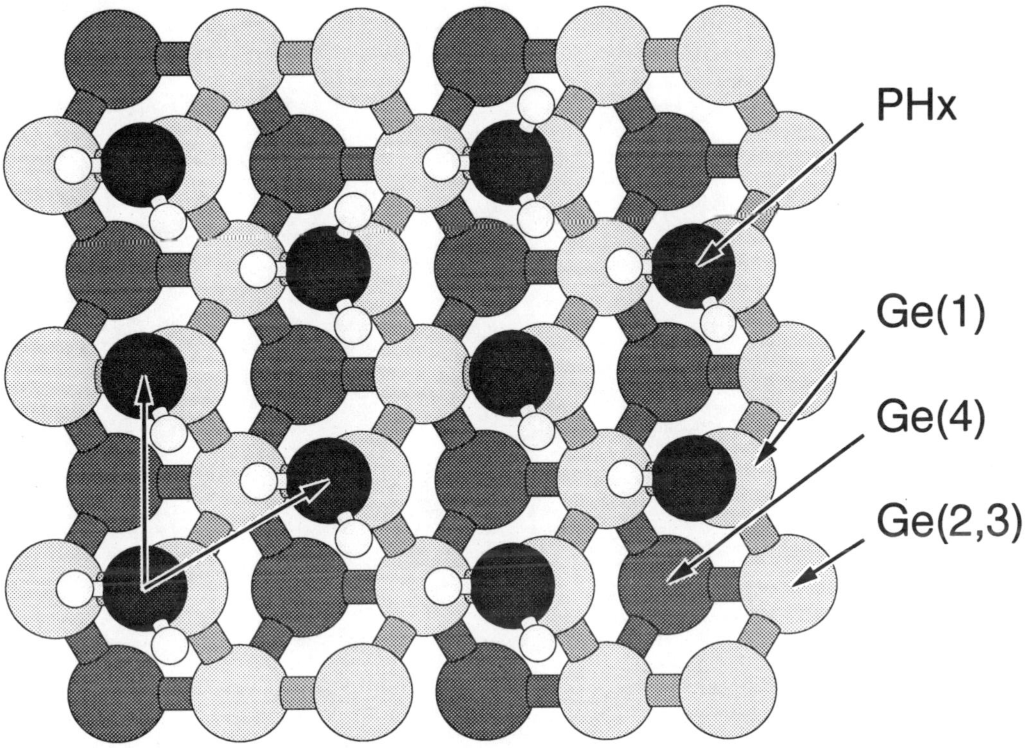

Fig. 110a : Ge(111)-(1x1)-PHx (top view)

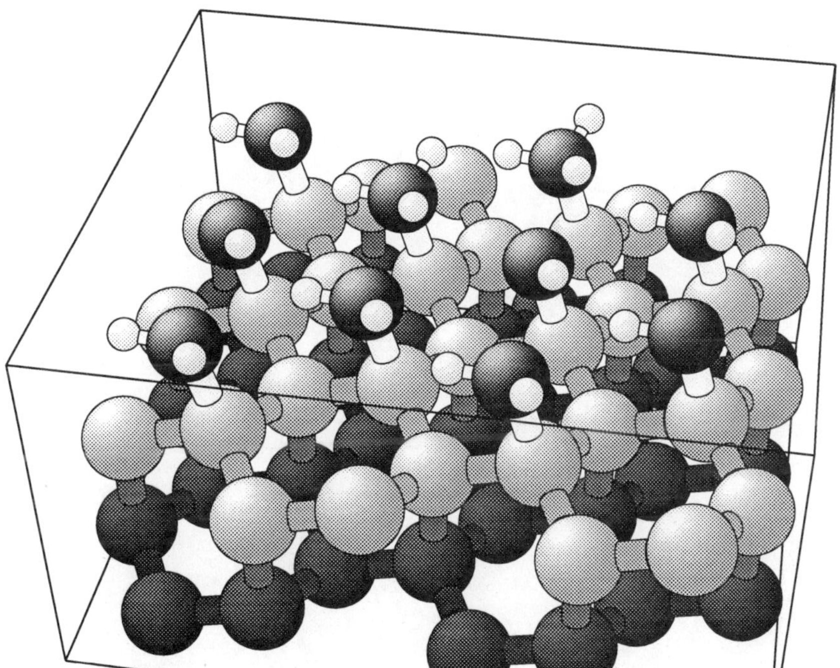

Fig. 110b : Ge(111)-(1x1)-PHx (perspective)

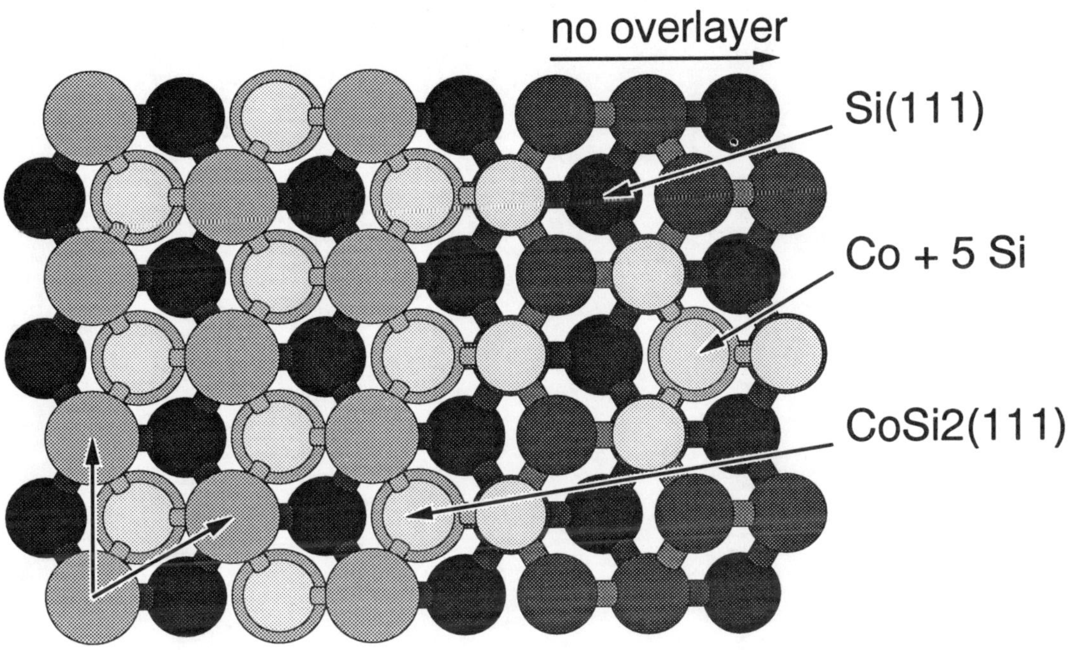

Fig. 111a : Si(111)-(1x1)-CoSi2(111) interface I (top view)

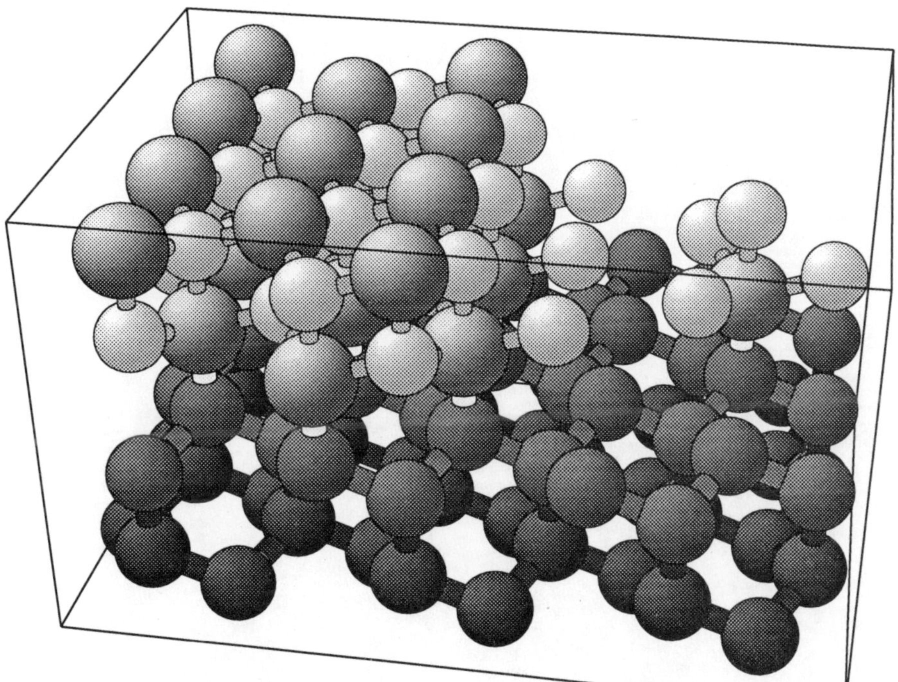

Fig. 111b : Si(111)-(1x1)-CoSi2(111) interface I (perspective)

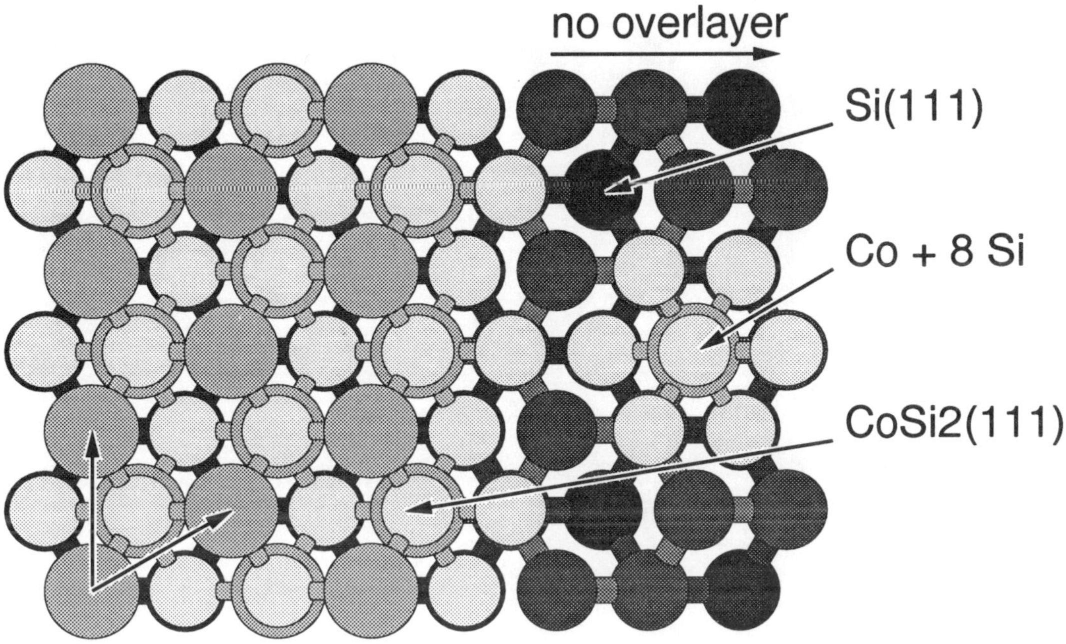

Fig. 112a : Si(111)-(1x1)-CoSi2(111) interface II (top view)

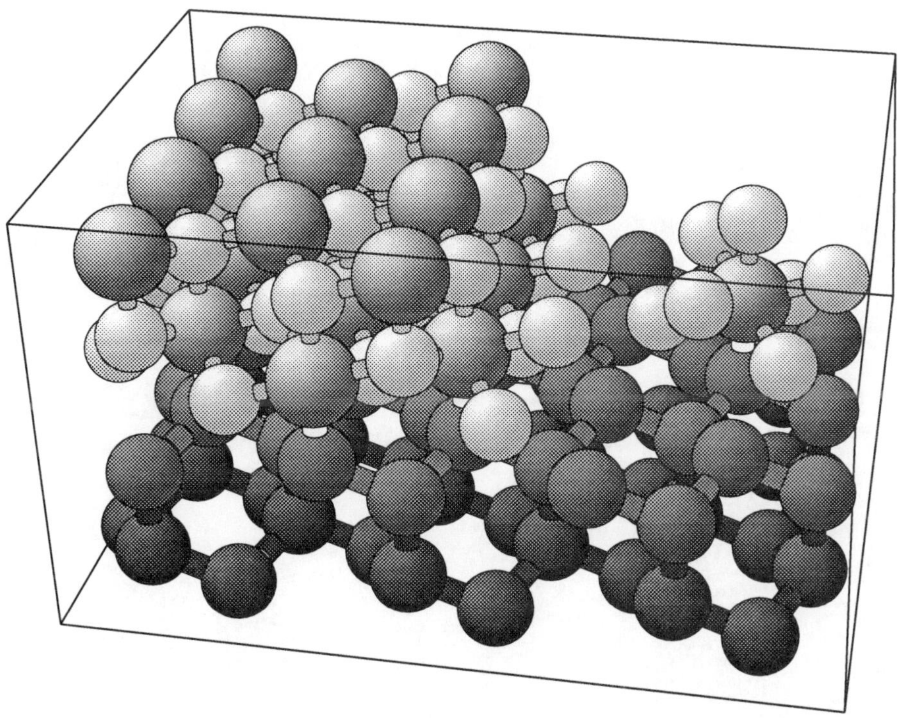

Fig. 112b : Si(111)-(1x1)-CoSi2(111) interface II (perspective)

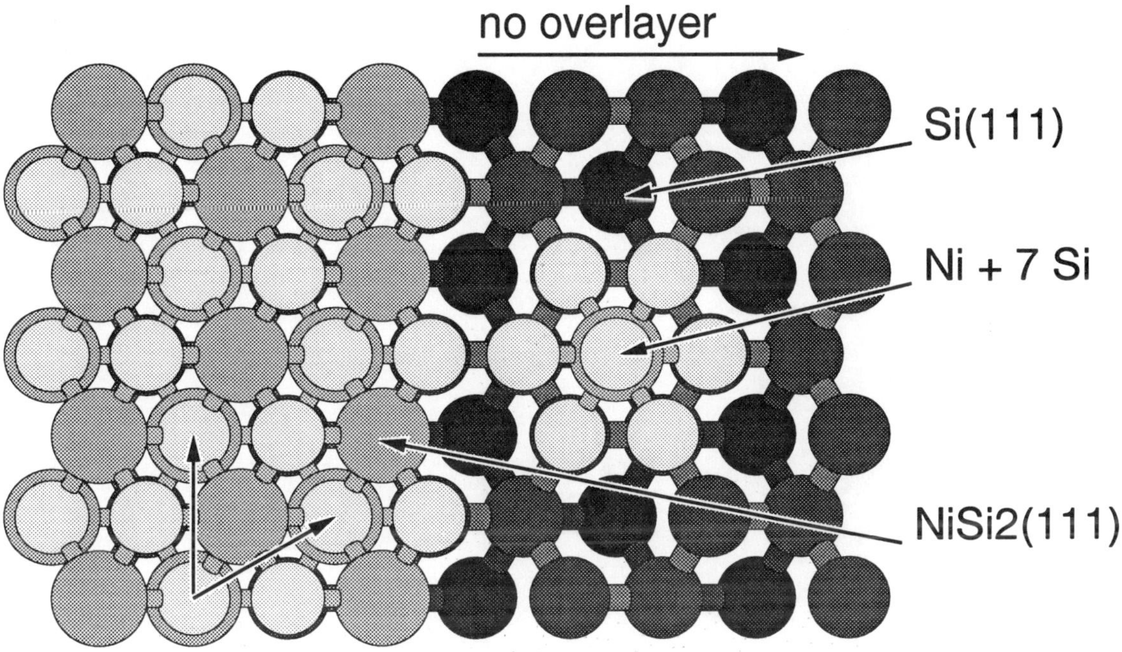

Fig. 113a : Si(111)-(1x1)-NiSi2(111) interface I (top view)

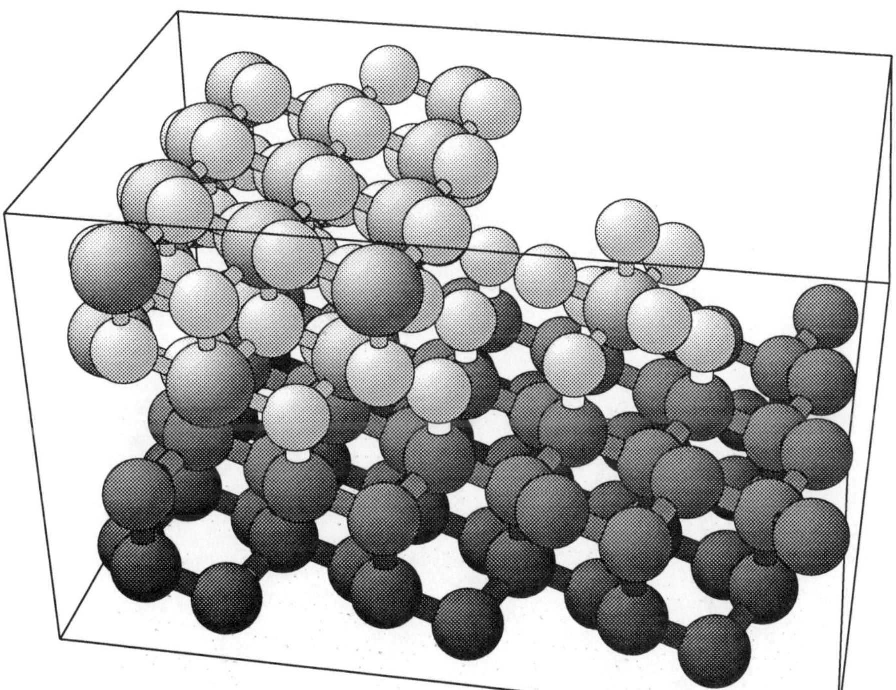

Fig. 113b : Si(111)-(1x1)-NiSi2(111) interface I (perspective)

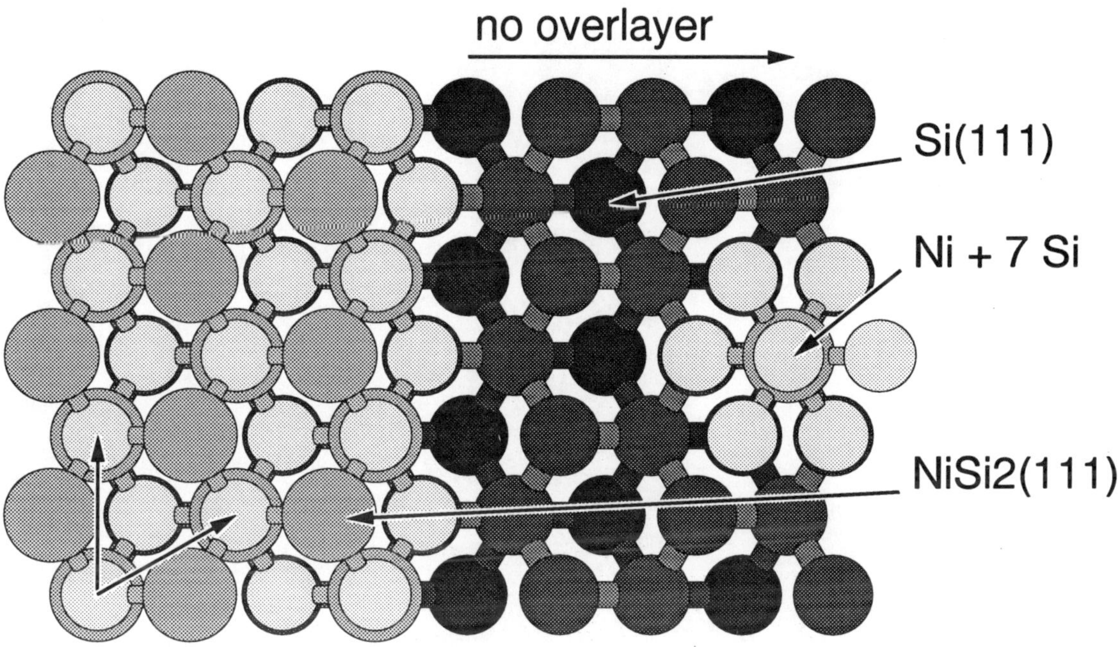

Fig. 114a : Si(111)-(1x1)-NiSi2(111) interface II (top view)

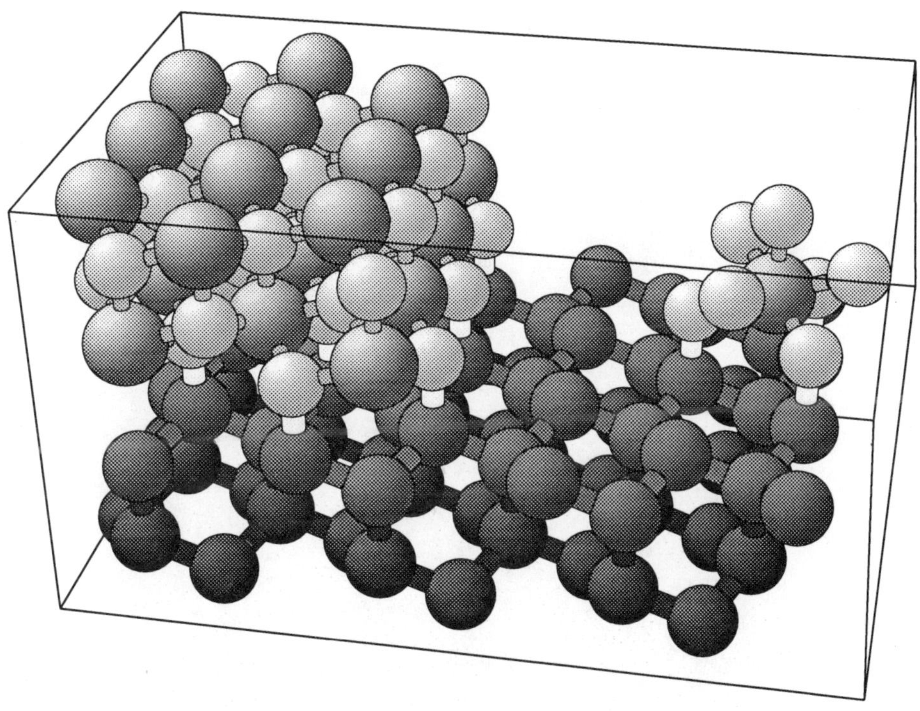

Fig. 114b : Si(111)-(1x1)-NiSi2(111) interface II (perspective)

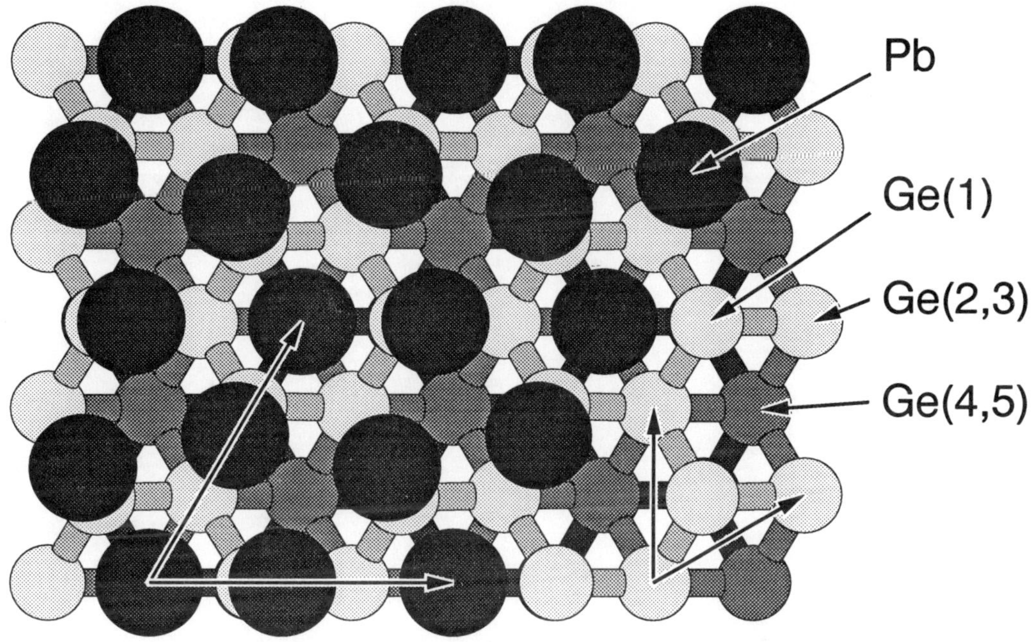

Fig. 115a : Ge(111)-(√3x√3)R30°-4Pb (4/3 ML) (top view)

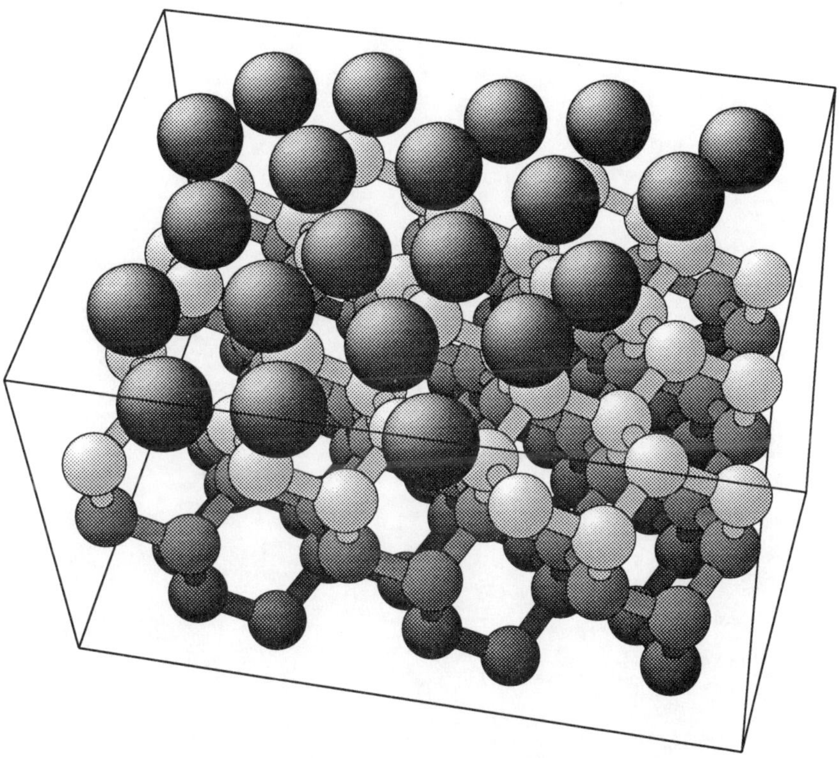

Fig. 115b : Ge(111)-(√3x√3)R30°-4Pb (4/3 ML) (perspective)

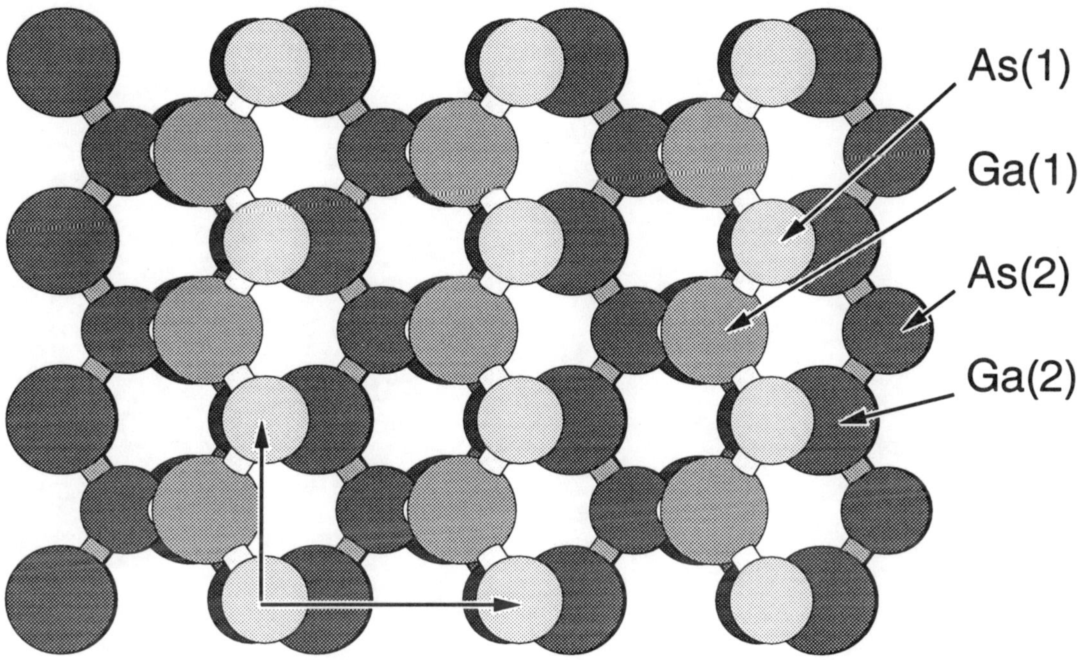

Fig. 116a : GaAs(110)-(1x1) (top view)

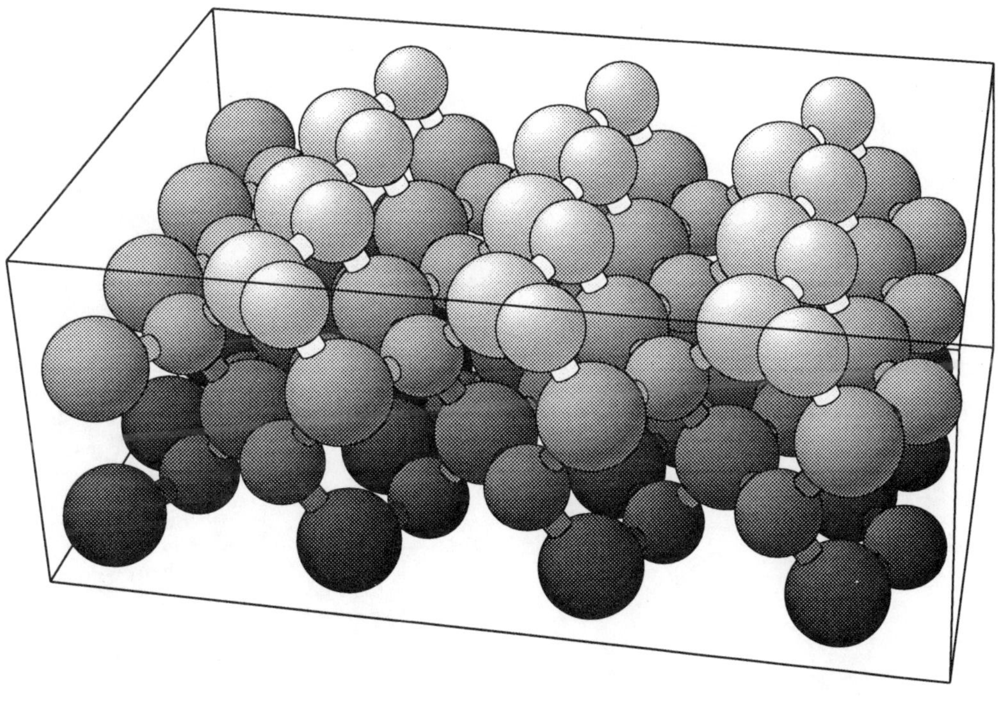

Fig. 116b : GaAs(110)-(1x1) (perspective)

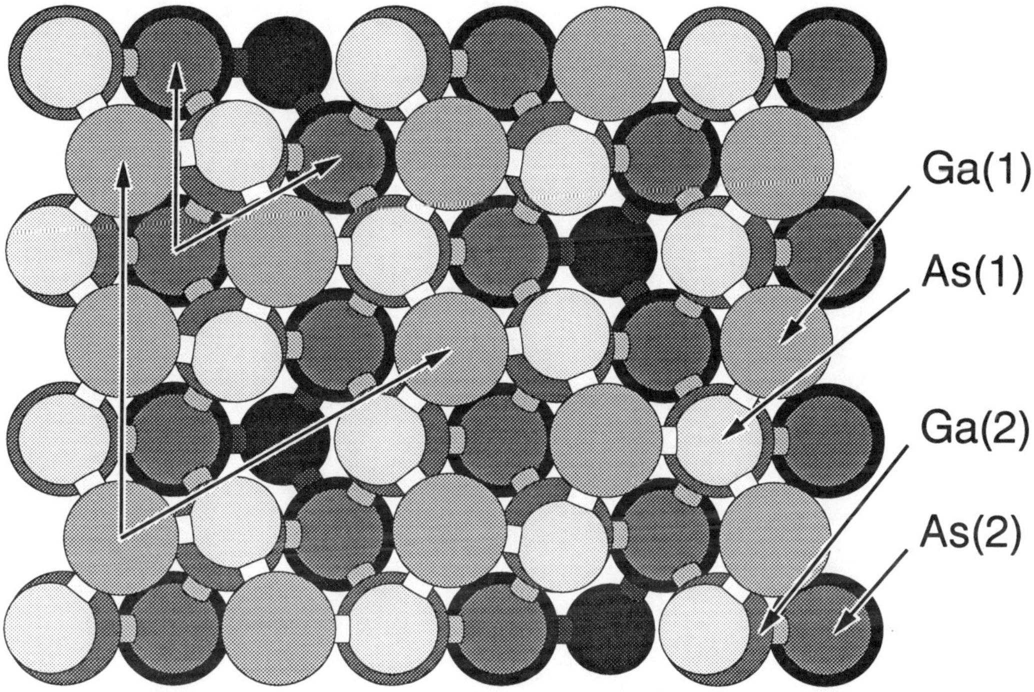

Fig. 117a : GaAs(111)-(2x2) (top view)

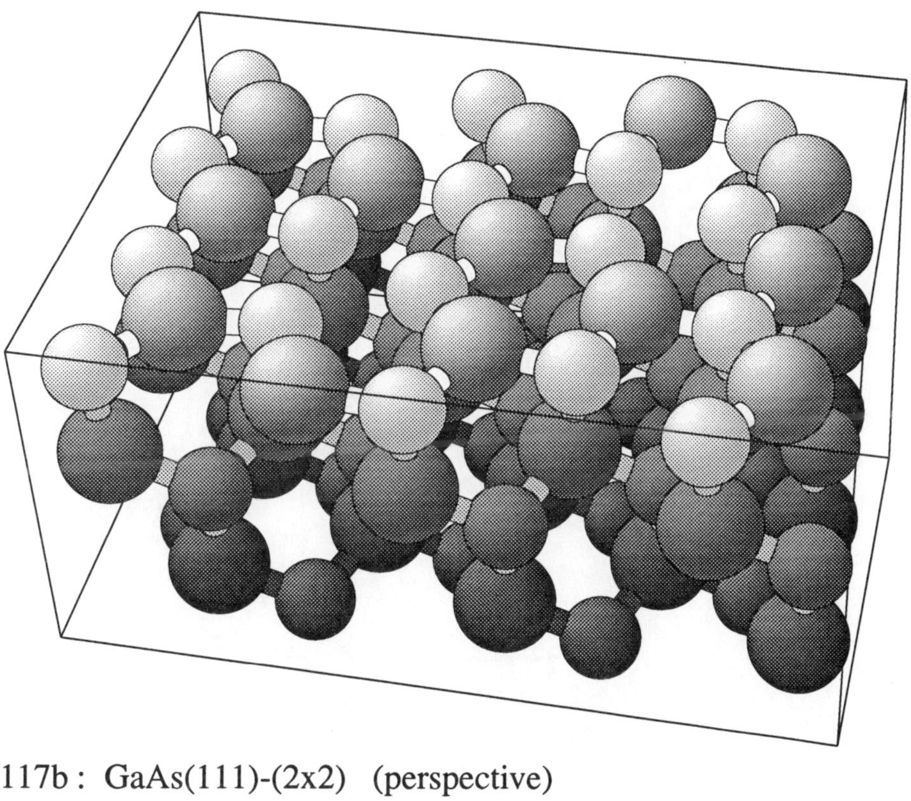

Fig. 117b : GaAs(111)-(2x2) (perspective)

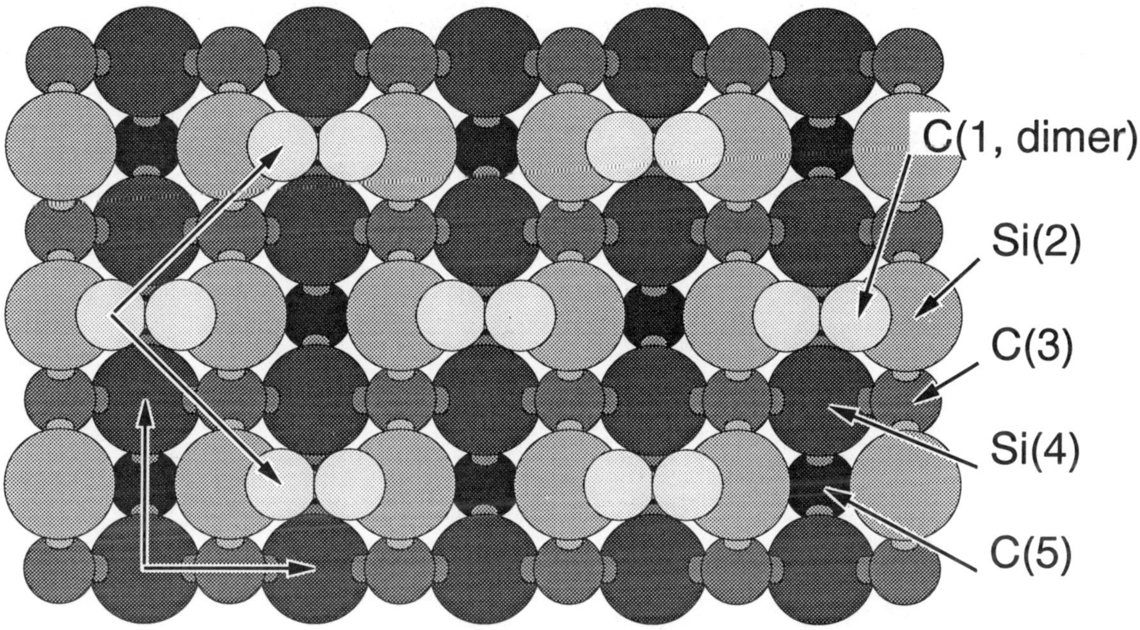

Fig. 118a : SiC(100)-c(2x2) (C2H4 exposed) (top view)

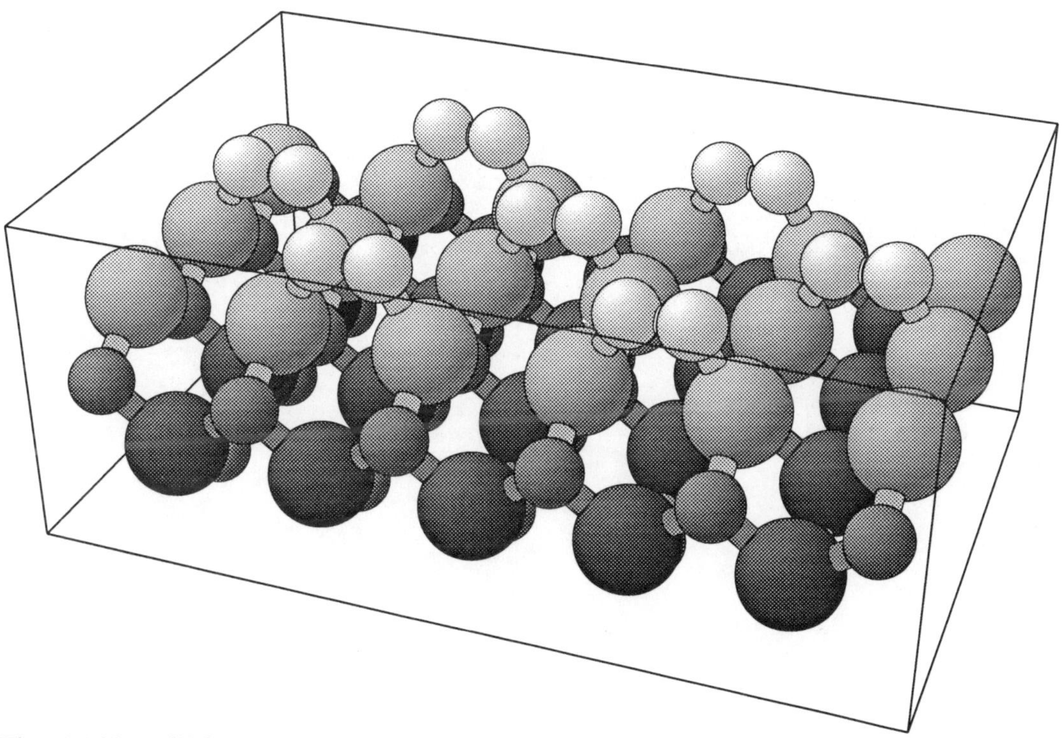

Fig. 118b : SiC(100)-c(2x2) (C2H4 exposed) (perspective)

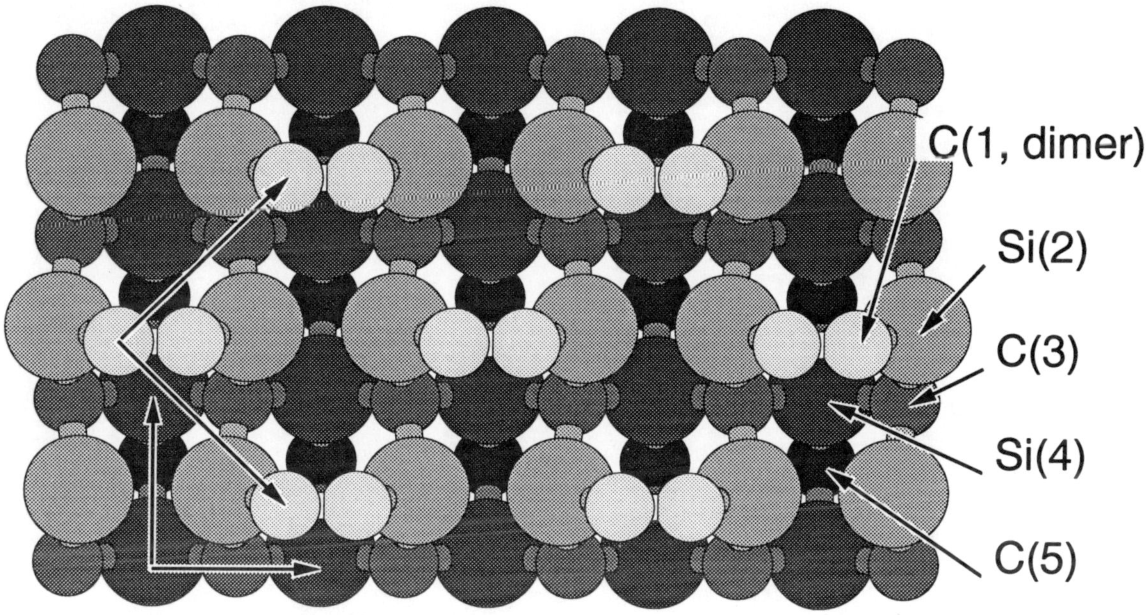

Fig. 119a : SiC(100)-c(2x2) (Si sublimation) (top view)

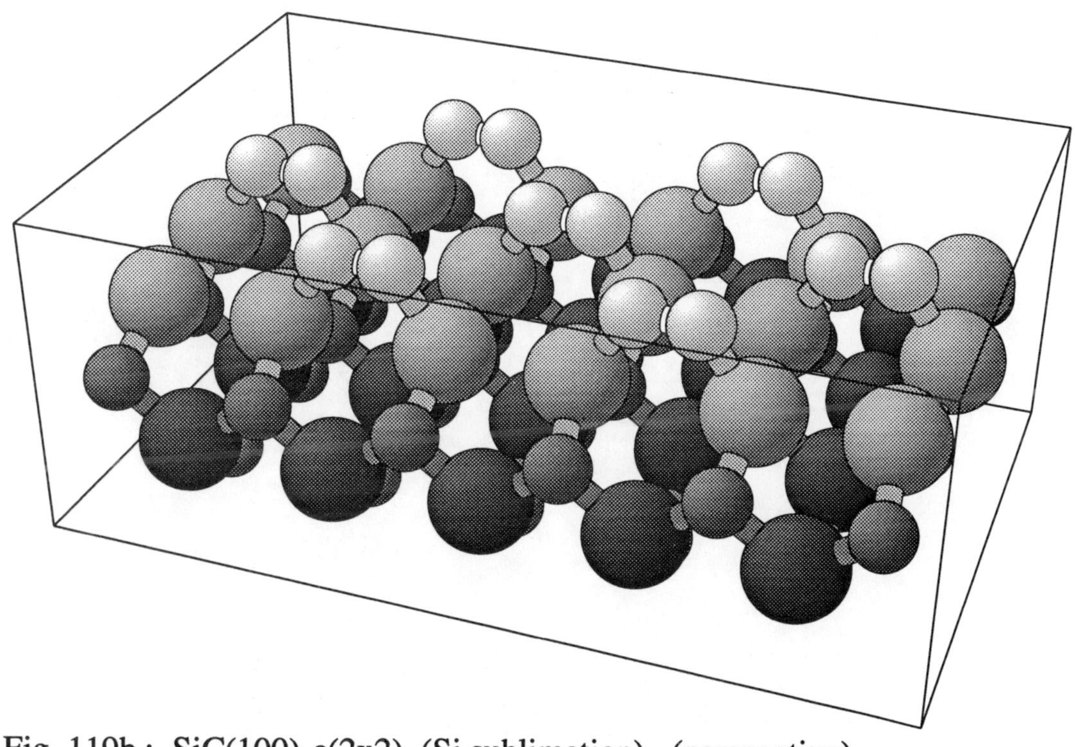

Fig. 119b : SiC(100)-c(2x2) (Si sublimation) (perspective)

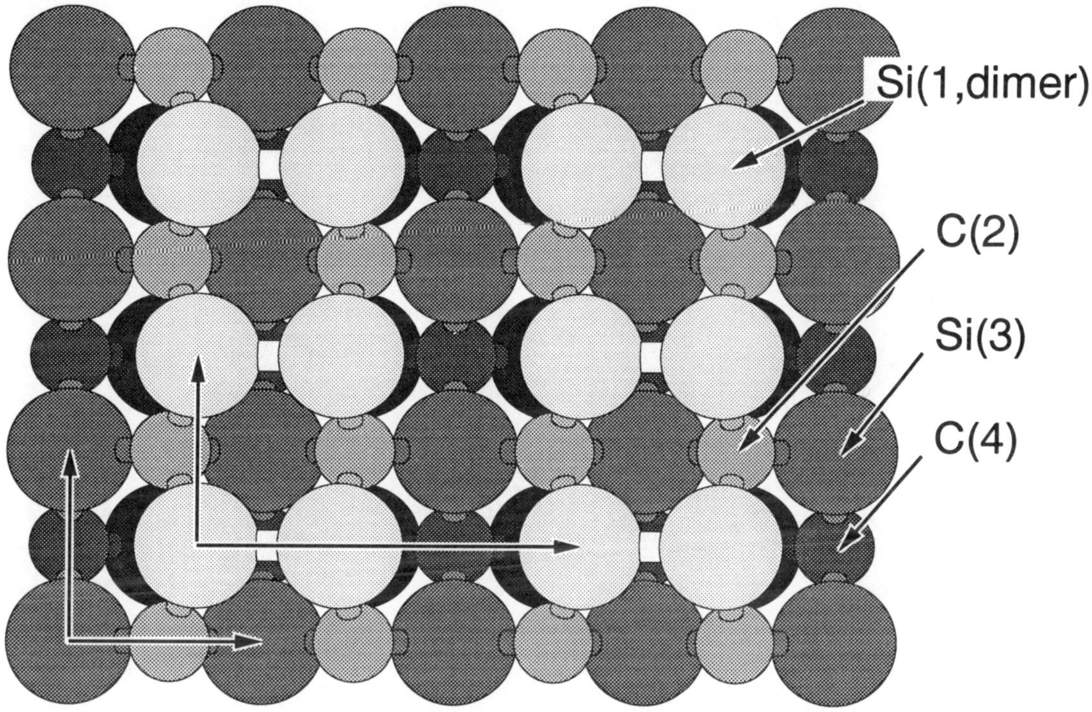

Fig. 120a : SiC(100)-p(2x1) (top view)

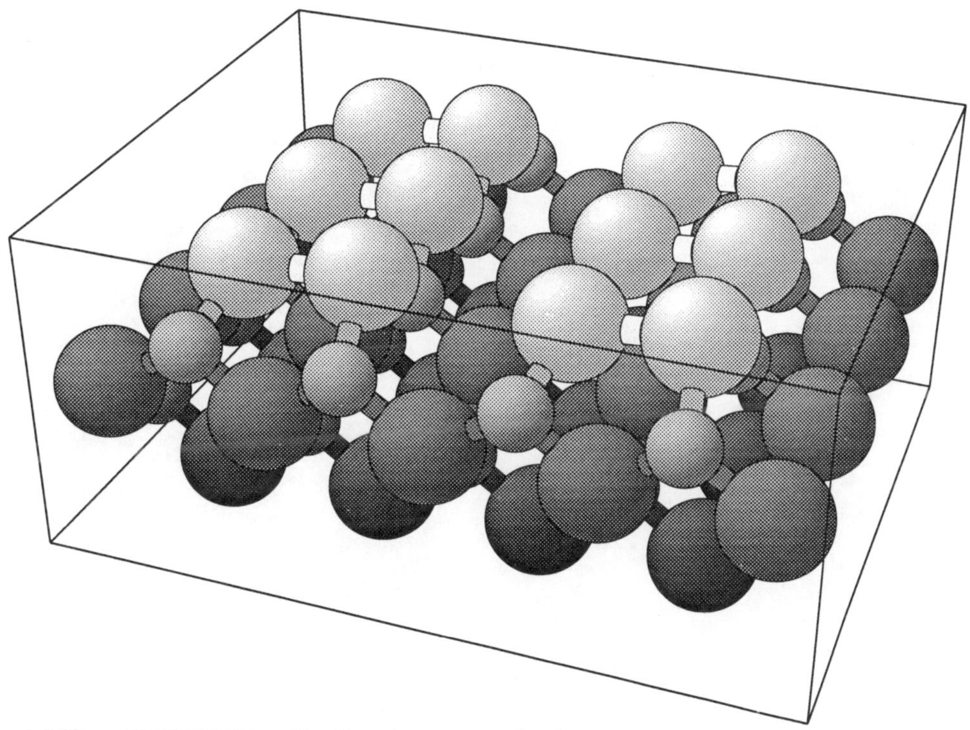

Fig. 120b : SiC(100)-p(2x1) (perspective)

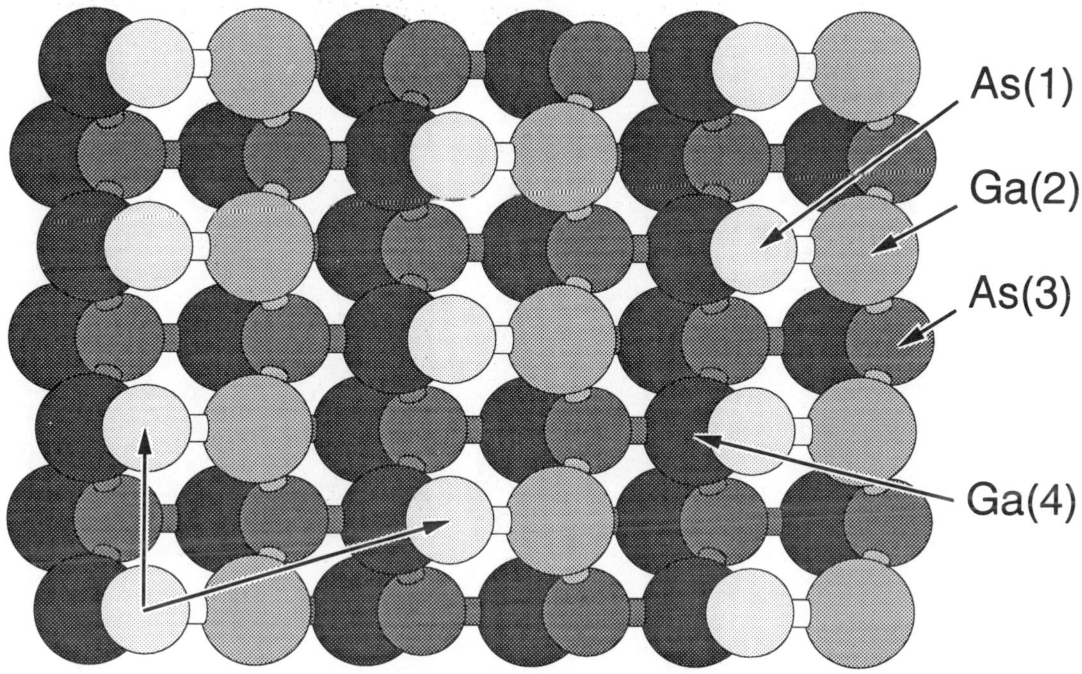

Fig. 121a : GaAs(311)-(1x1) As termination (top view)

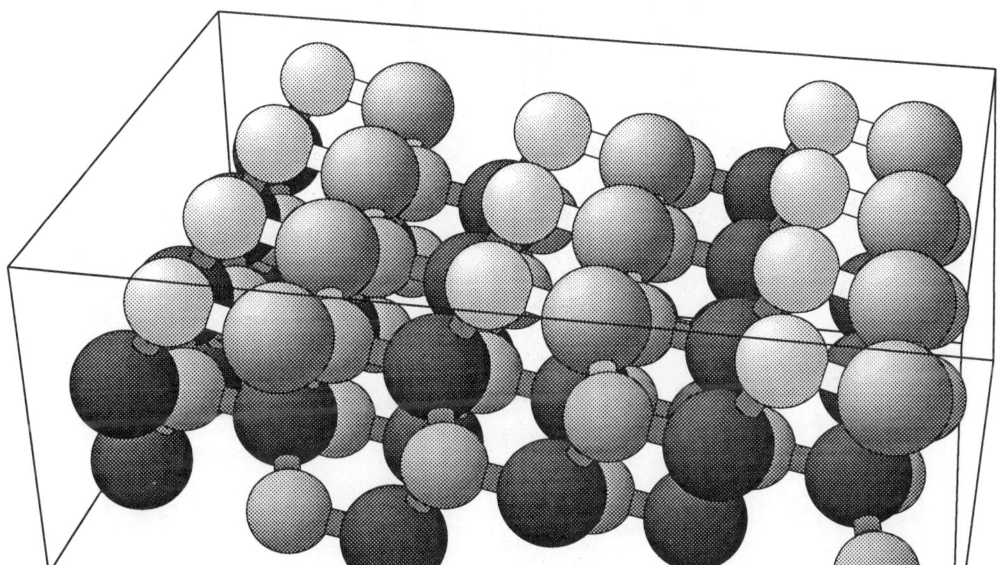

Fig. 121b : GaAs(311)-(1x1) As termination (perspective)

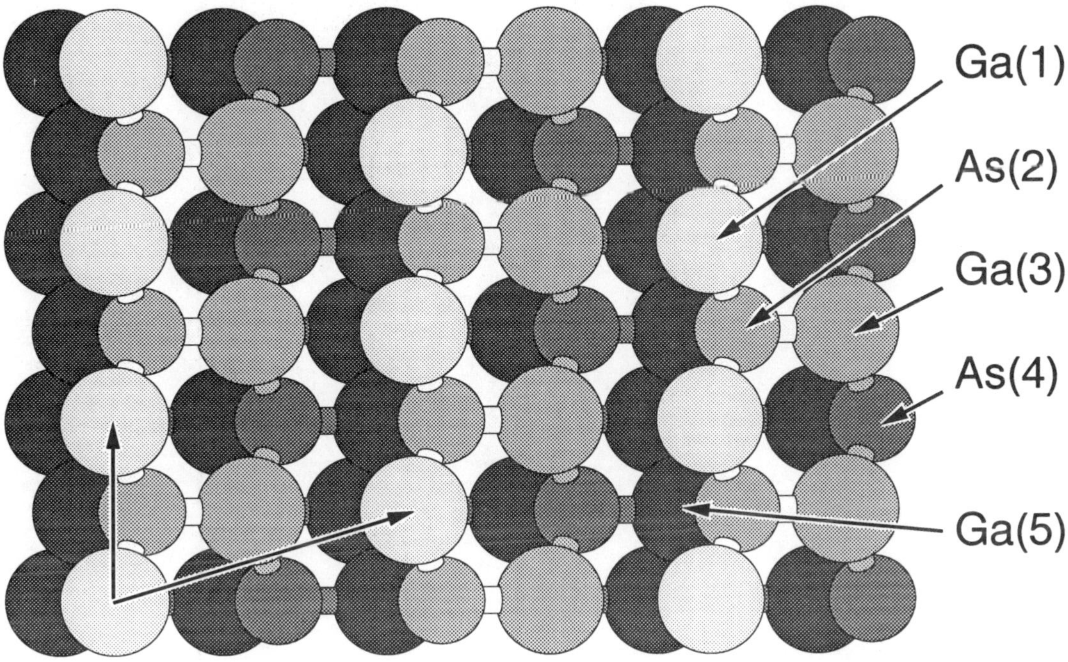

Fig. 122a : GaAs(311)-(1x1) Ga termination (top view)

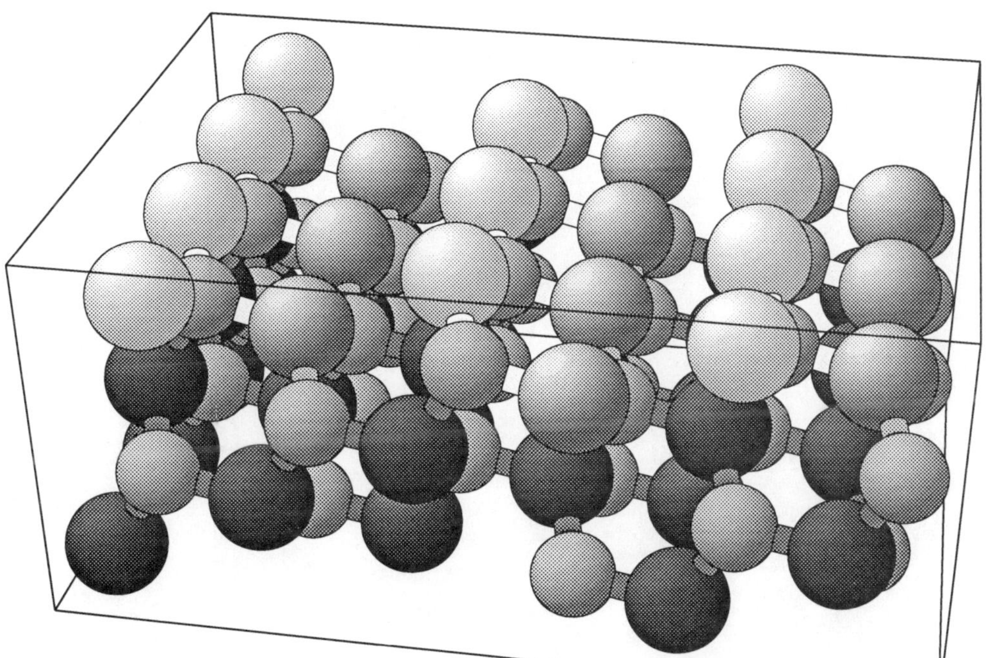

Fig. 122b : GaAs(311)-(1x1) Ga termination (perspective)

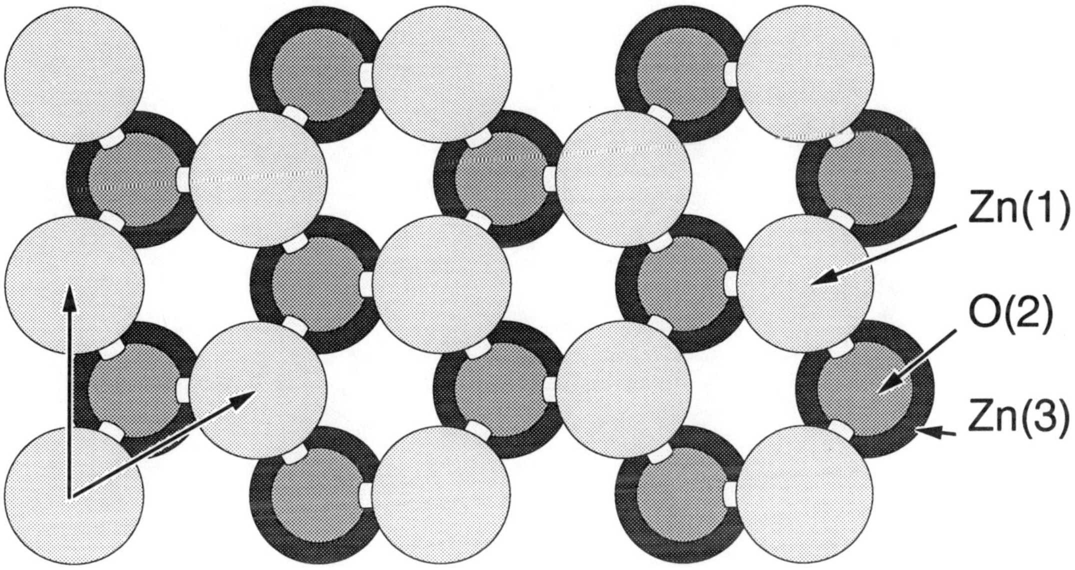

Fig. 123a : ZnO(0001)-(1x1) (top view)

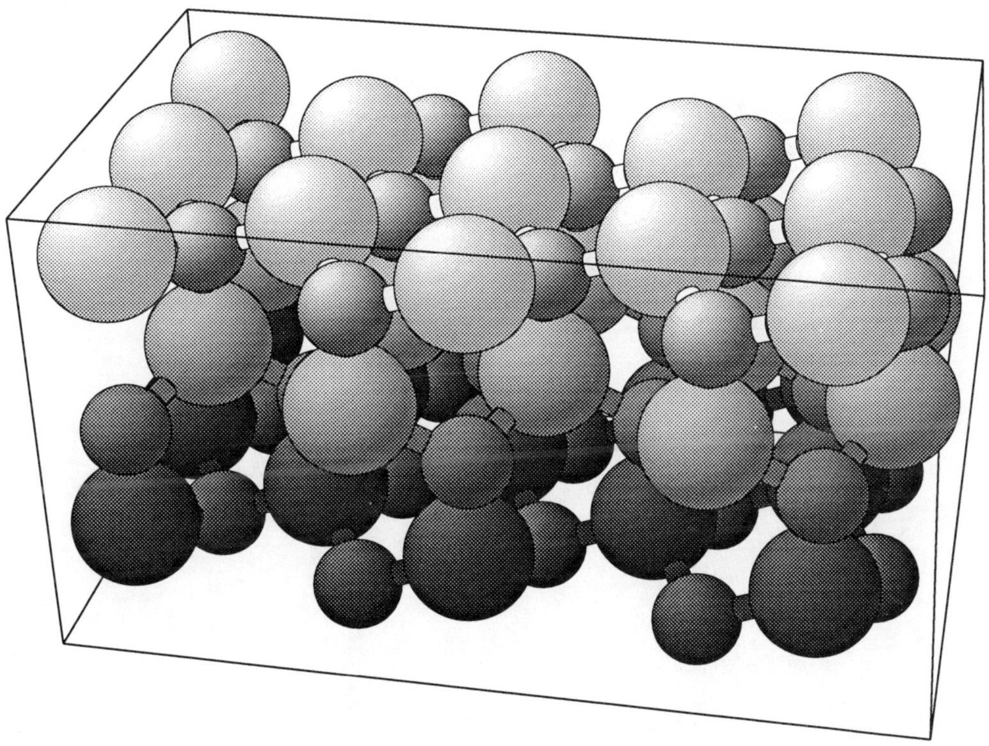

Fig. 123b : ZnO(0001)-(1x1) (perspective)

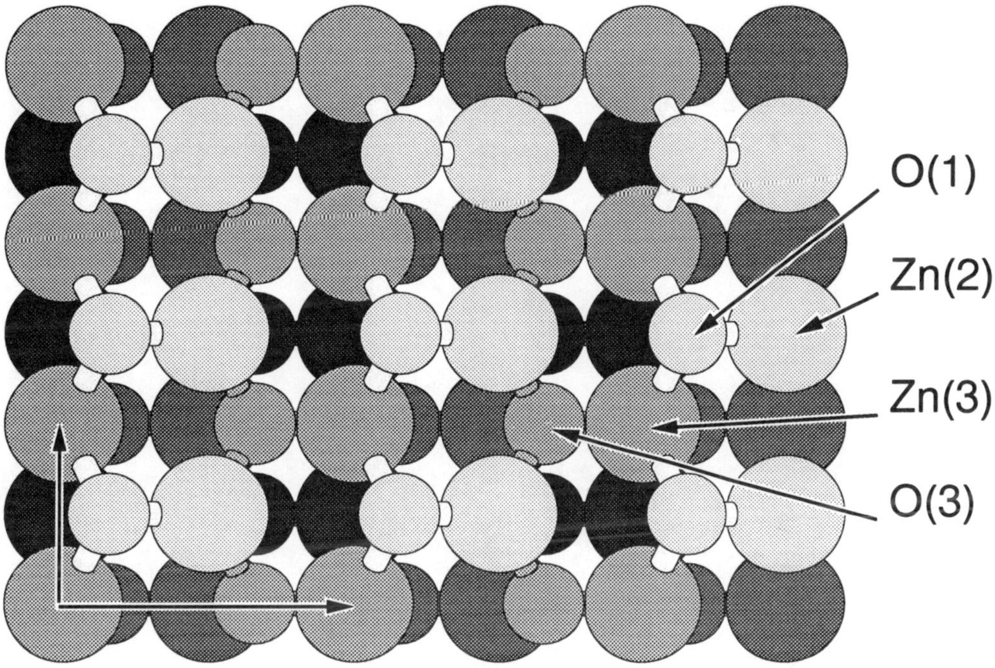

Fig. 124a : ZnO(10-10)-(1x1) (top view)

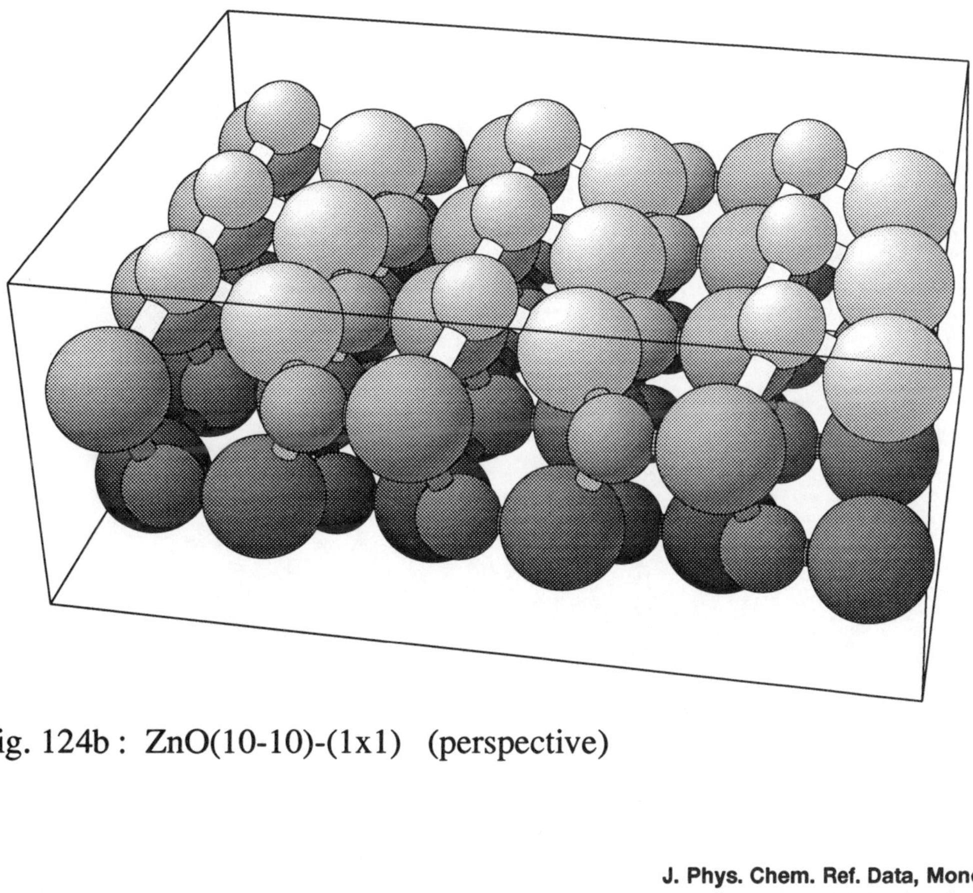

Fig. 124b : ZnO(10-10)-(1x1) (perspective)

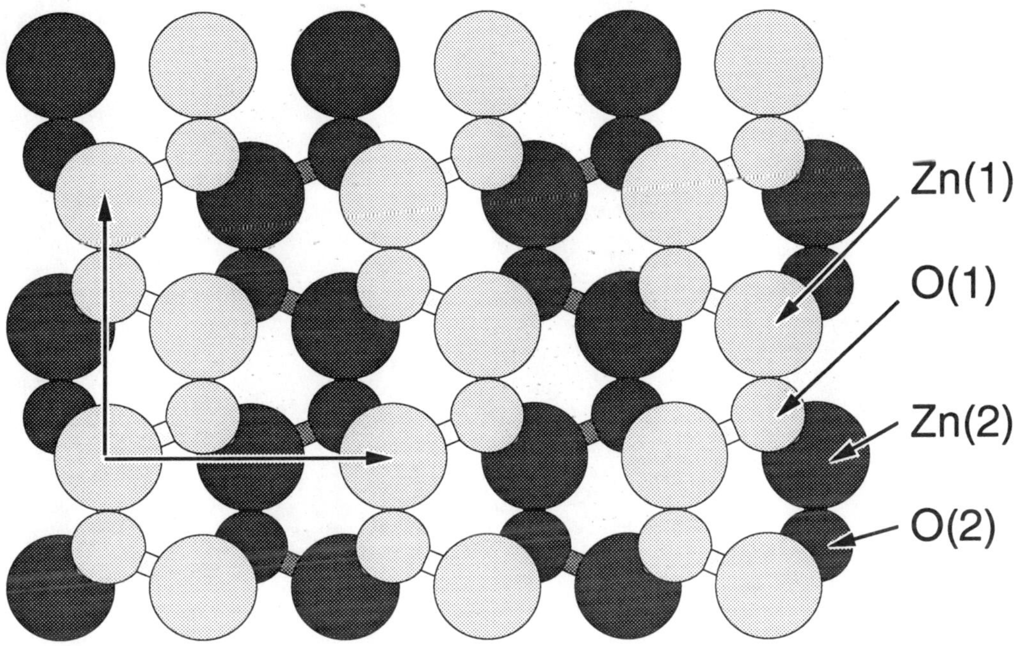

Fig. 125a : ZnO(11-20)-(1x1) (top view)

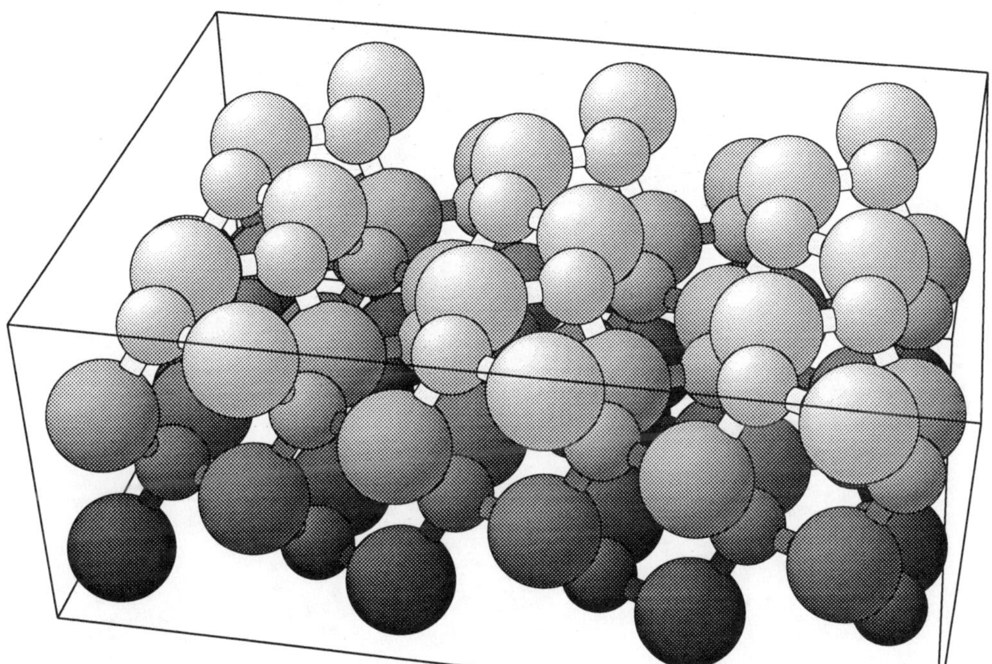

Fig. 125b : ZnO(11-20)-(1x1) (perspective)

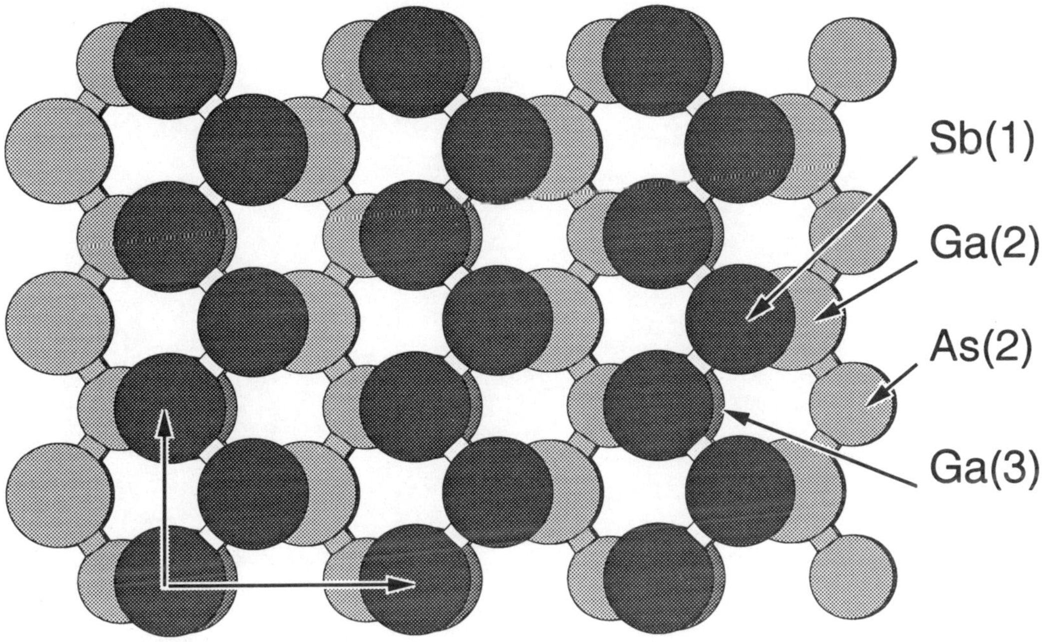

Fig. 126a : GaAs(110)-(1x1)-2Sb (top view)

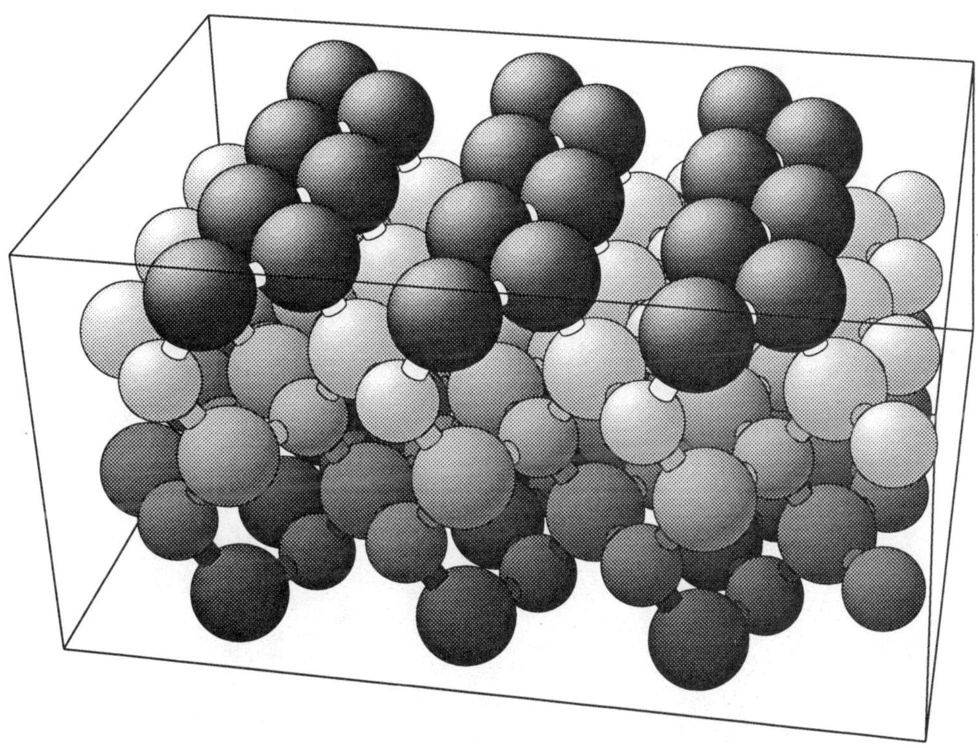

Fig. 126b : GaAs(110)-(1x1)-2Sb (perspective)

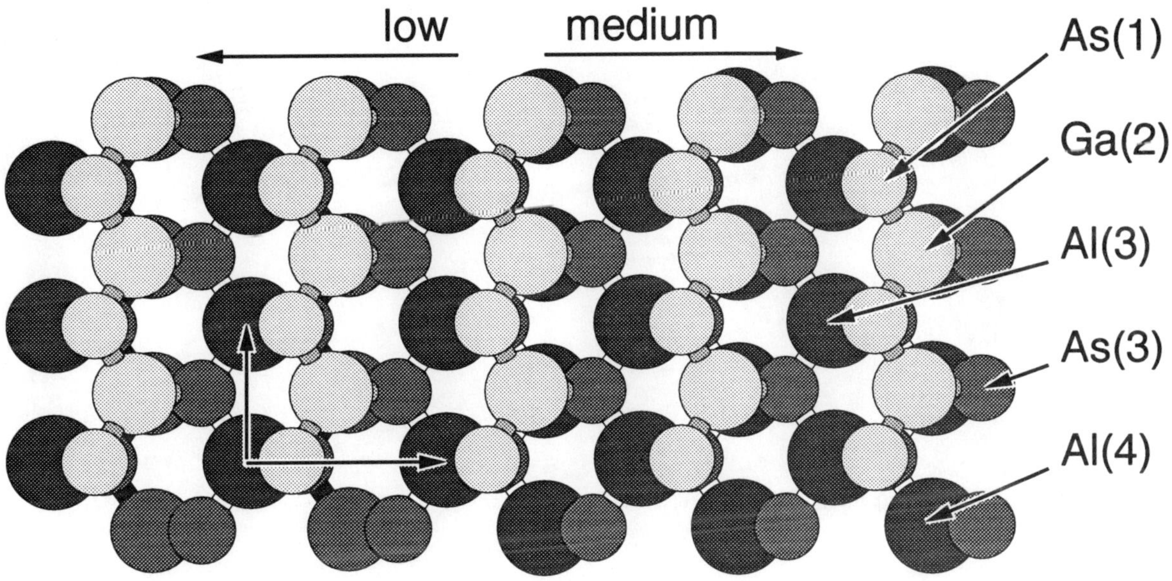

Fig. 127a : GaAs(110)-(1x1)-Al/2Al (low/medium coverage) (top view)

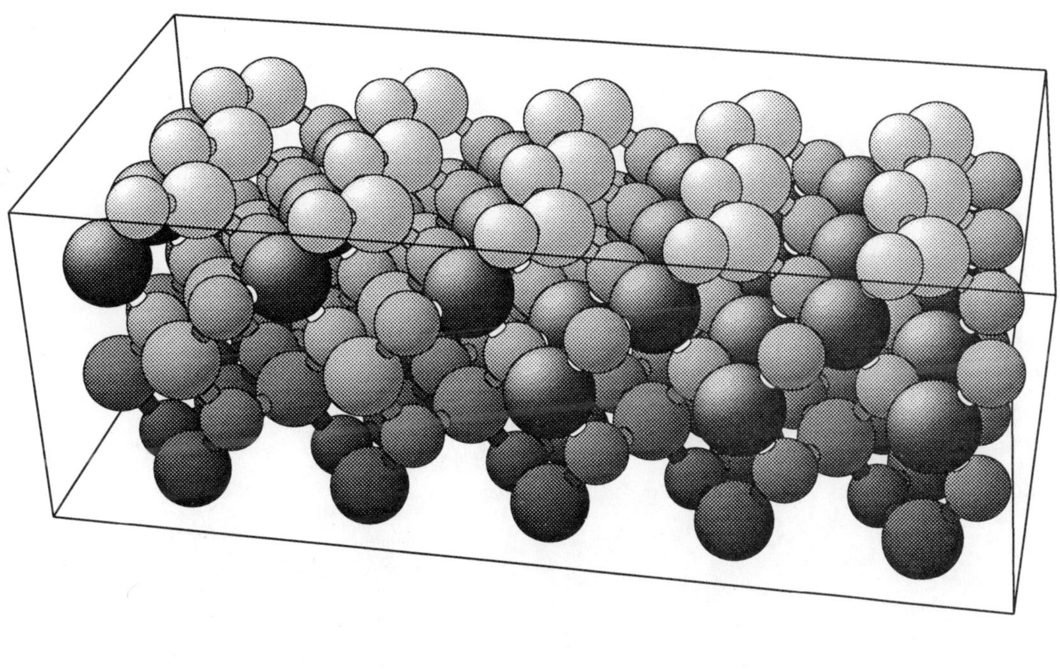

Fig. 127b : GaAs(110)-(1x1)-Al/2Al (low/medium coverage) (perspective)

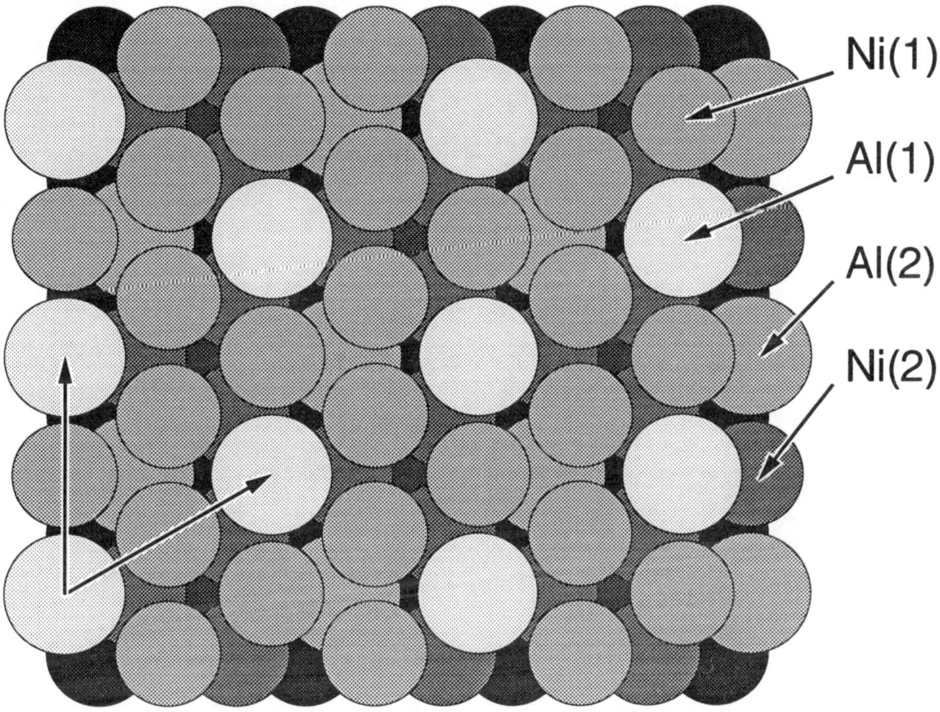

Fig. 128a : Ni3Al(111)-(1x1) (top view)

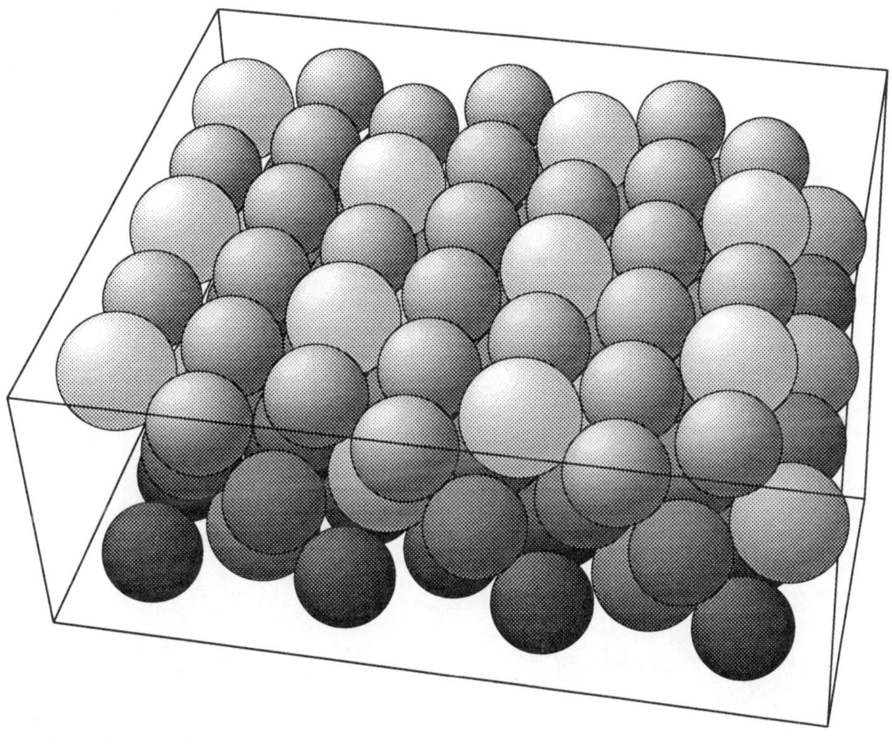

Fig. 128b : Ni3Al(111)-(1x1) (perspective)

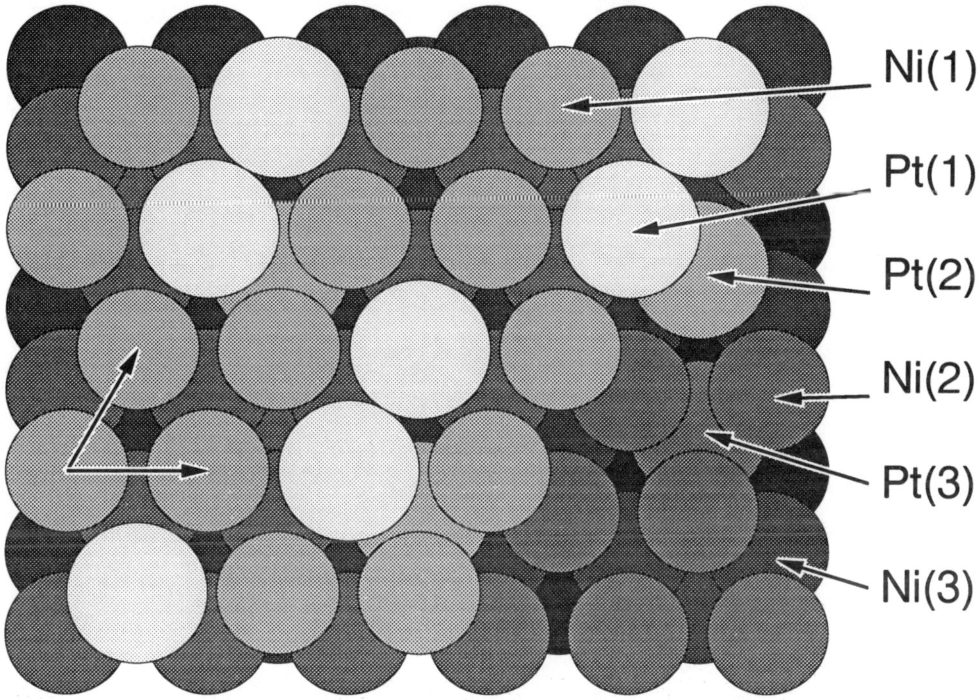

Fig. 129a : Pt0.1Ni0.9(111)-(1x1) (top view)

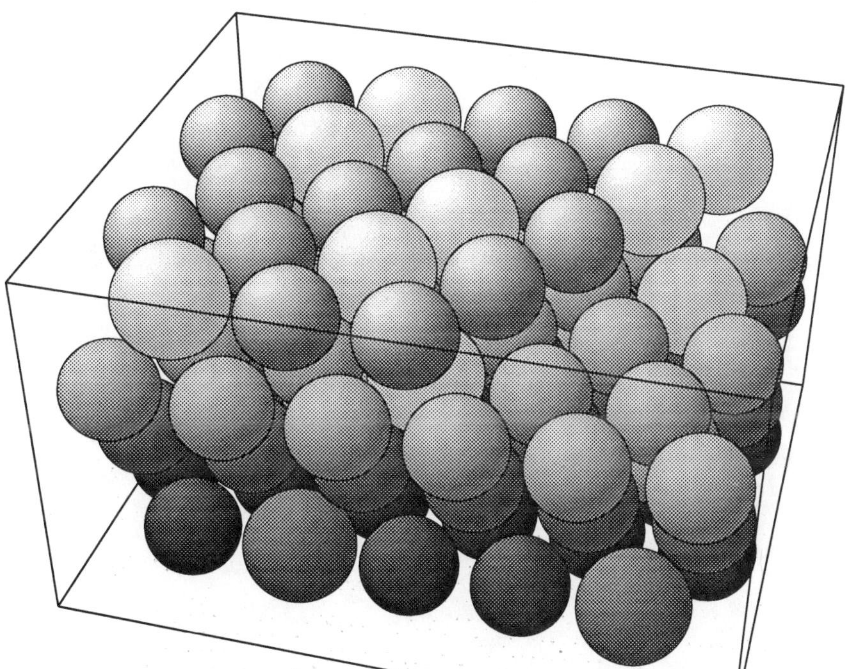

Fig. 129b : Pt0.1Ni0.9(111)-(1x1) (perspective)

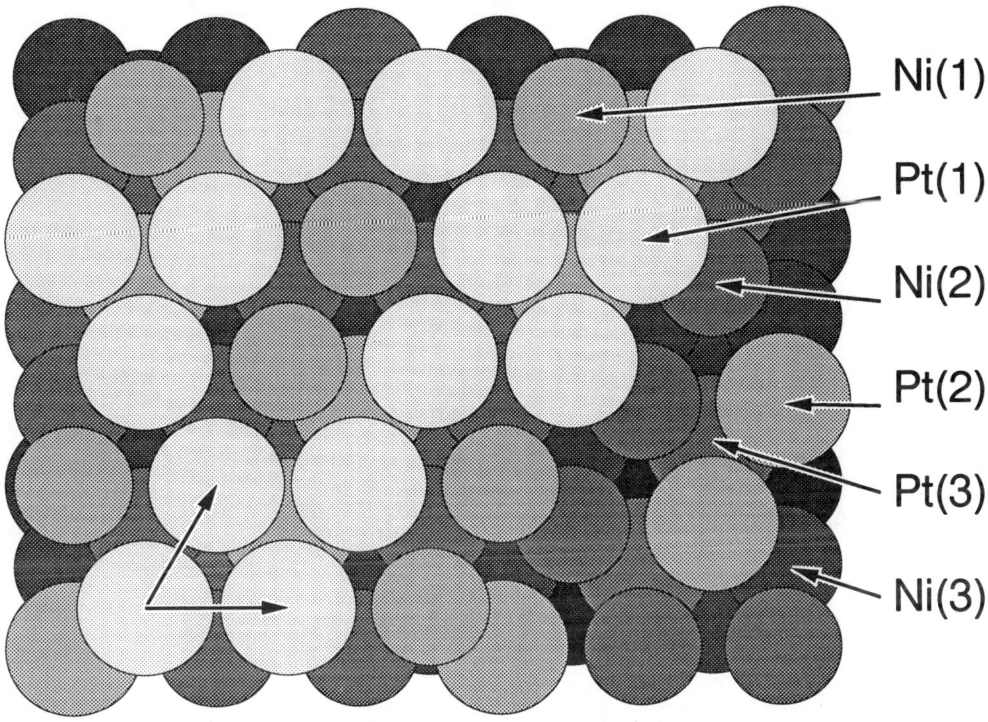

Fig. 130a : Pt0.5Ni0.5(111)-(1x1) (top view)

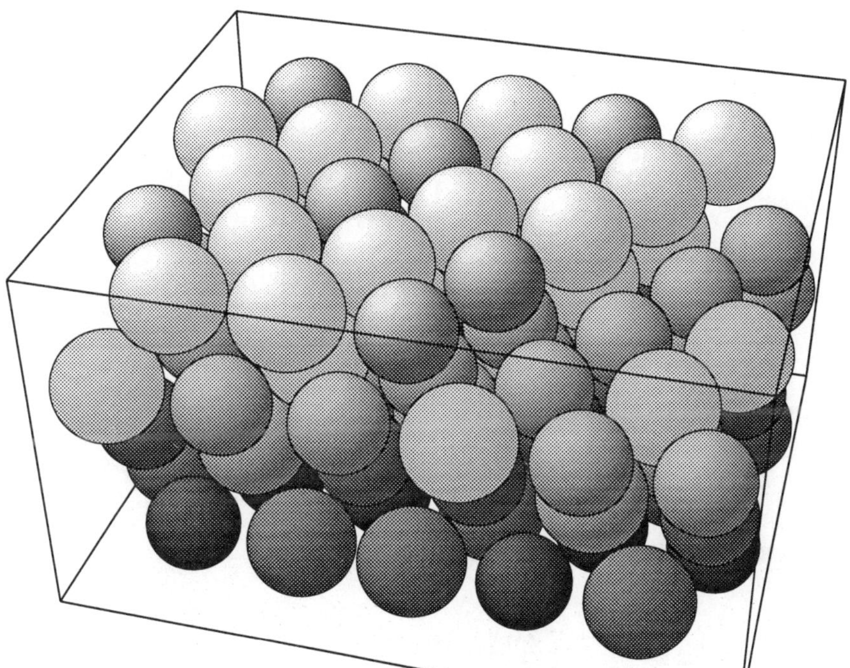

Fig. 130b : Pt0.5Ni0.5(111)-(1x1) (perspective)

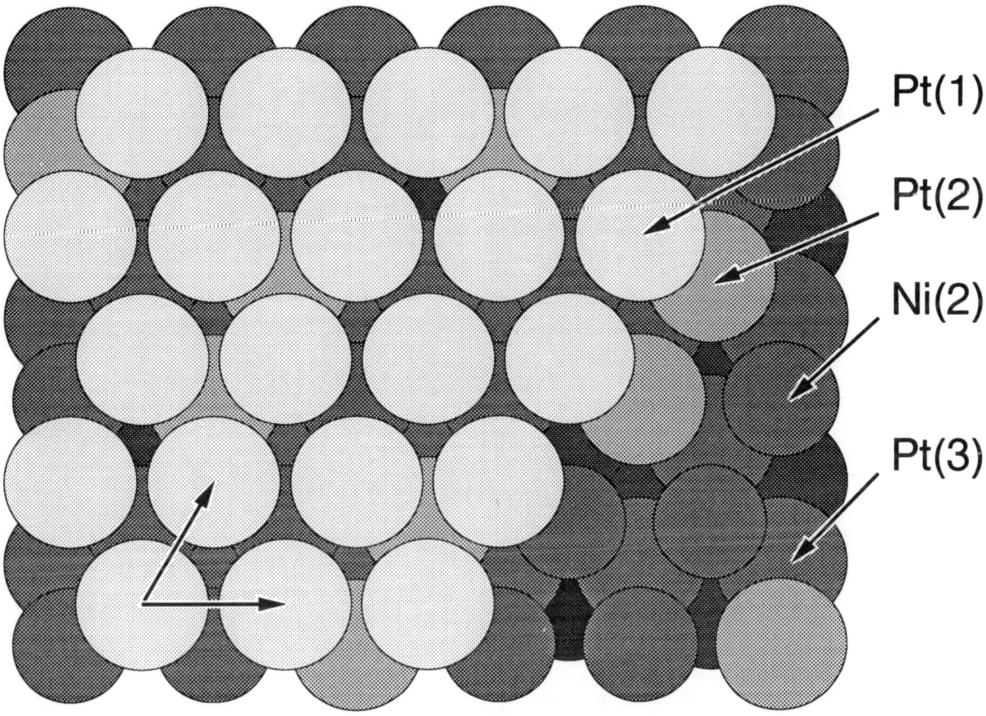

Fig. 131a : Pt0.78Ni0.22(111)-(1x1) (top view)

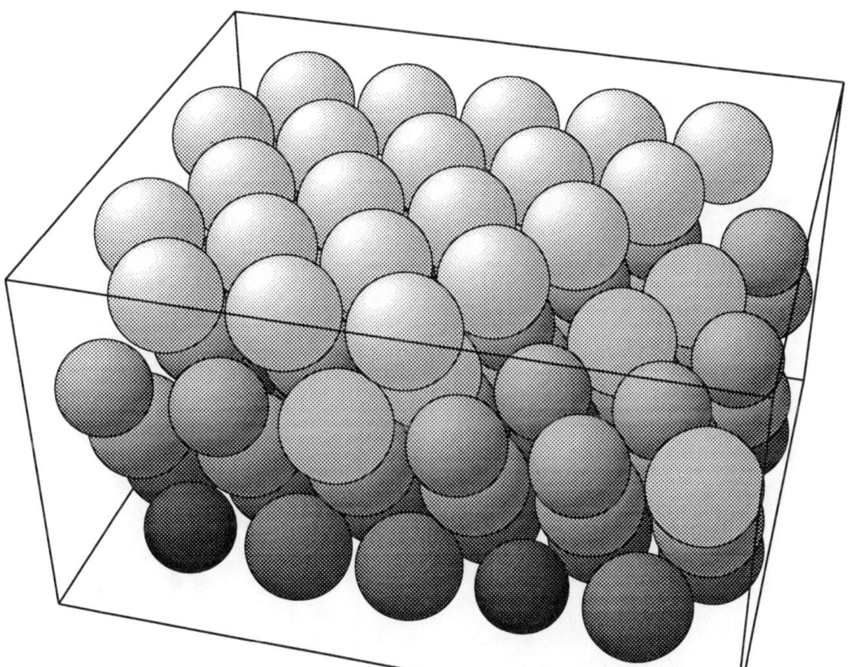

Fig. 131b : Pt0.78Ni0.22(111)-(1x1) (perspective)

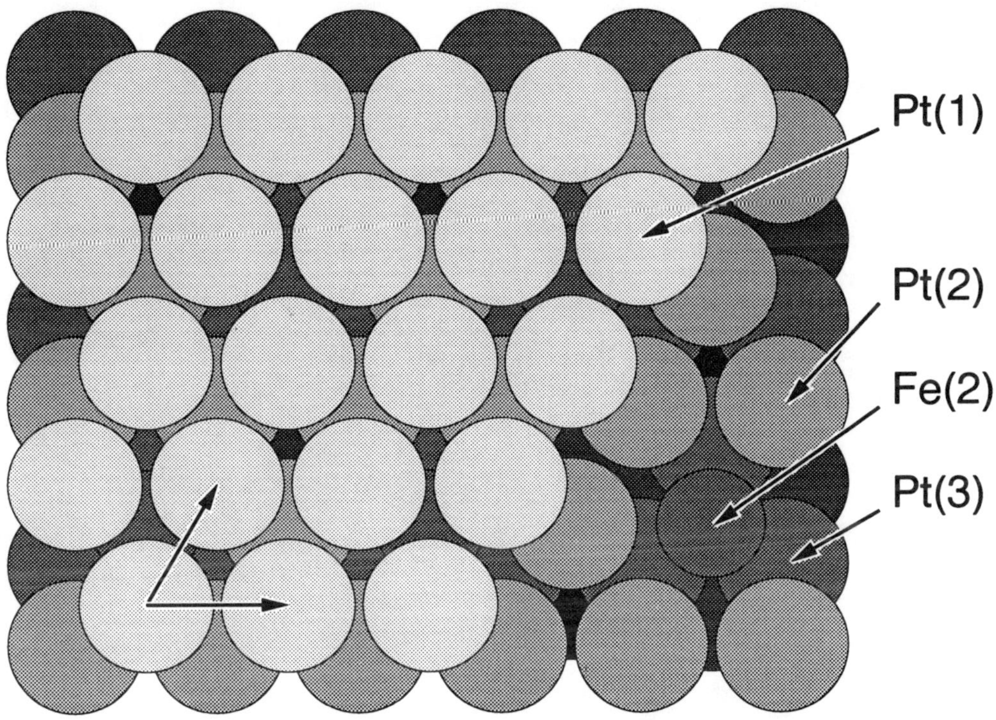

Fig. 132a : Pt0.8Fe0.2(111)-(1x1) (top view)

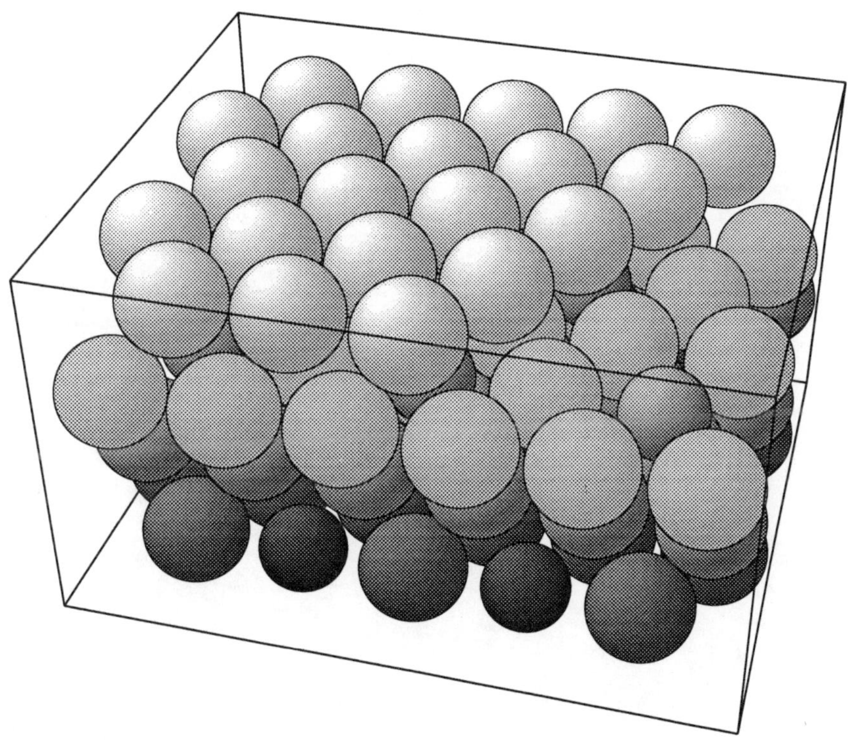

Fig. 132b : Pt0.8Fe0.2(111)-(1x1) (perspective)

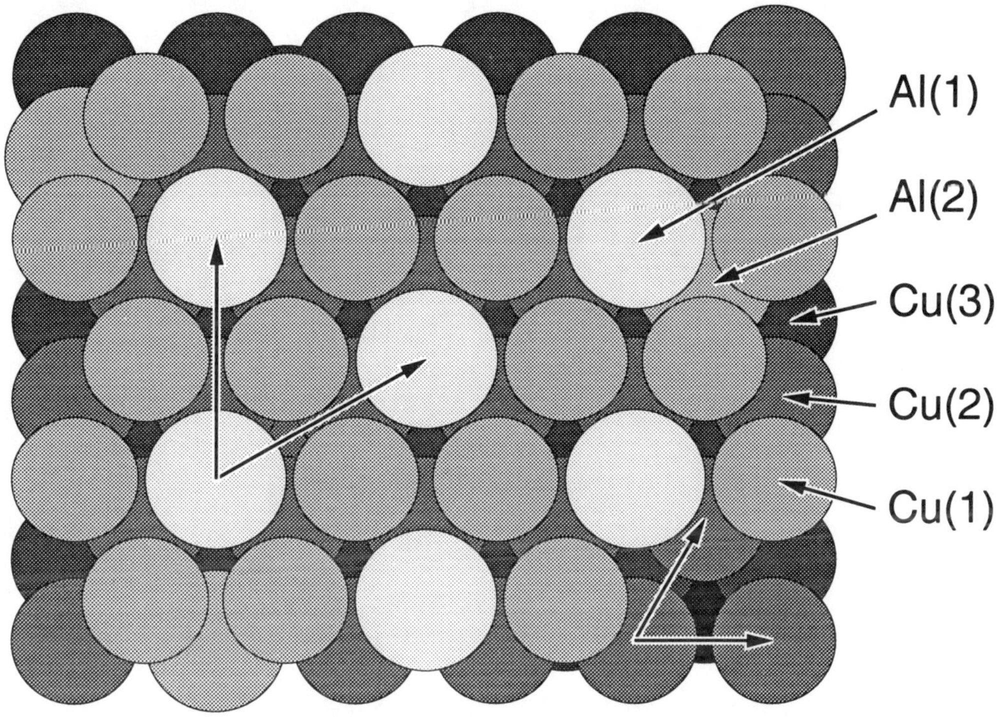

Fig. 133a : α-Cu(111)-16%Al-(√3x√3)R30° (top view)

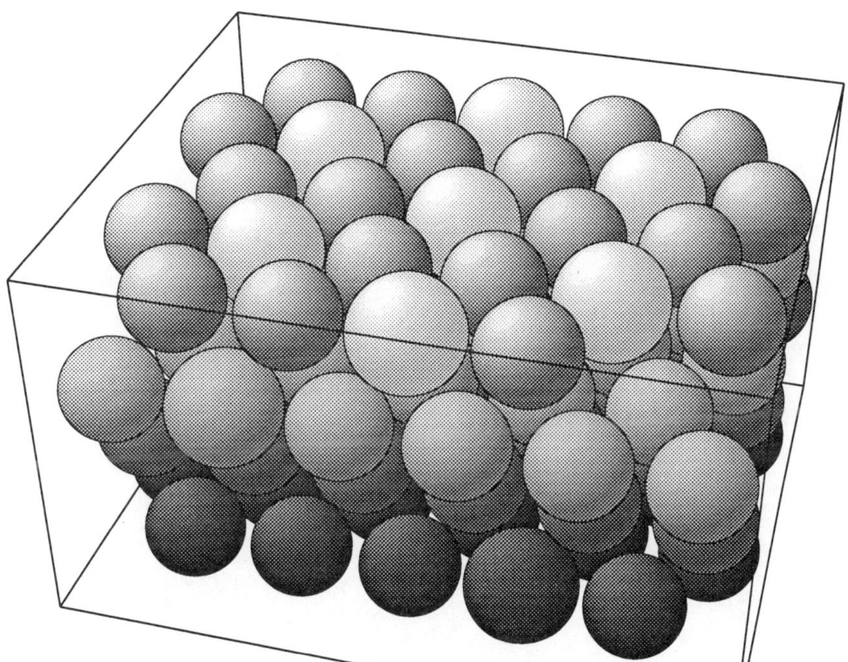

Fig. 133b : α-Cu(111)-16%Al-(√3x√3)R30° (perspective)

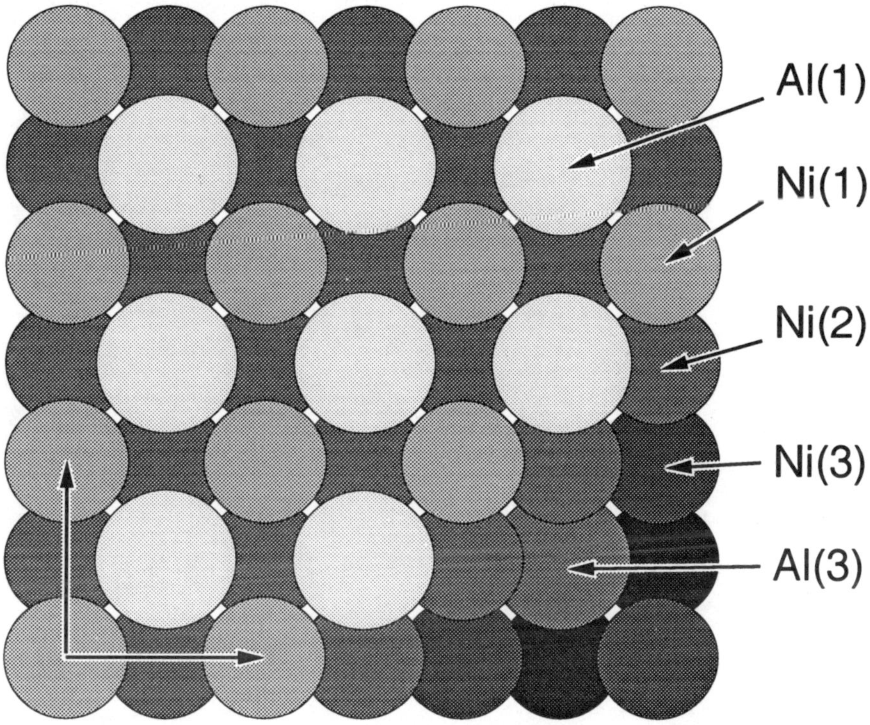

Fig. 134a : Ni3Al(100)-(1x1) (top view)

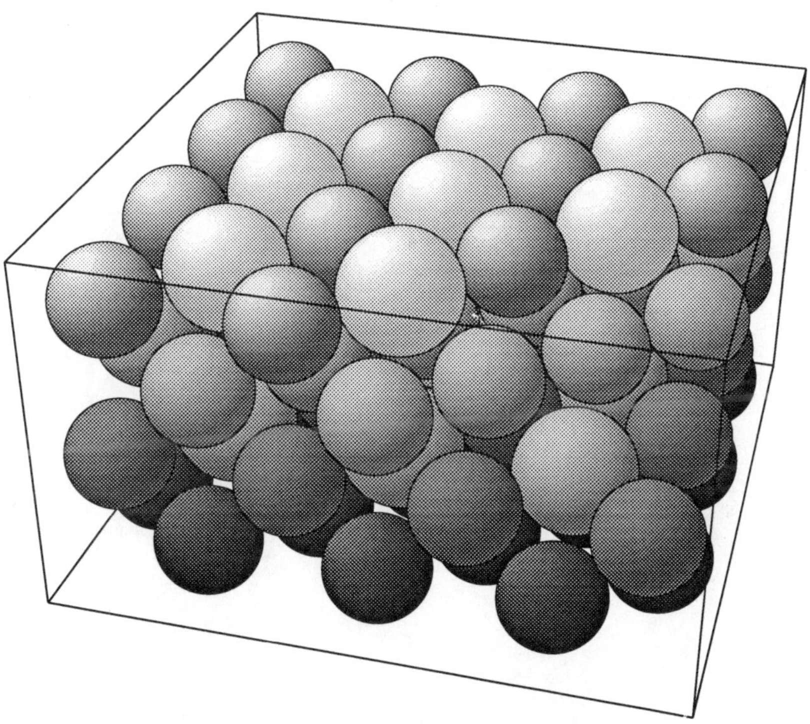

Fig. 134b : Ni3Al(100)-(1x1) (perspective)

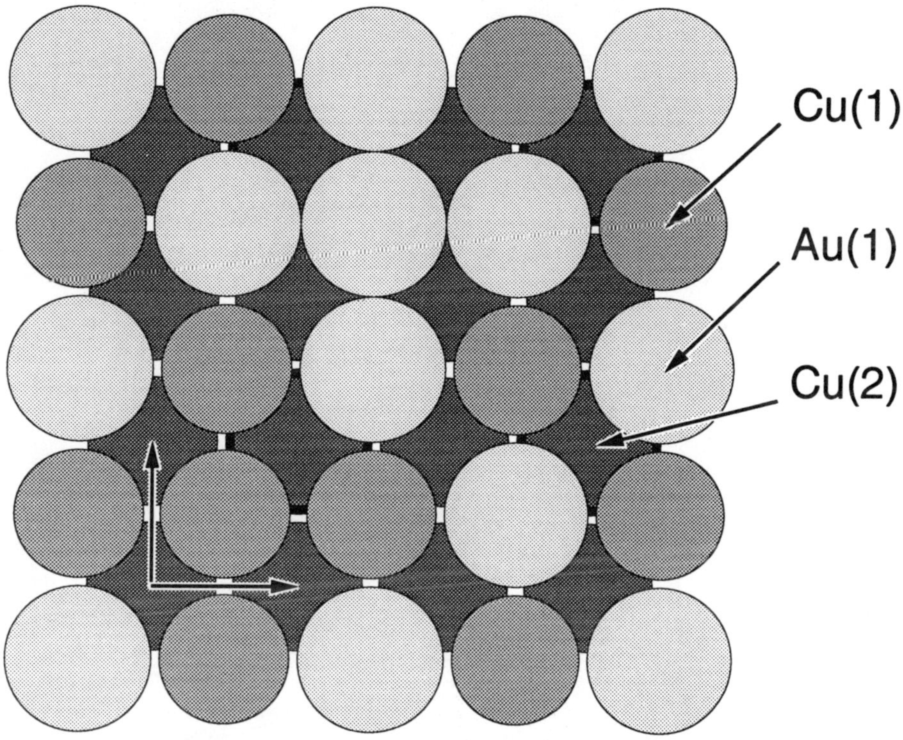

Fig. 135a : AuCu3(100) disordered (top view)

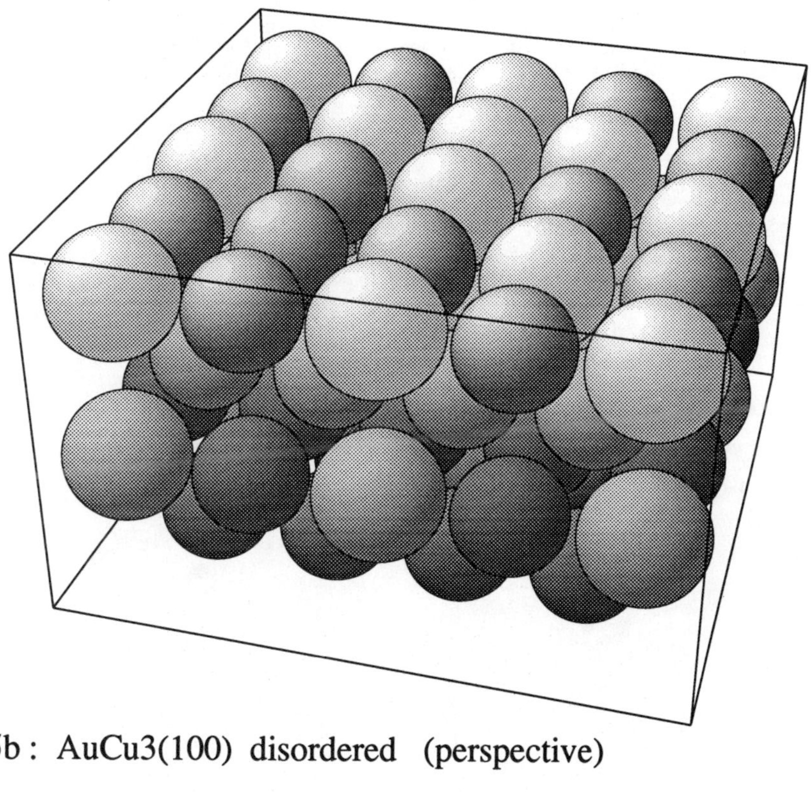

Fig. 135b : AuCu3(100) disordered (perspective)

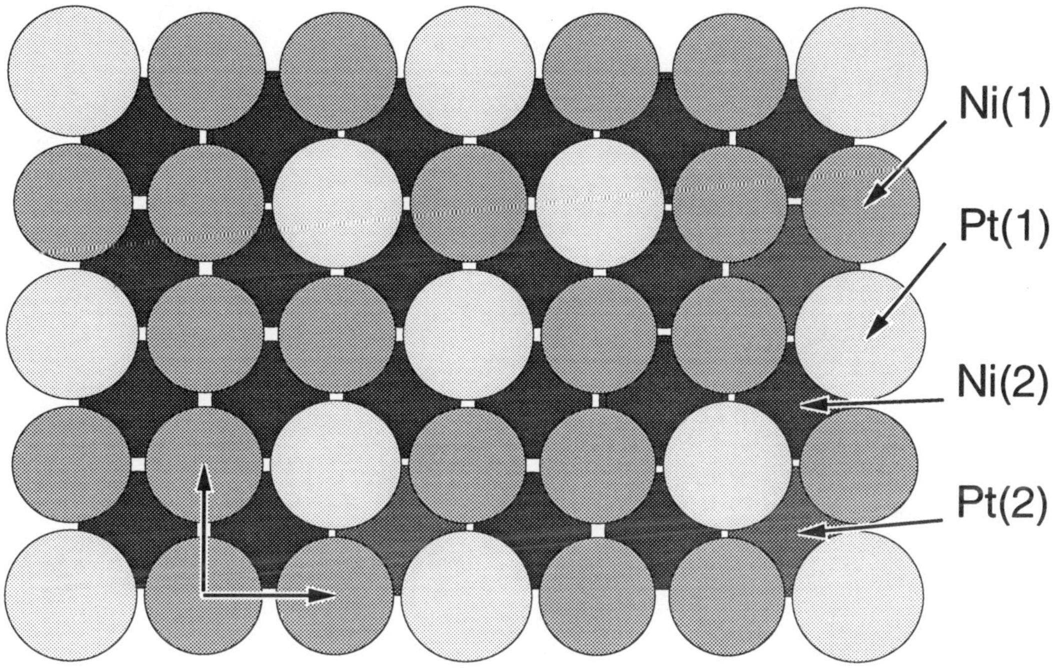

Fig. 136a : Pt0.1Ni0.9(100)-(1x1) (top view)

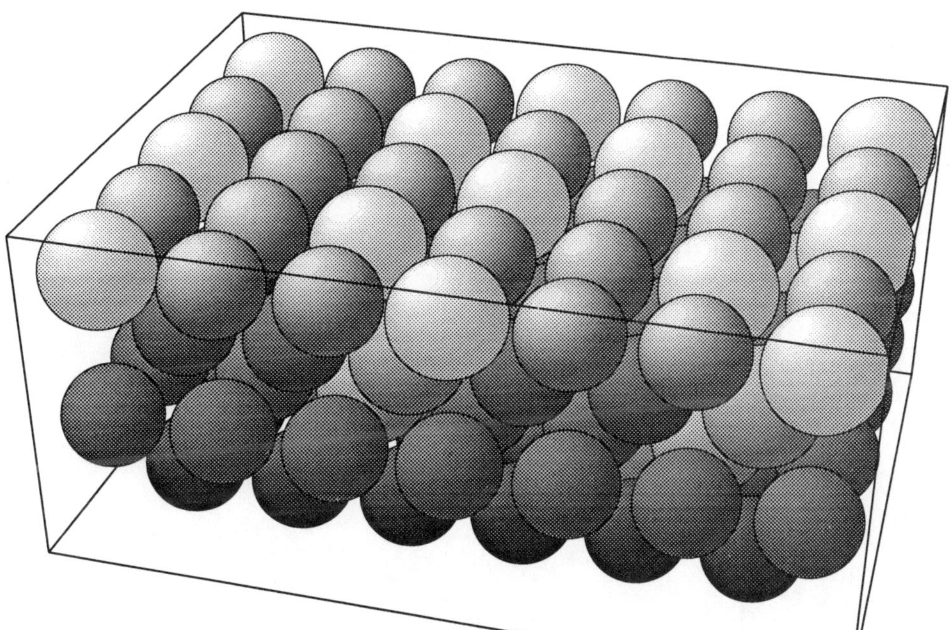

Fig. 136b : Pt0.1Ni0.9(100)-(1x1) (perspective)

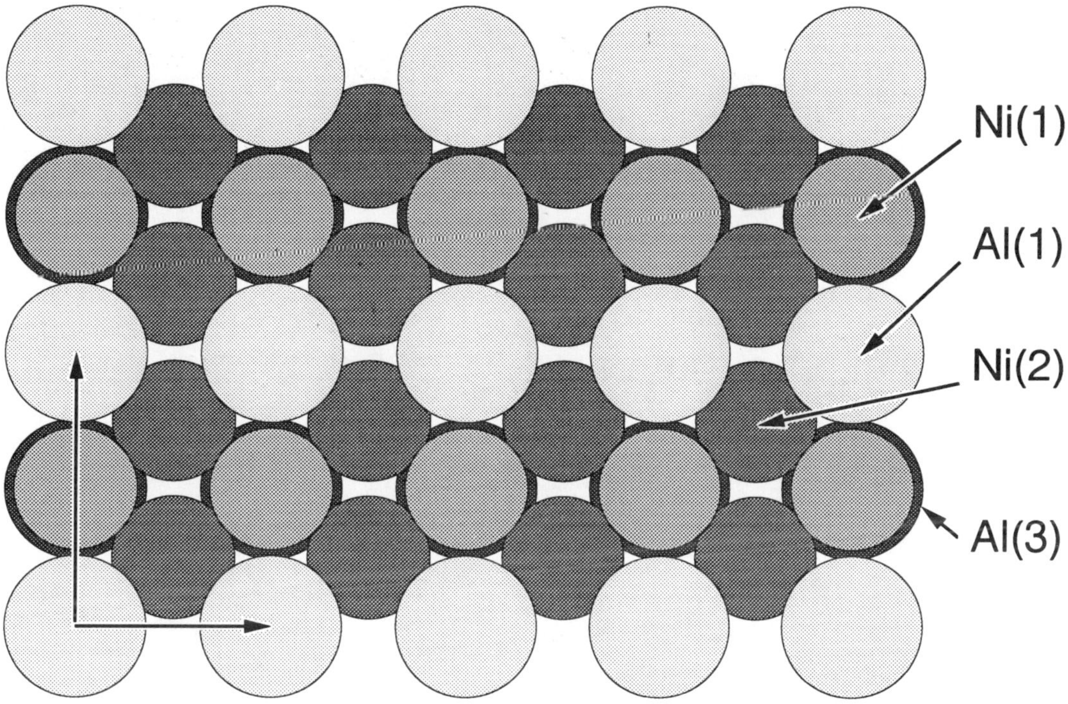

Fig. 137a : Ni3Al(110)-(1x1) (top view)

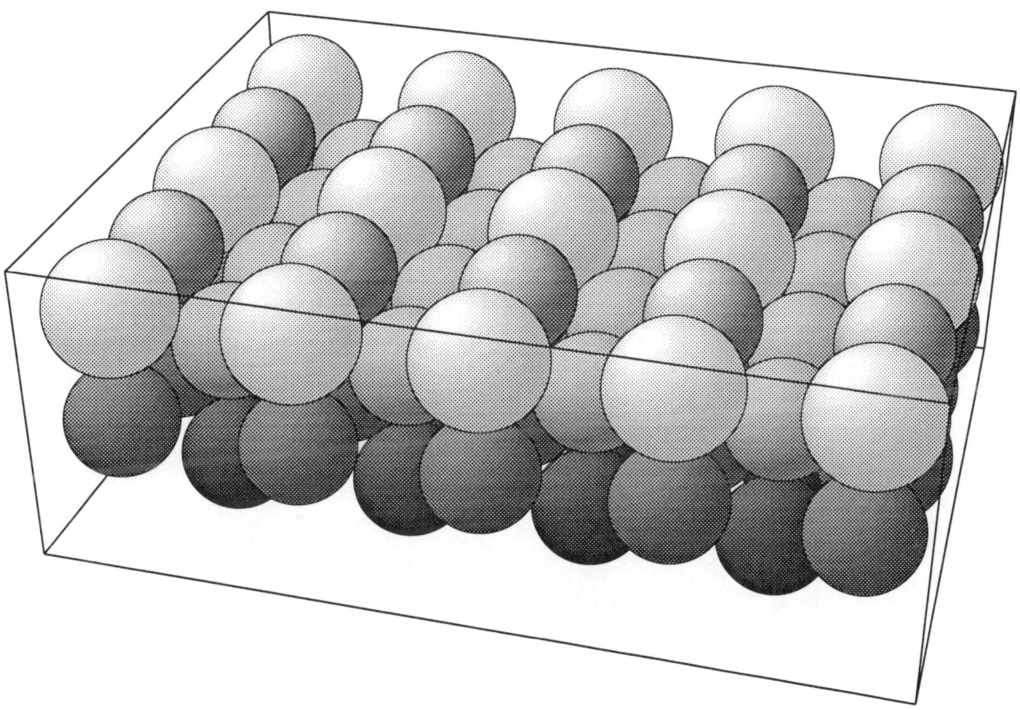

Fig. 137b : Ni3Al(110)-(1x1) (perspective)

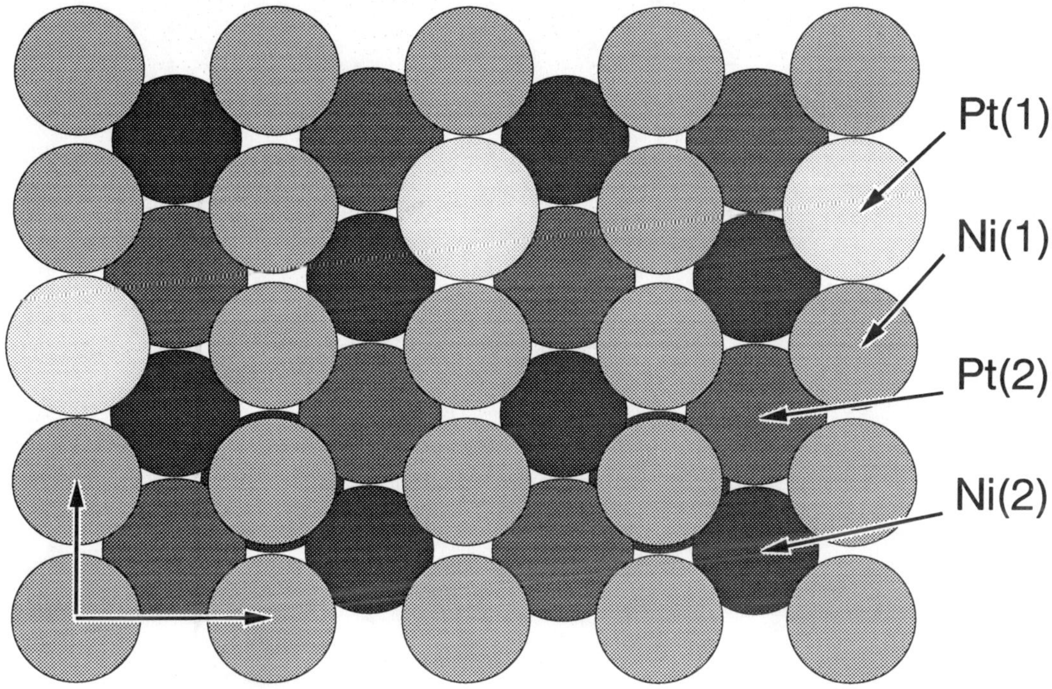

Fig. 138a : Pt0.1Ni0.9(110)-(1x1) (top view)

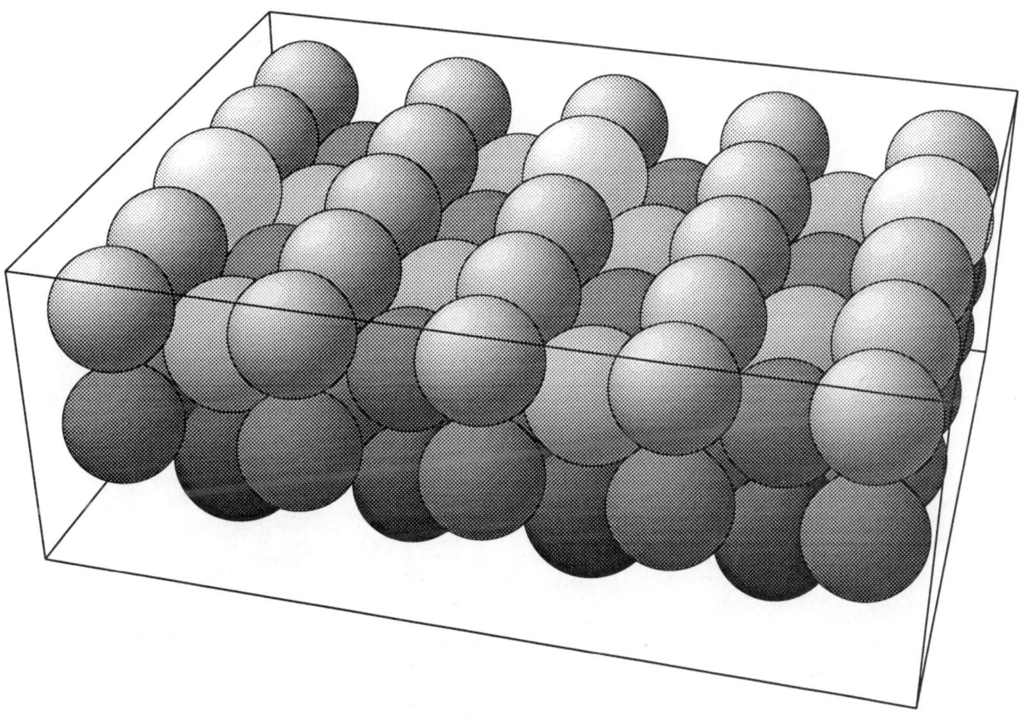

Fig. 138b : Pt0.1Ni0.9(110)-(1x1) (perspective)

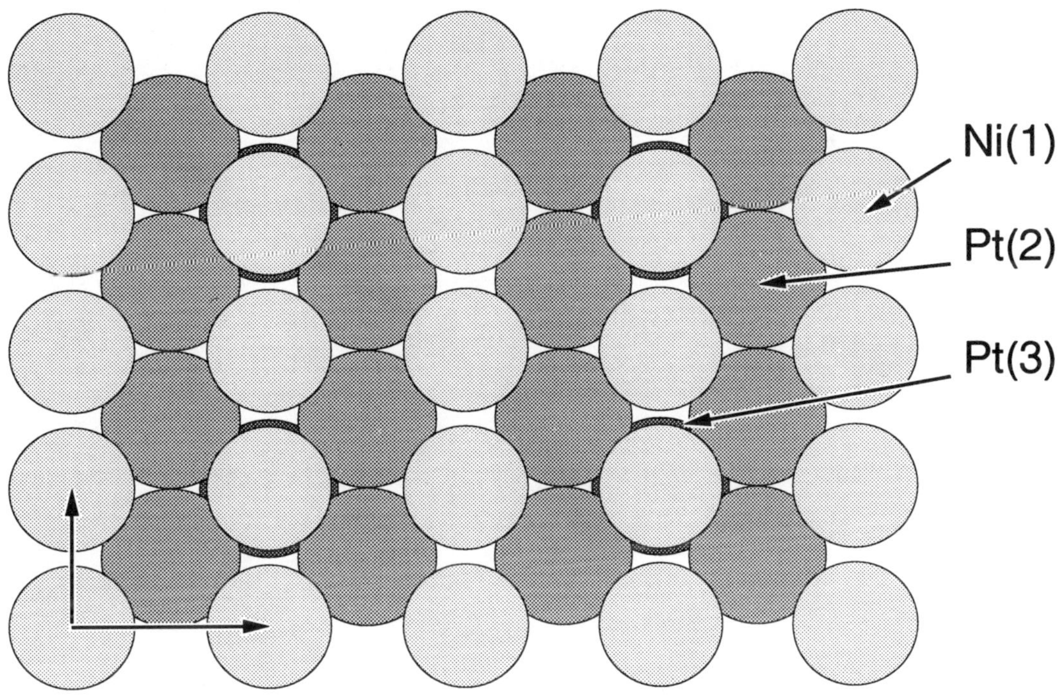

Fig. 139a : Pt0.5Ni0.5(110)-(1x1) (top view)

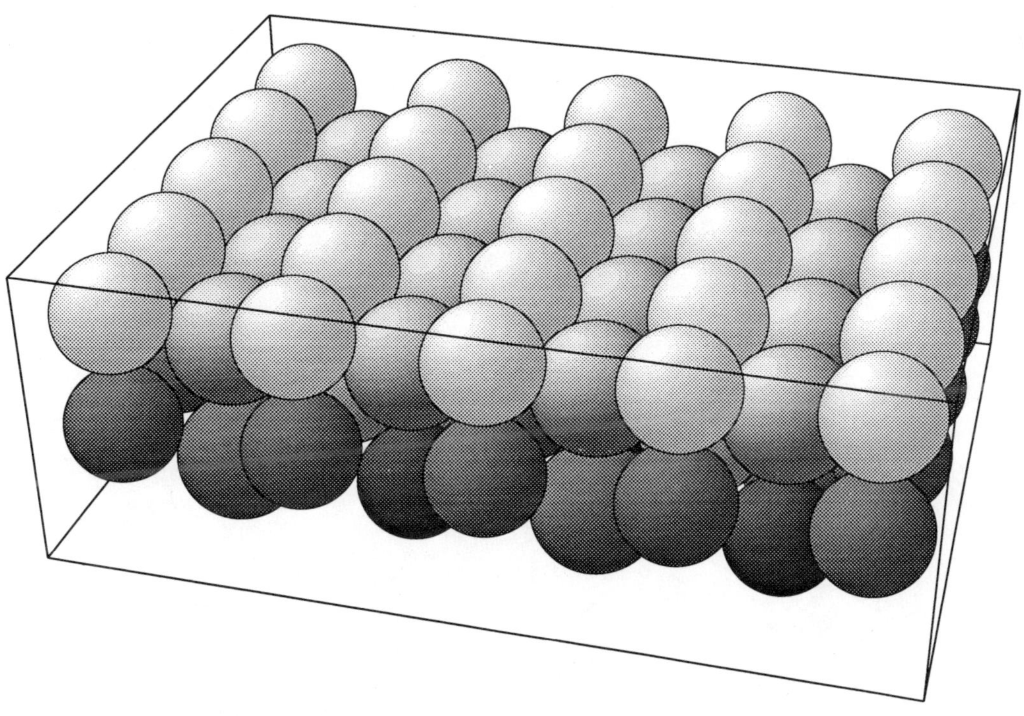

Fig. 139b : Pt0.5Ni0.5(110)-(1x1) (perspective)

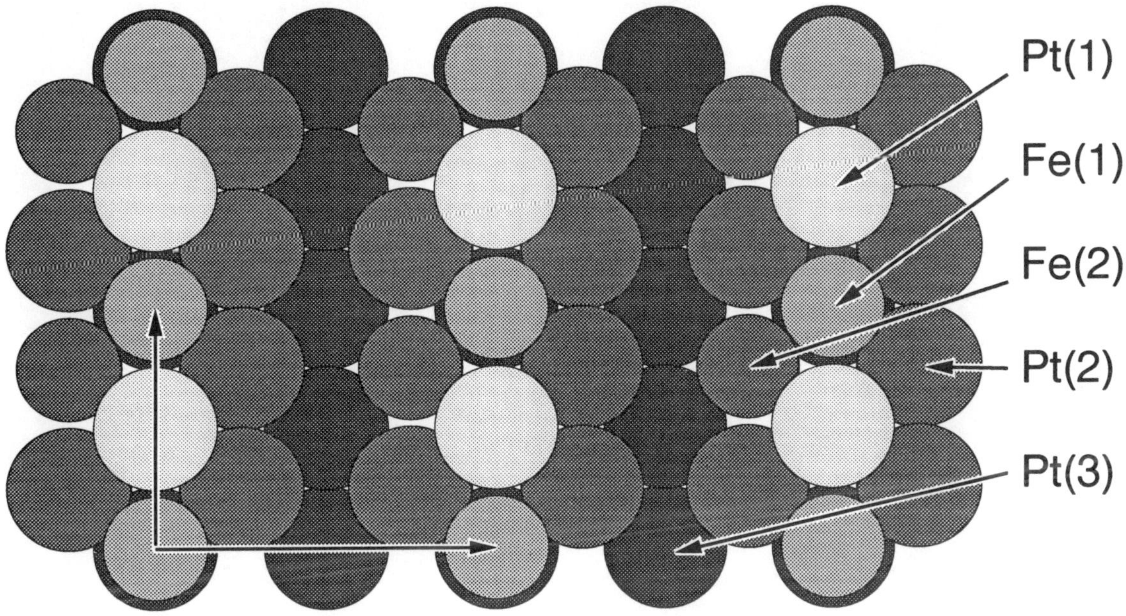

Fig. 140a : Pt0.8Fe0.2(110)-(1x2) (top view)

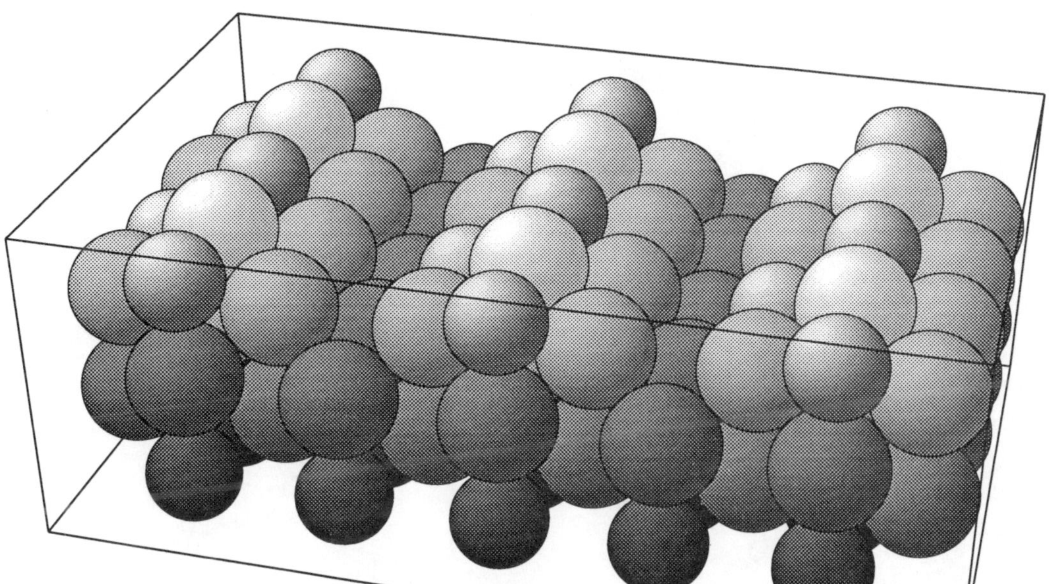

Fig. 140b : Pt0.8Fe0.2(110)-(1x2) (perspective)

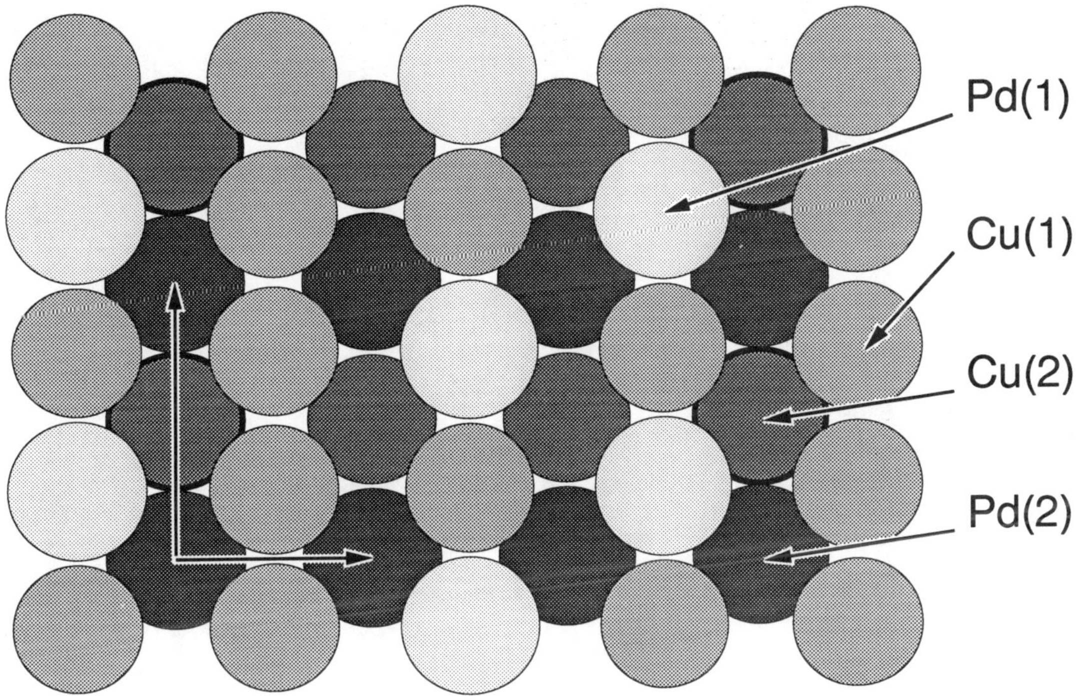

Fig. 141a : Cu0.85Pd0.15(110)-(2x1) (top view)

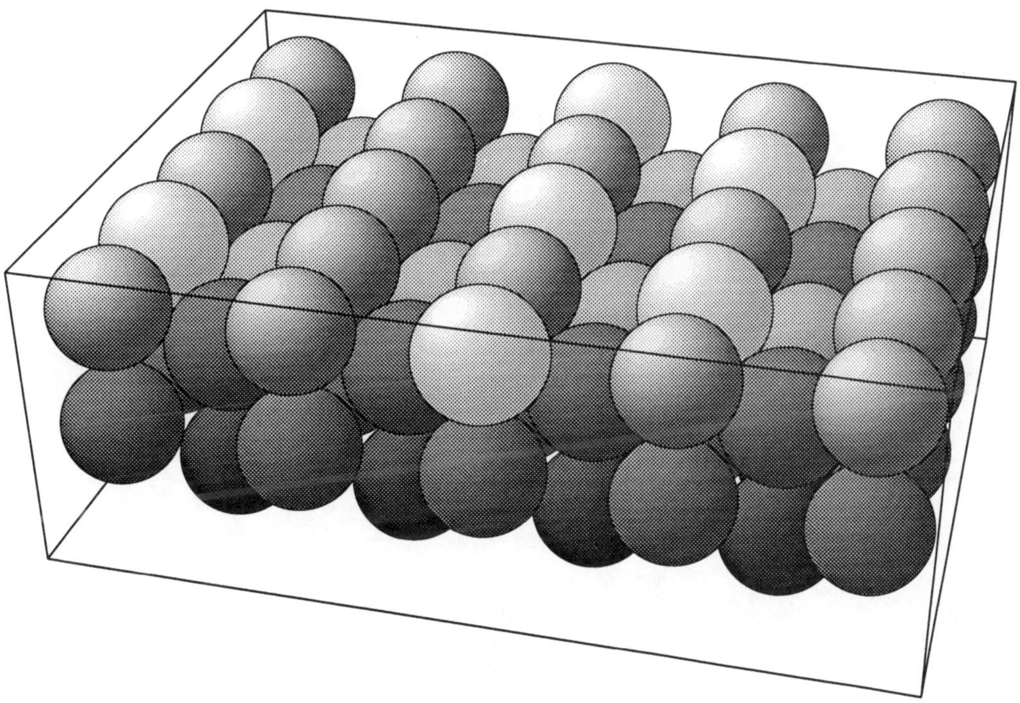

Fig. 141b : Cu0.85Pd0.15(110)-(2x1) (perspective)

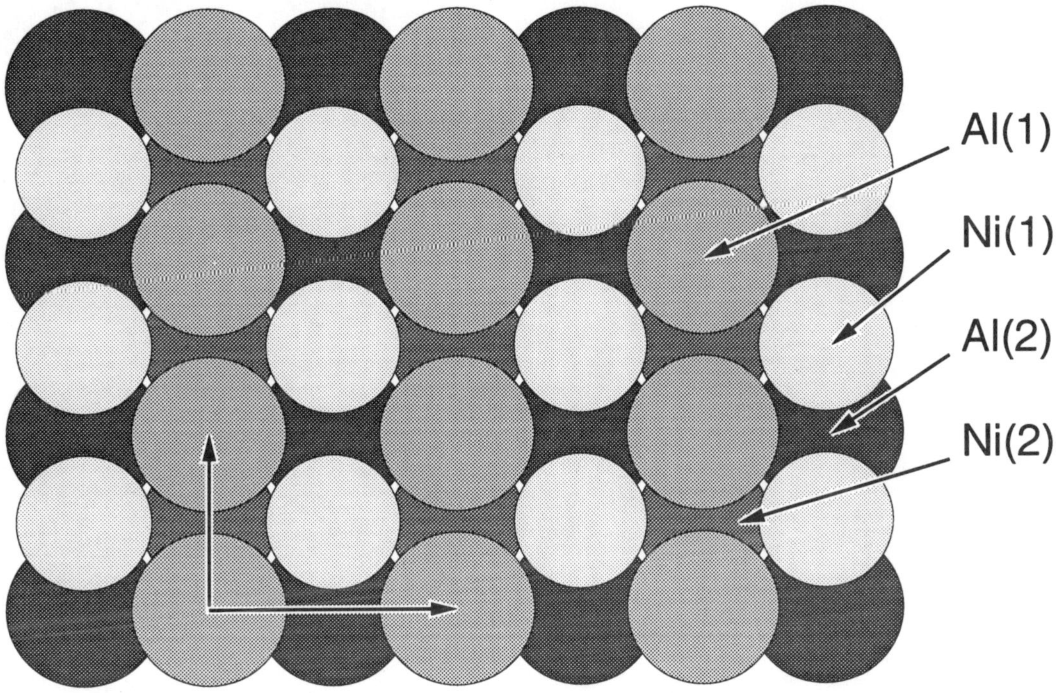

Fig. 142a : NiAl(110)-(1x1) (top view)

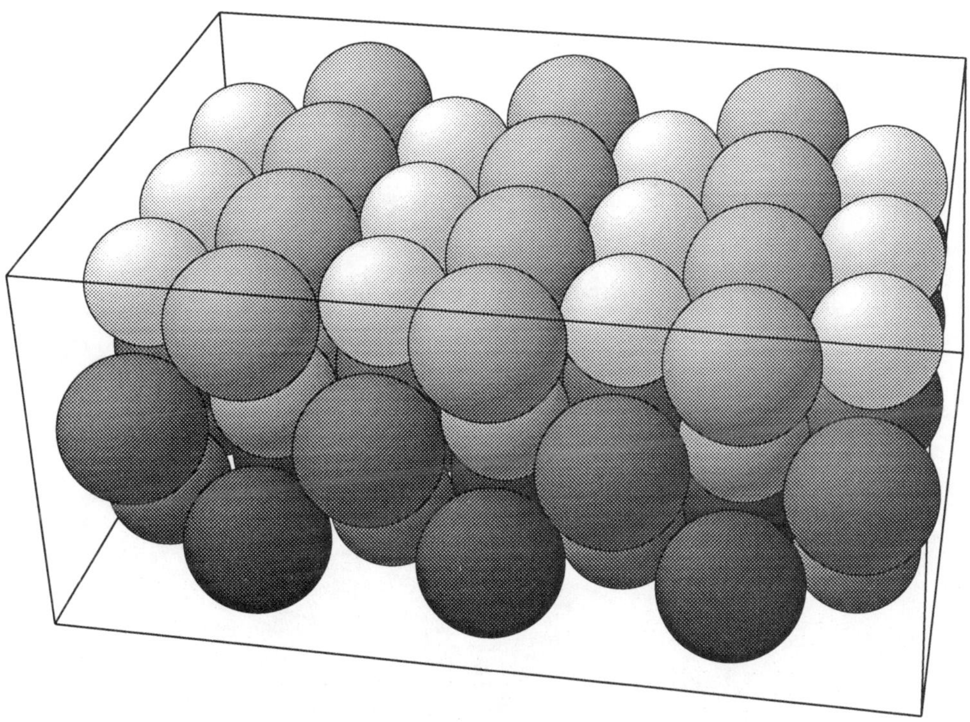

Fig. 142b : NiAl(110)-(1x1) (perspective)

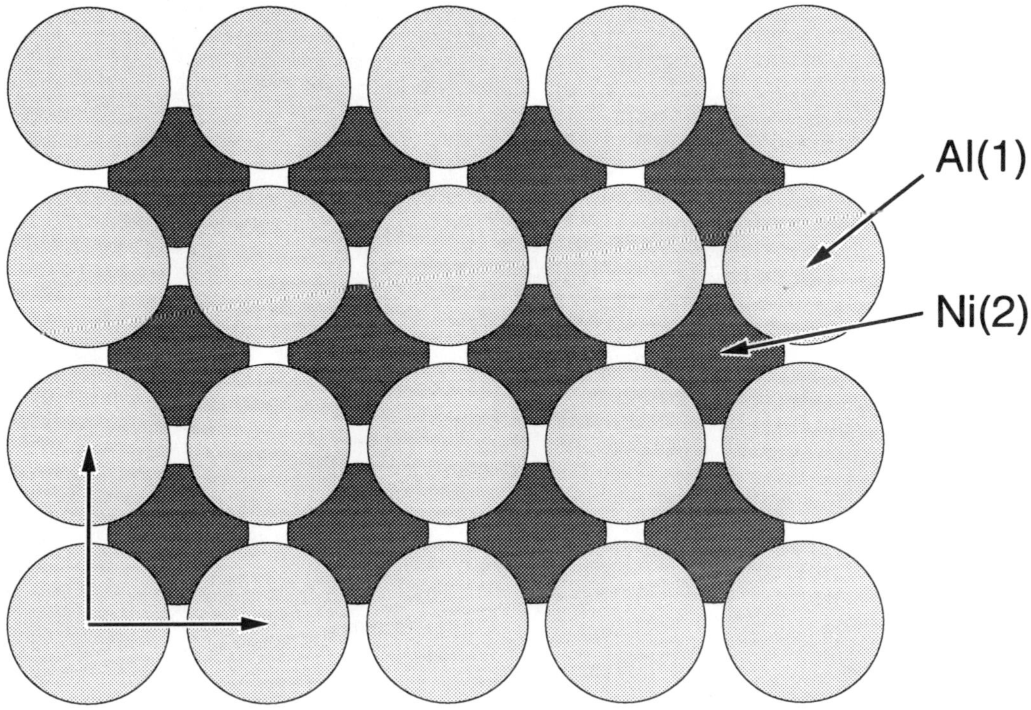

Fig. 143a : NiAl(100)-(1x1) (top view)

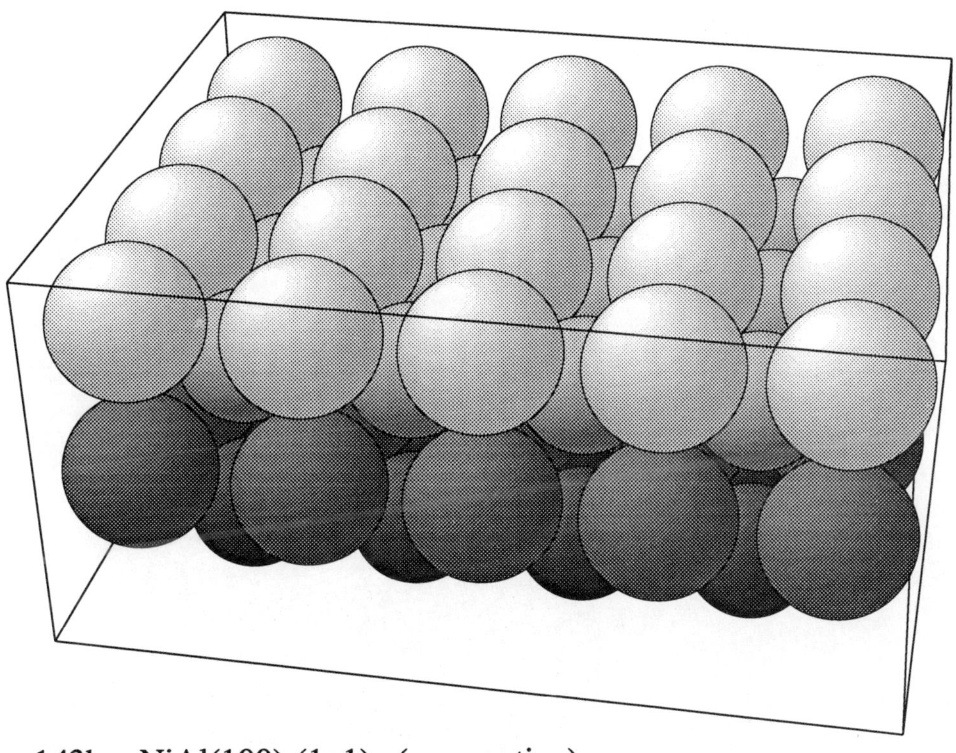

Fig. 143b : NiAl(100)-(1x1) (perspective)

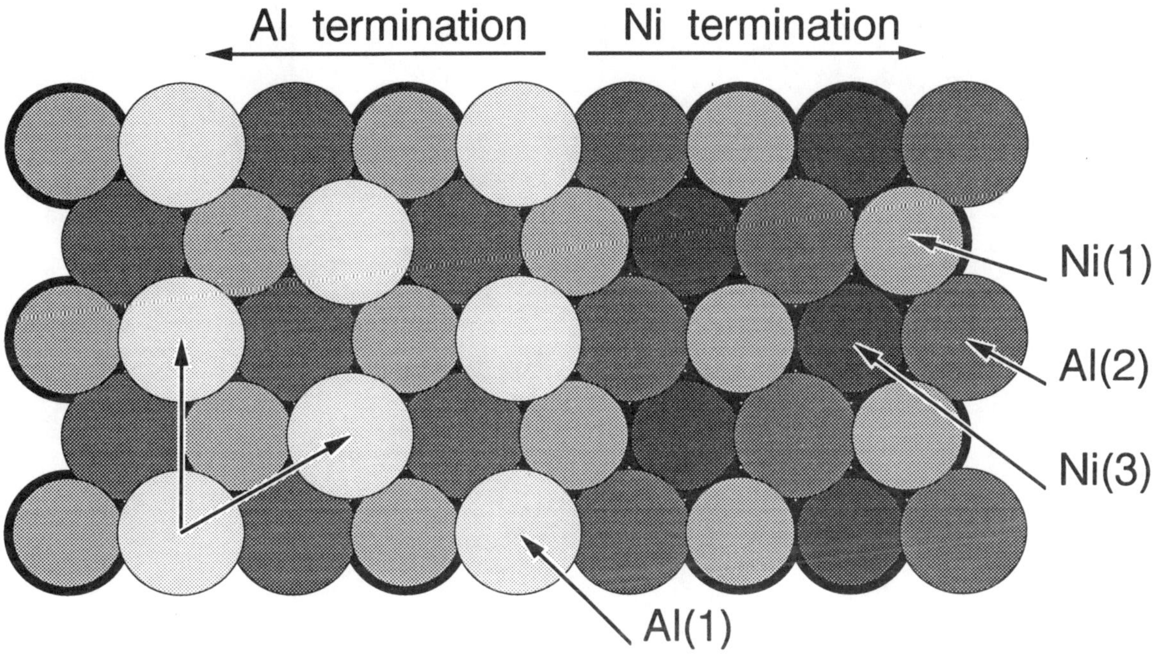

Fig. 144a : NiAl(111)-(1x1) Al/Ni terminated (top view)

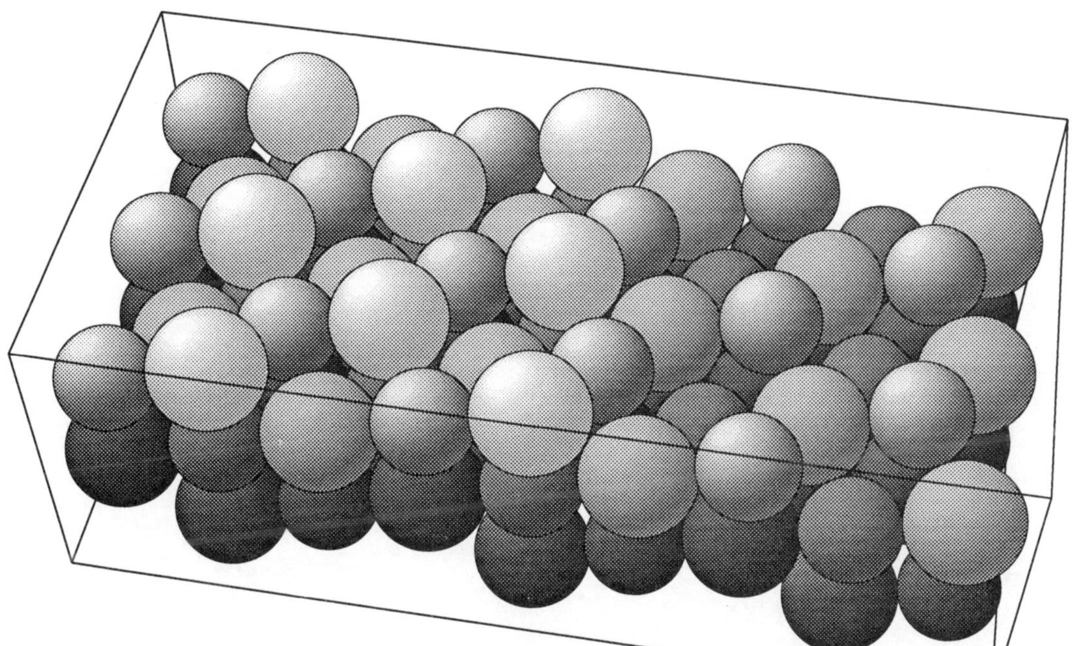

Fig. 144b : NiAl(111)-(1x1) Al/Ni terminated (perspective)

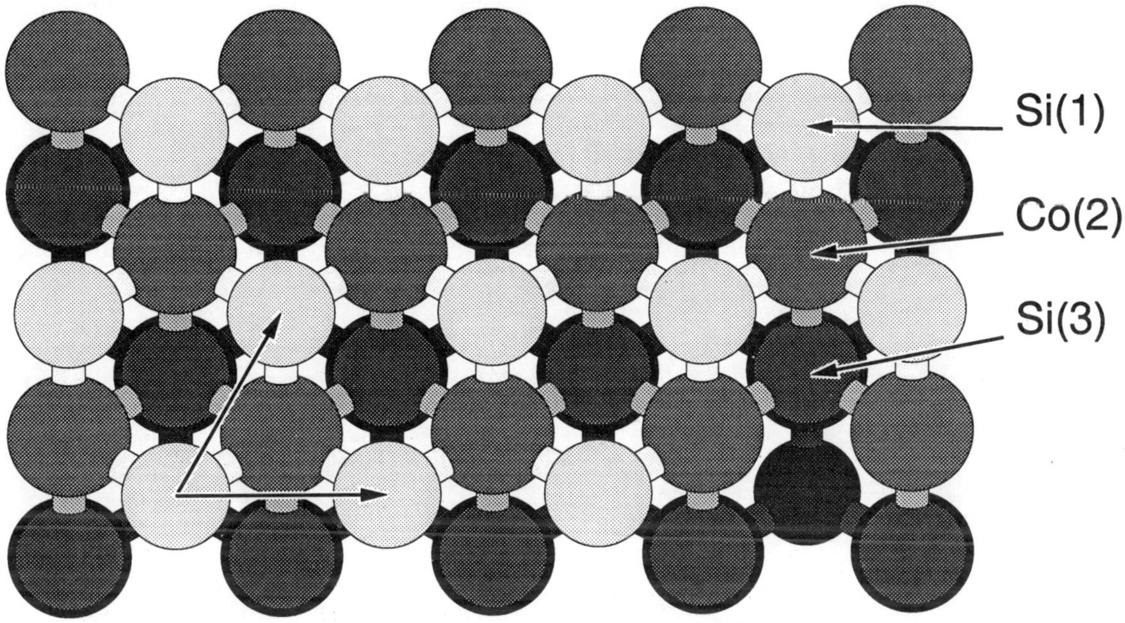

Fig. 145a : CoSi2(111)-(1x1) (top view)

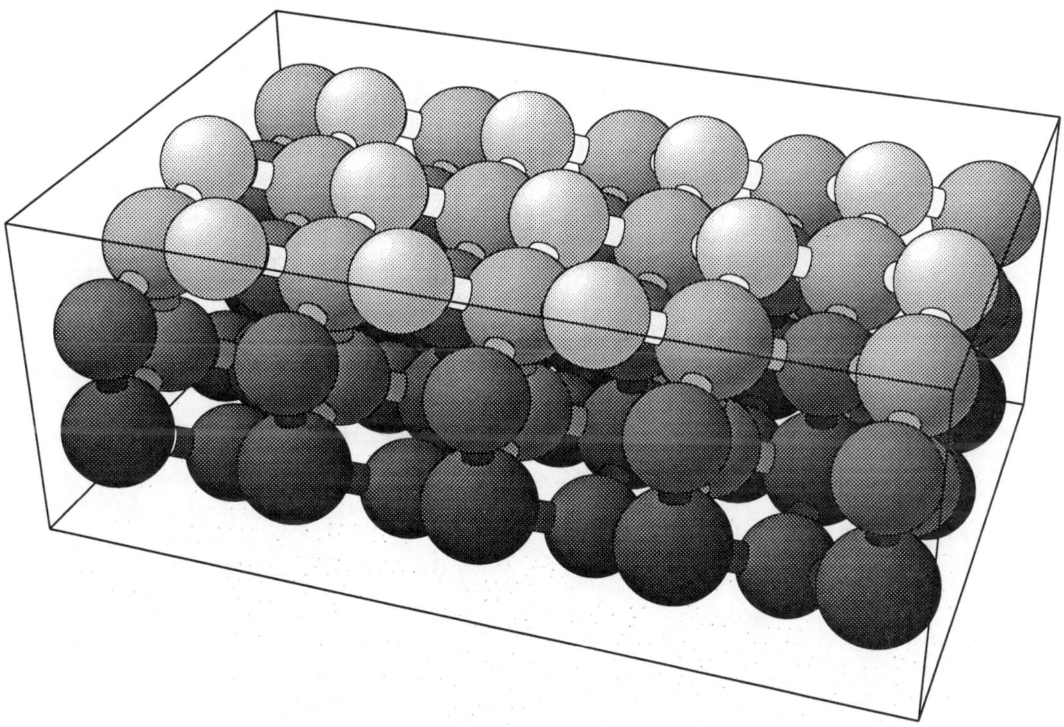

Fig. 145b : CoSi2(111)-(1x1) (perspective)

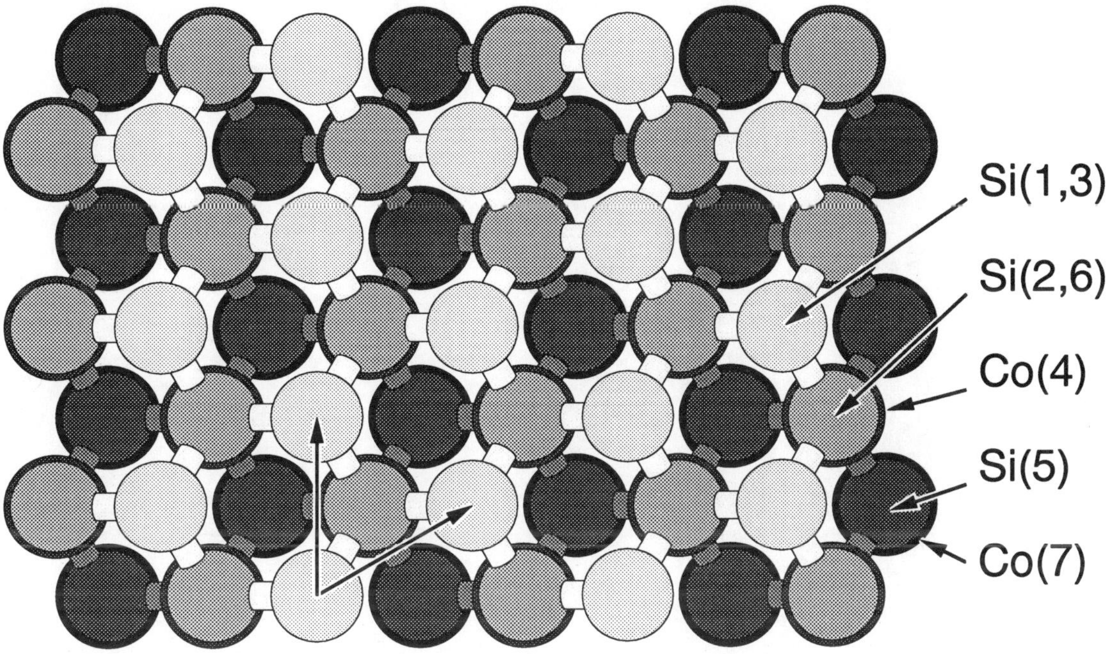

Fig. 146a : CoSi2(111)-(1x1) Si-rich (top view)

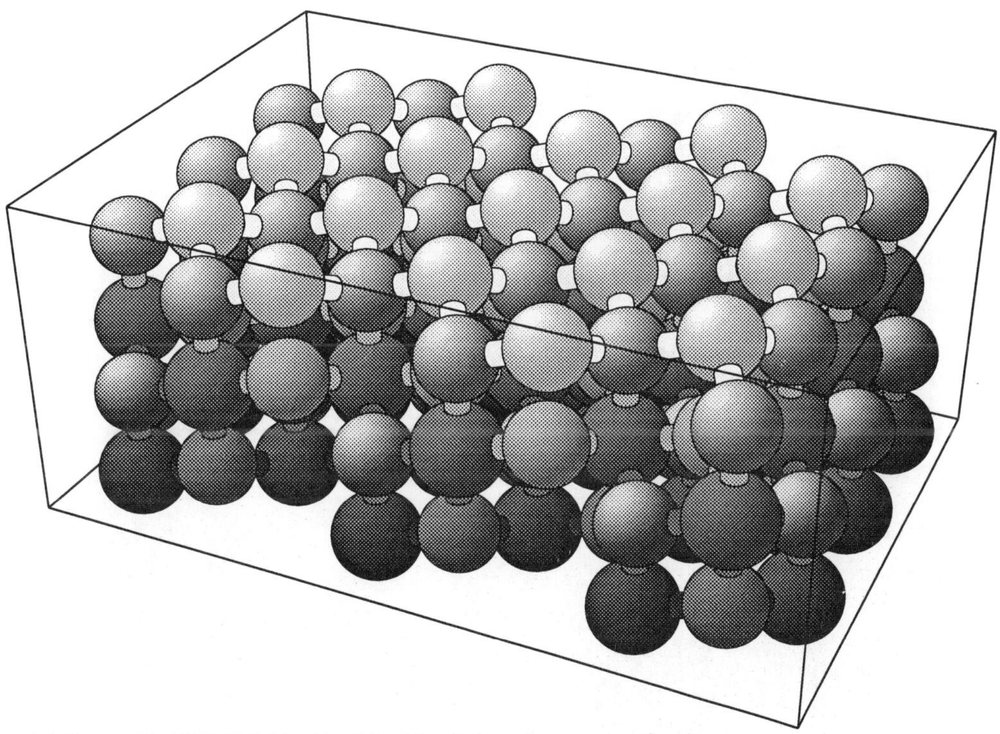

Fig. 146b : CoSi2(111)-(1x1) Si-rich (perspective)

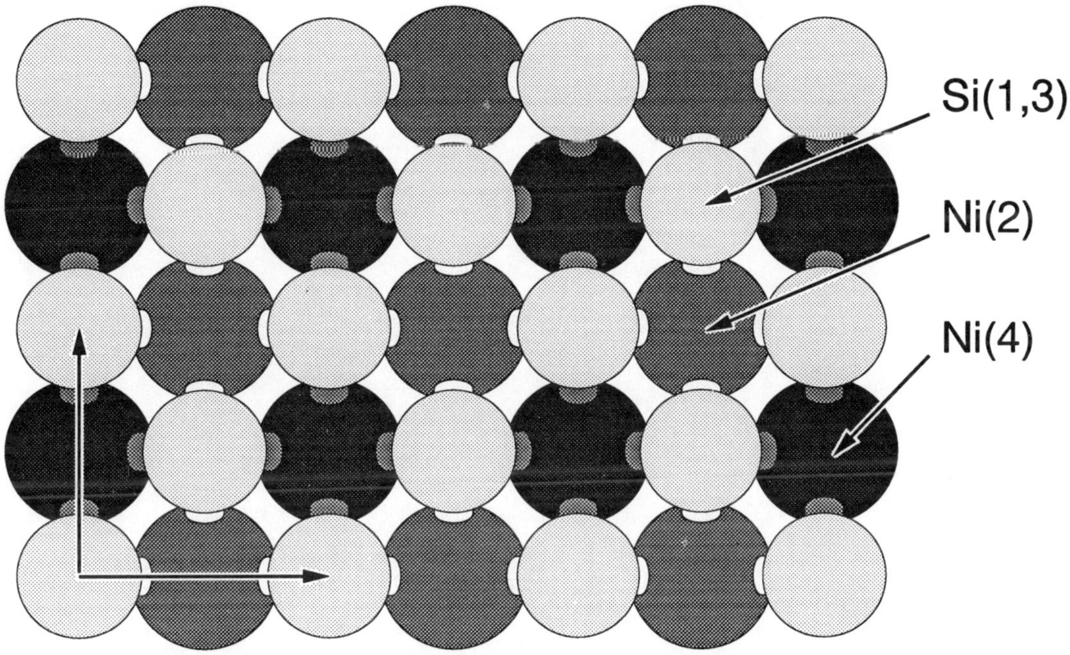

Fig. 147a : NiSi2(100)-(1x1) (top view)

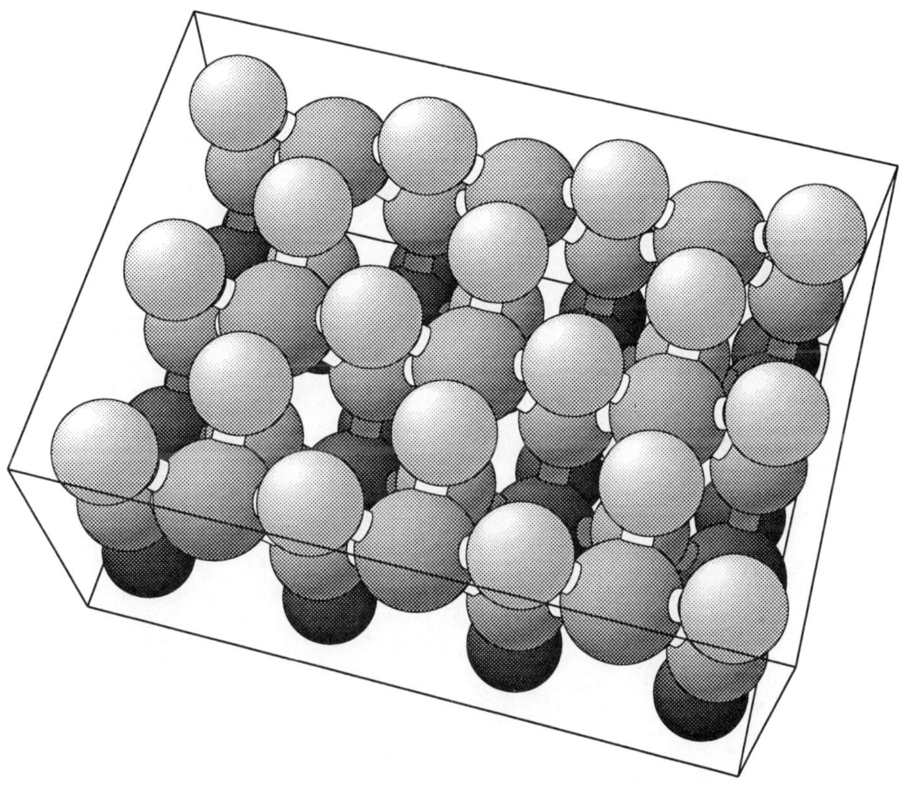

Fig. 147b : NiSi2(100)-(1x1) (perspective)

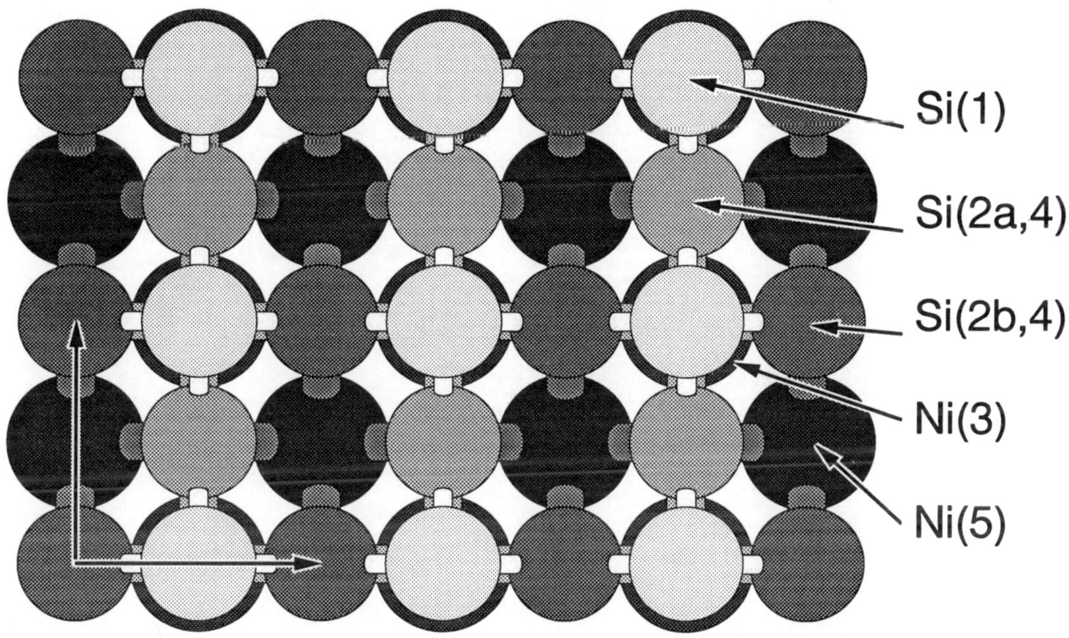

Fig. 148a : NiSi2(100)-(1x1) Si-rich (top view)

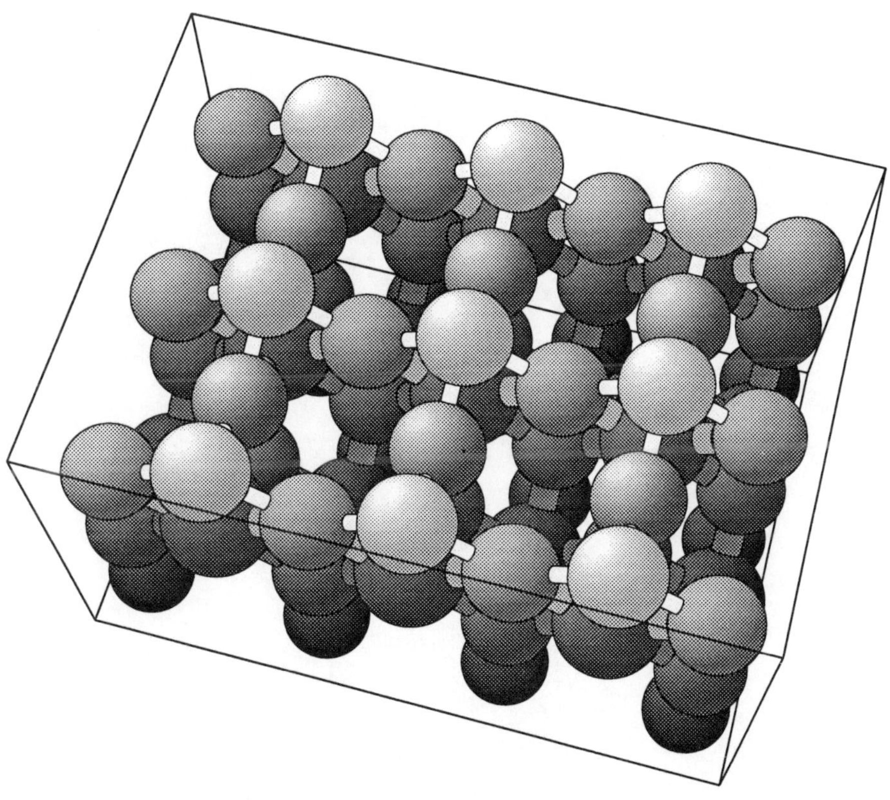

Fig. 148b : NiSi2(100)-(1x1) Si-rich (perspective)

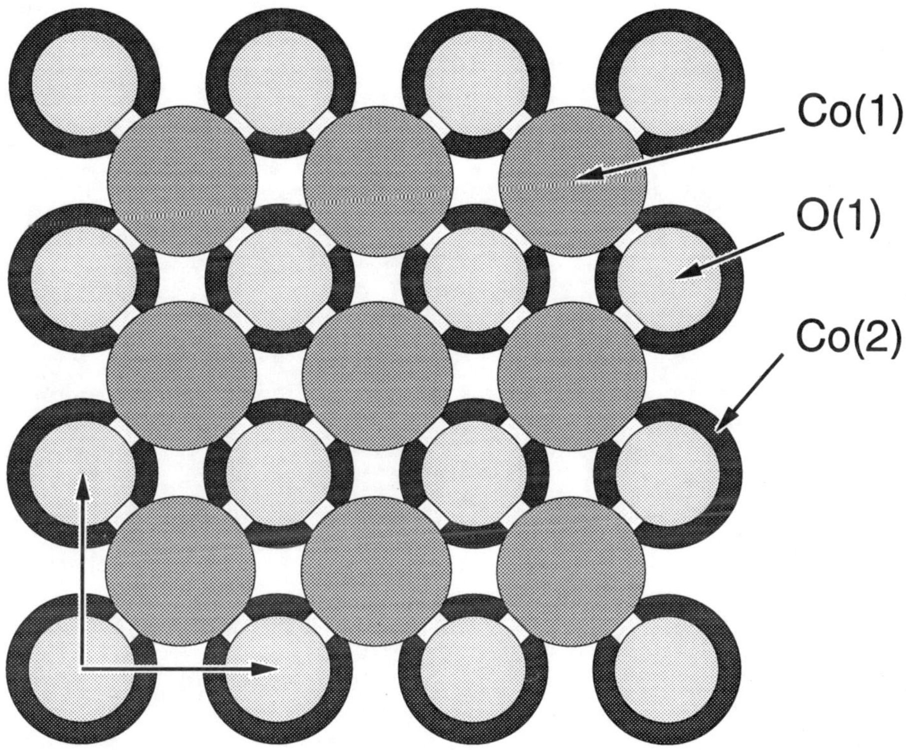

Fig. 149a : CoO(100)-(1x1) (top view)

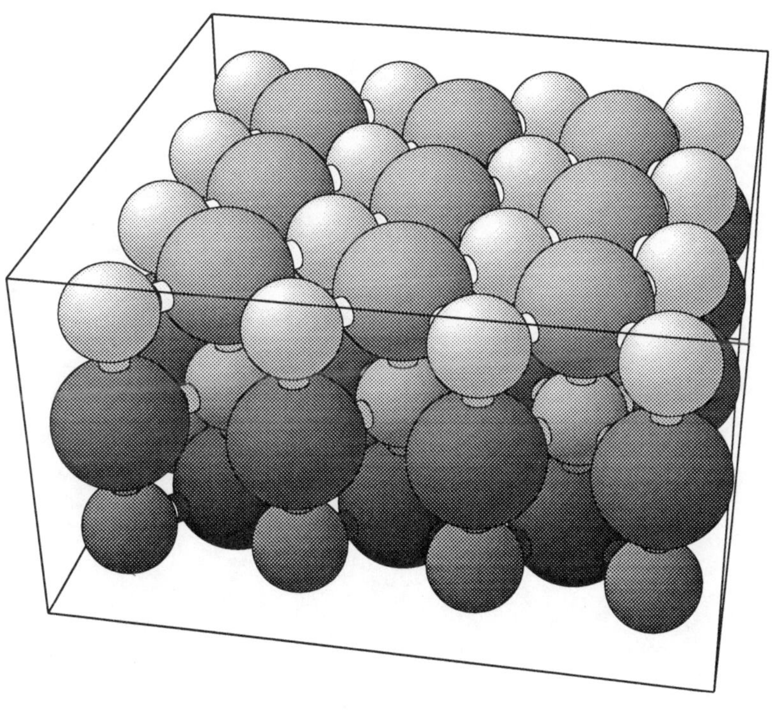

Fig. 149b : CoO(100)-(1x1) (perspective)

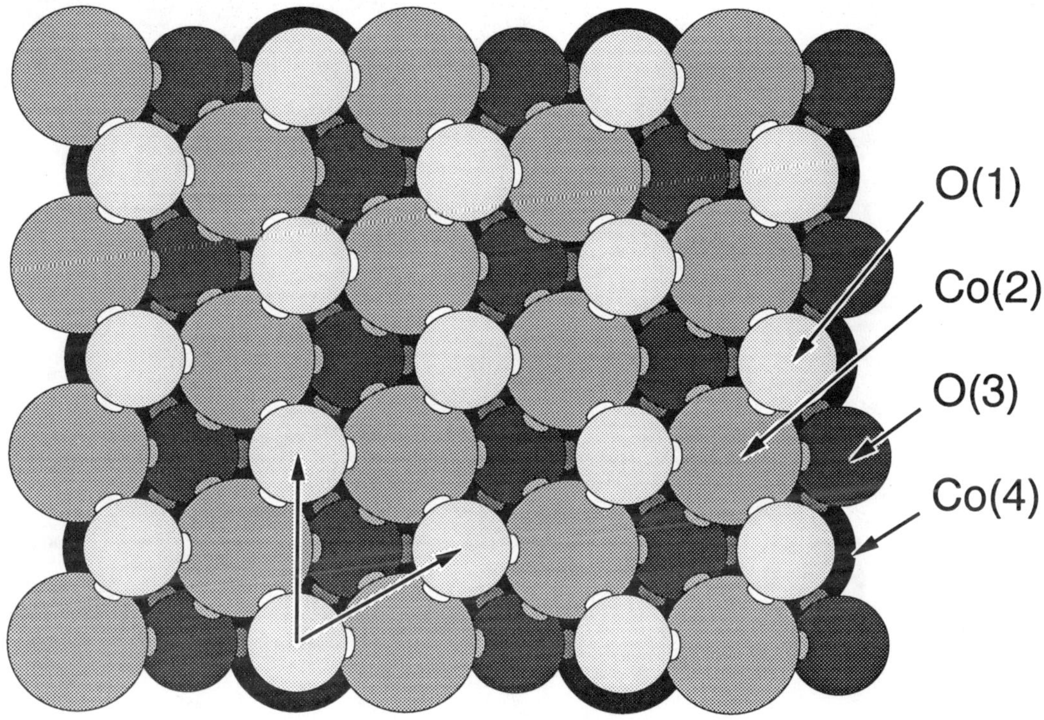

Fig. 150a : CoO(111)-(1x1) (top view)

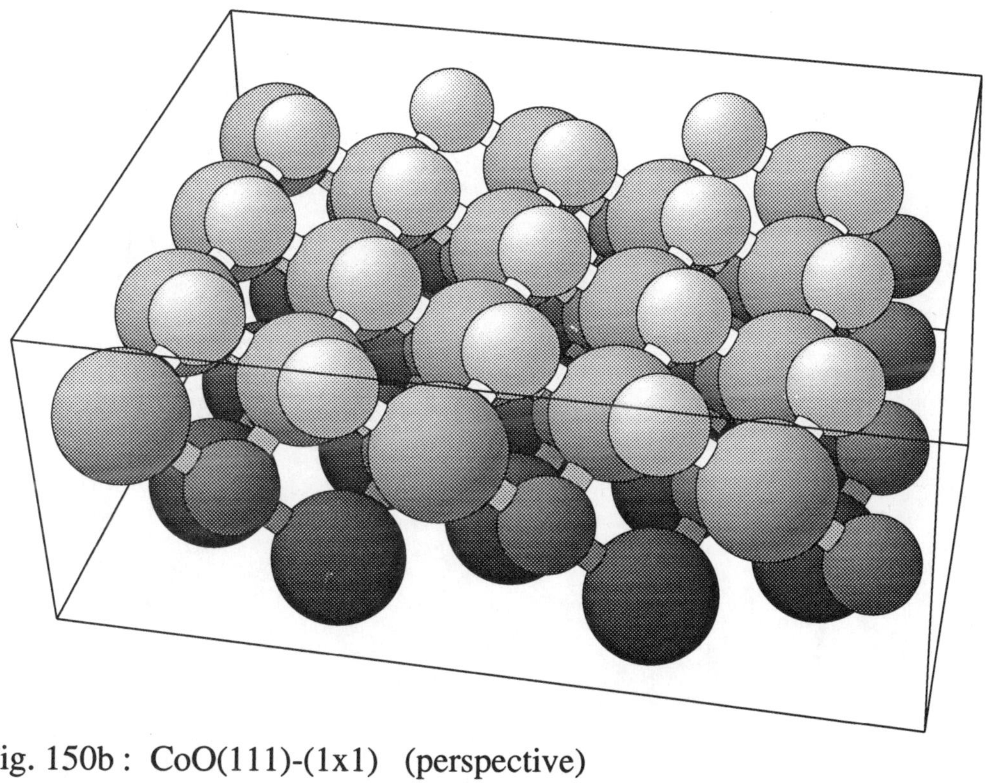

Fig. 150b : CoO(111)-(1x1) (perspective)

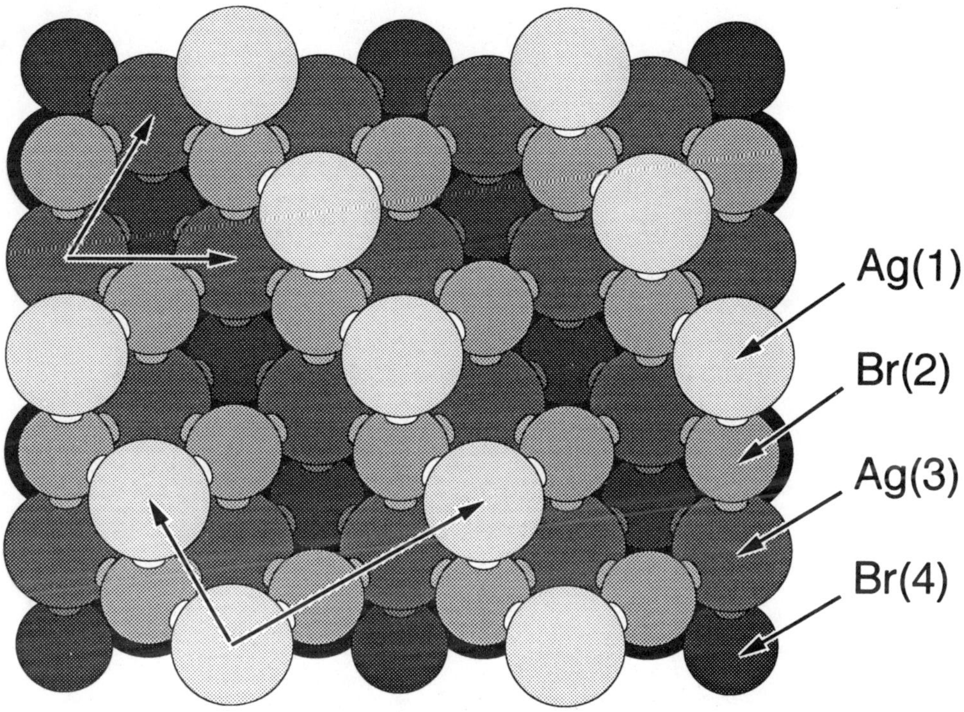

Fig. 151a : AgBr(111)-(2x1) (top view)

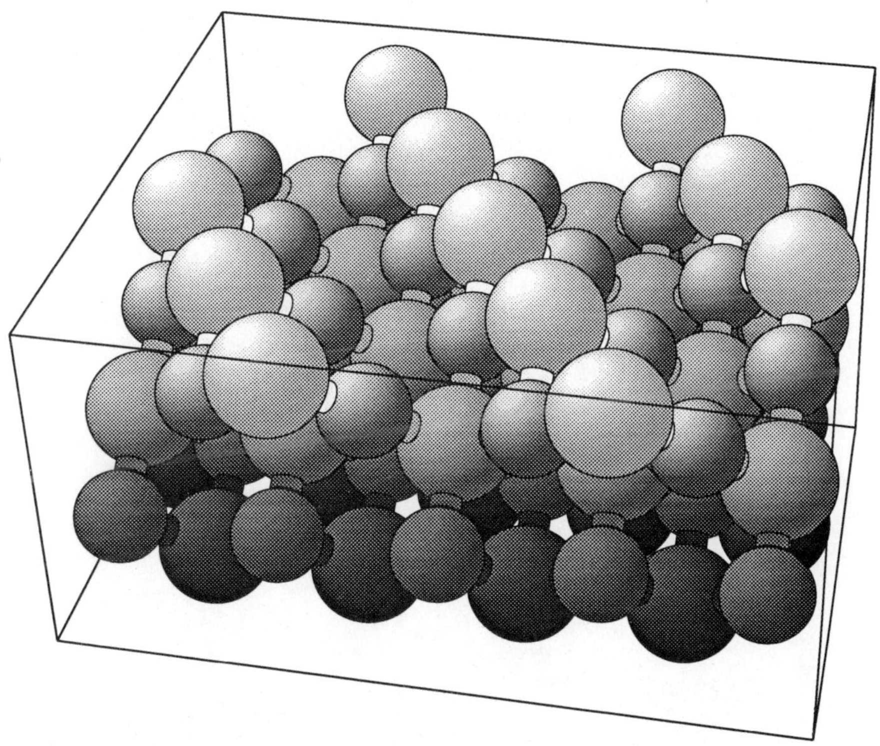

Fig. 151b : AgBr(111)-(2x1) (perspective)

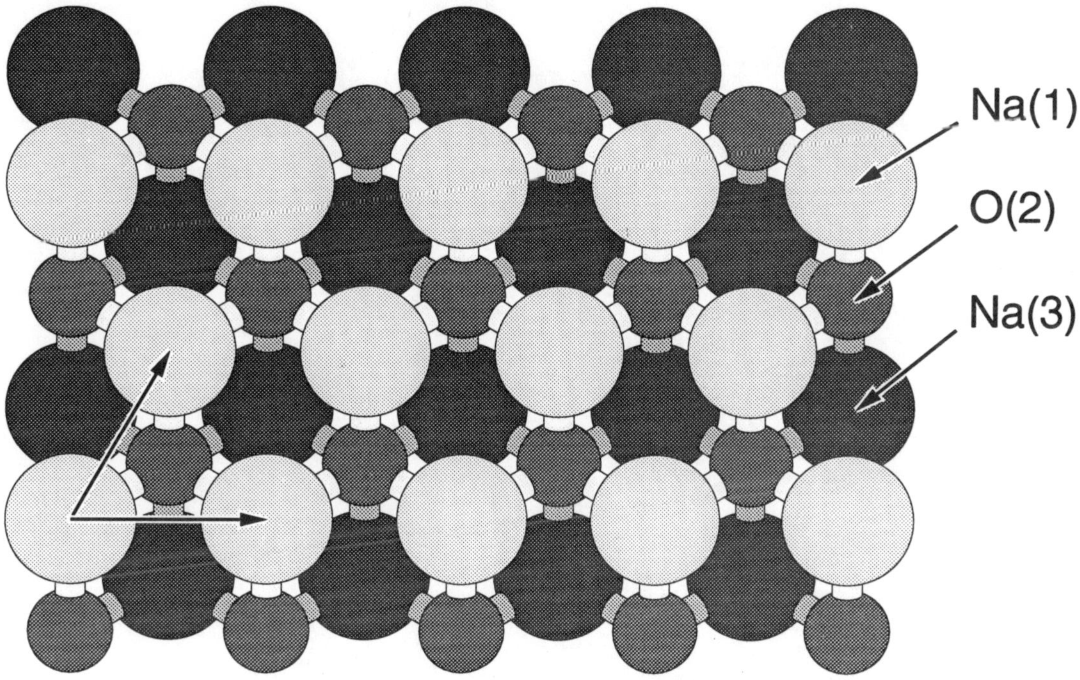

Fig. 152a : Na2O(111)-(1x1) (top view)

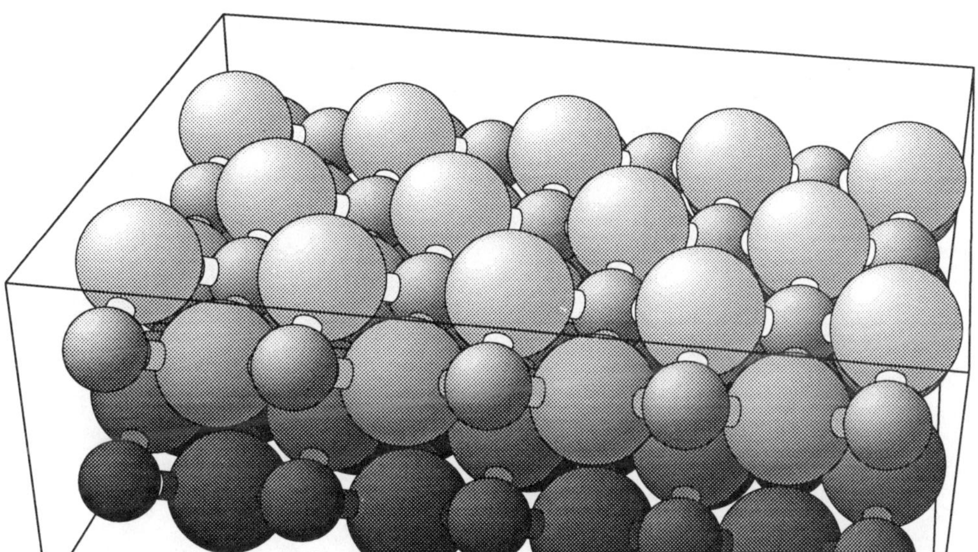

Fig. 152b : Na2O(111)-(1x1) (perspective)

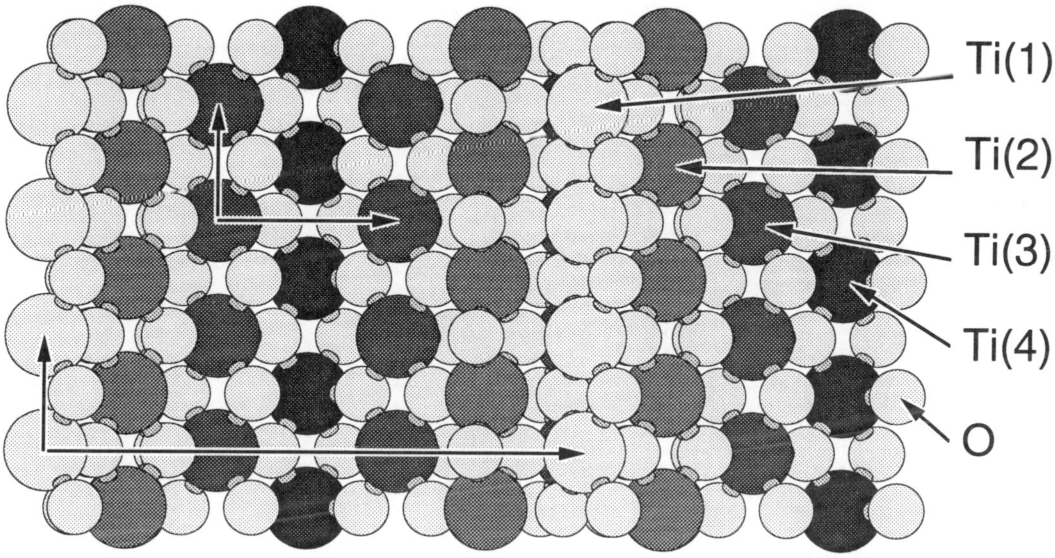

Fig. 153a : TiO2(100)-(3x1) (top view)

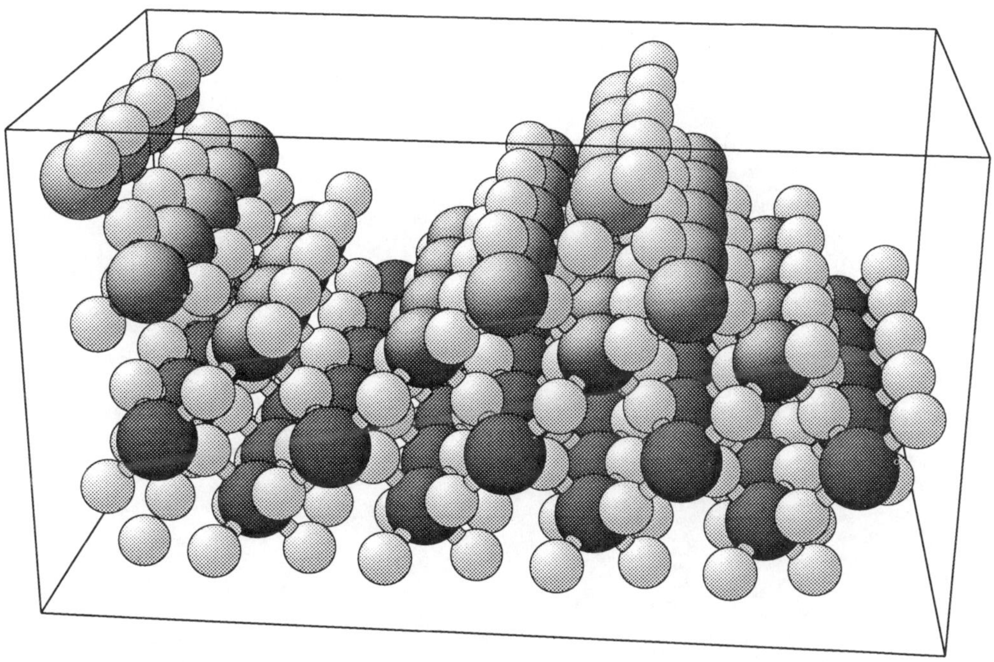

Fig. 153b : TiO2(100)-(3x1) (perspective)

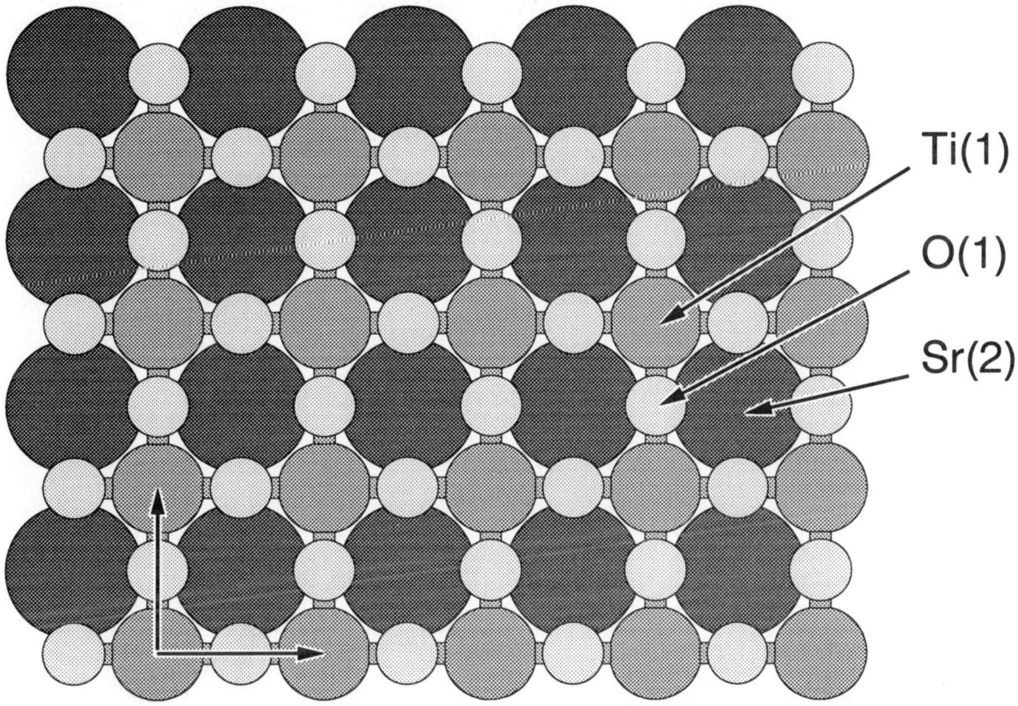

Fig. 154a : SrTiO3(100)-(1x1) O-Ti-O termination (top view)

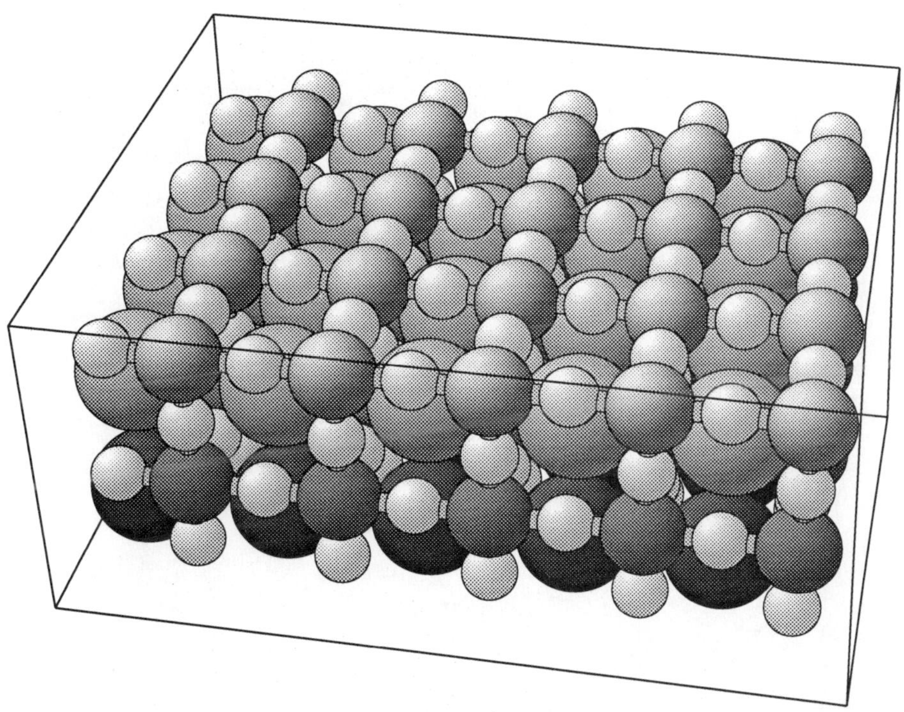

Fig. 154b : SrTiO3(100)-(1x1) O-Ti-O termination (perspective)

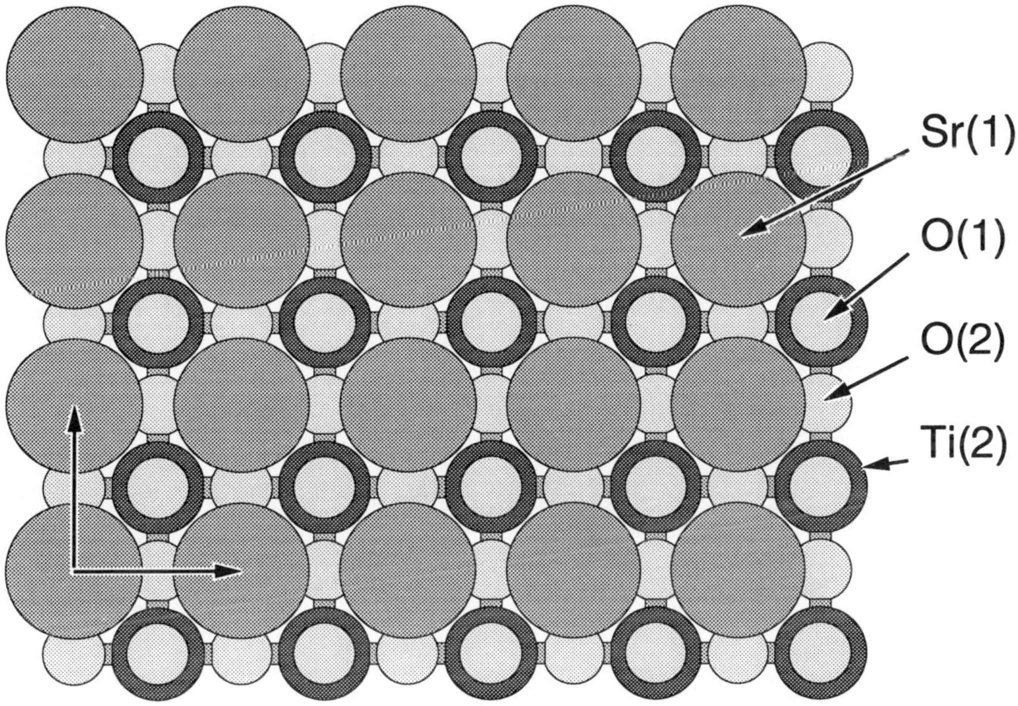

Fig. 155a : SrTiO3(100)-(1x1) Sr-O termination (top view)

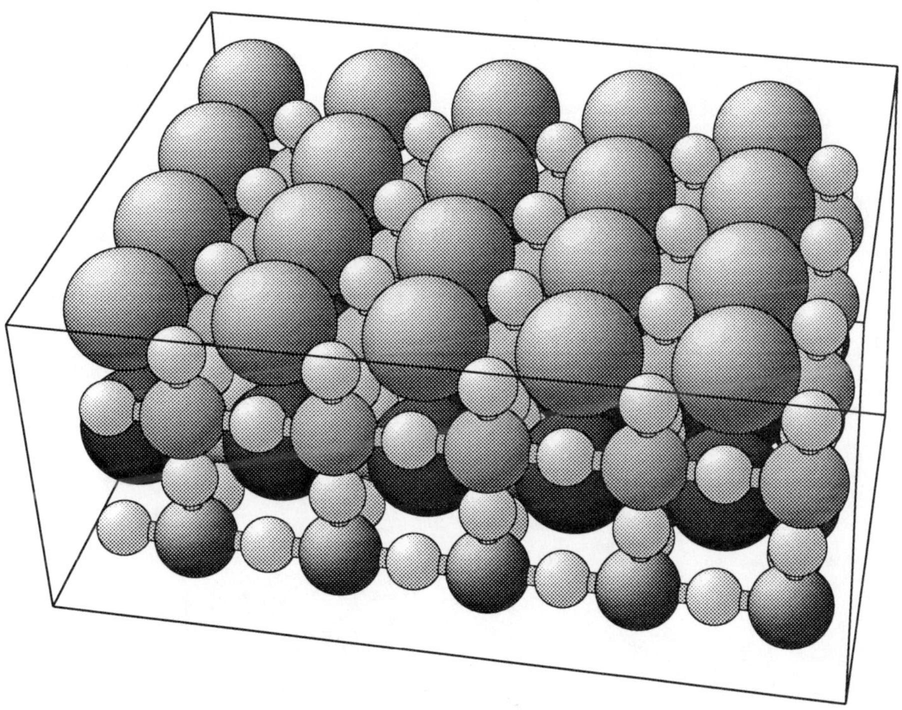

Fig. 155b : SrTiO3(100)-(1x1) Sr-O termination (perspective)

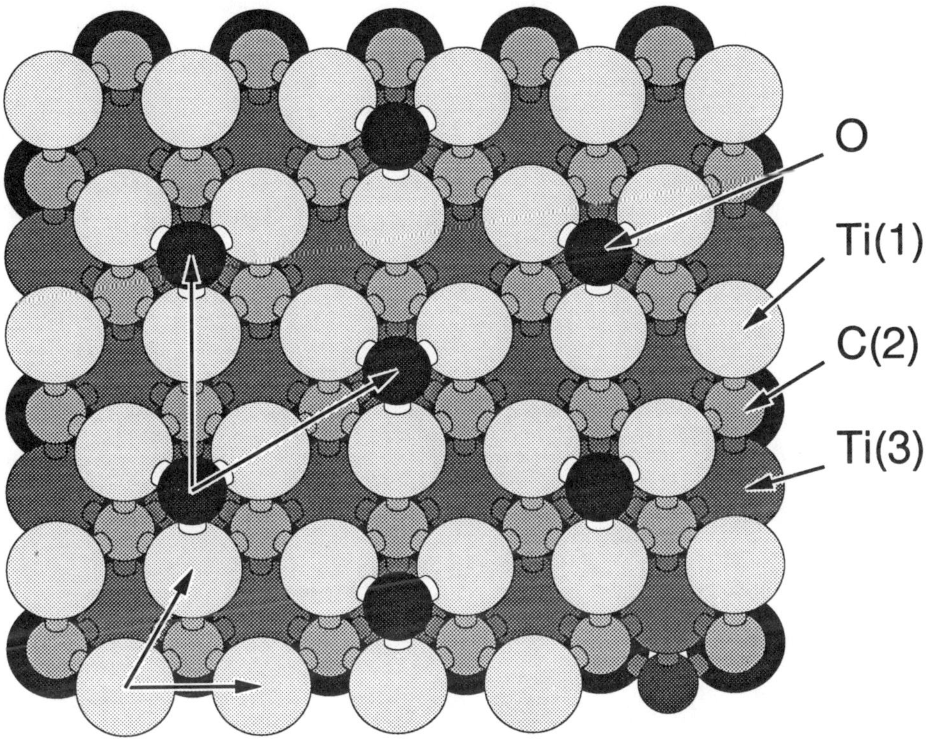

Fig. 156a : TiC(111)-(√3x√3)R30°-O (top view)

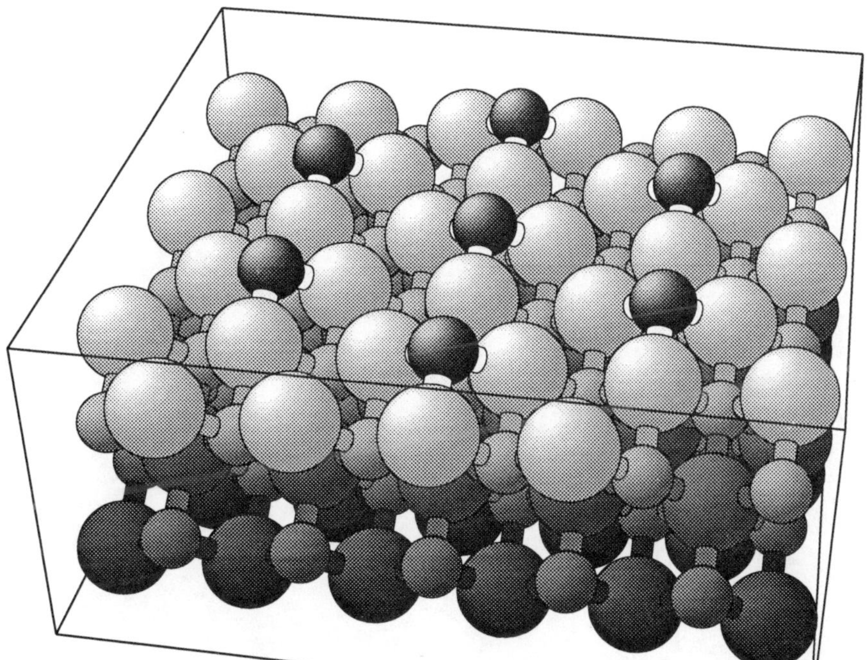

Fig. 156b : TiC(111)-(√3x√3)R30°-O (perspective)

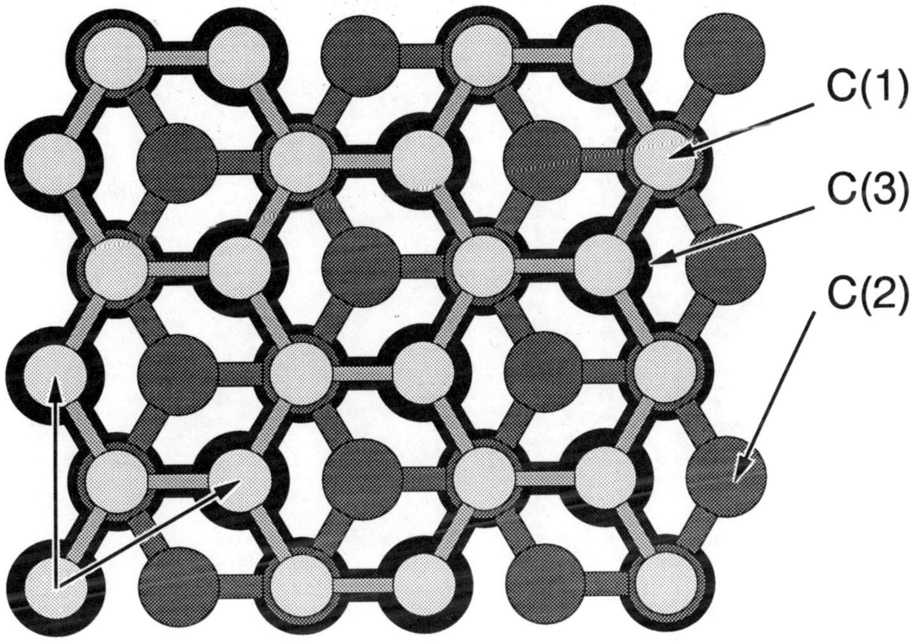

Fig. 157a : C(0001)-(1x1) graphite (top view)

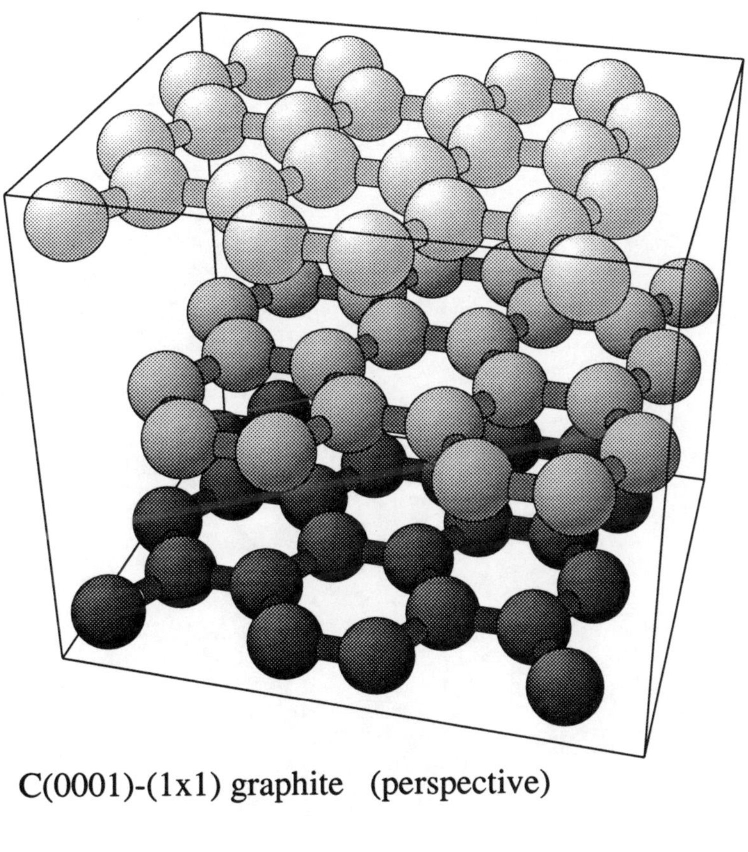

Fig. 157b : C(0001)-(1x1) graphite (perspective)

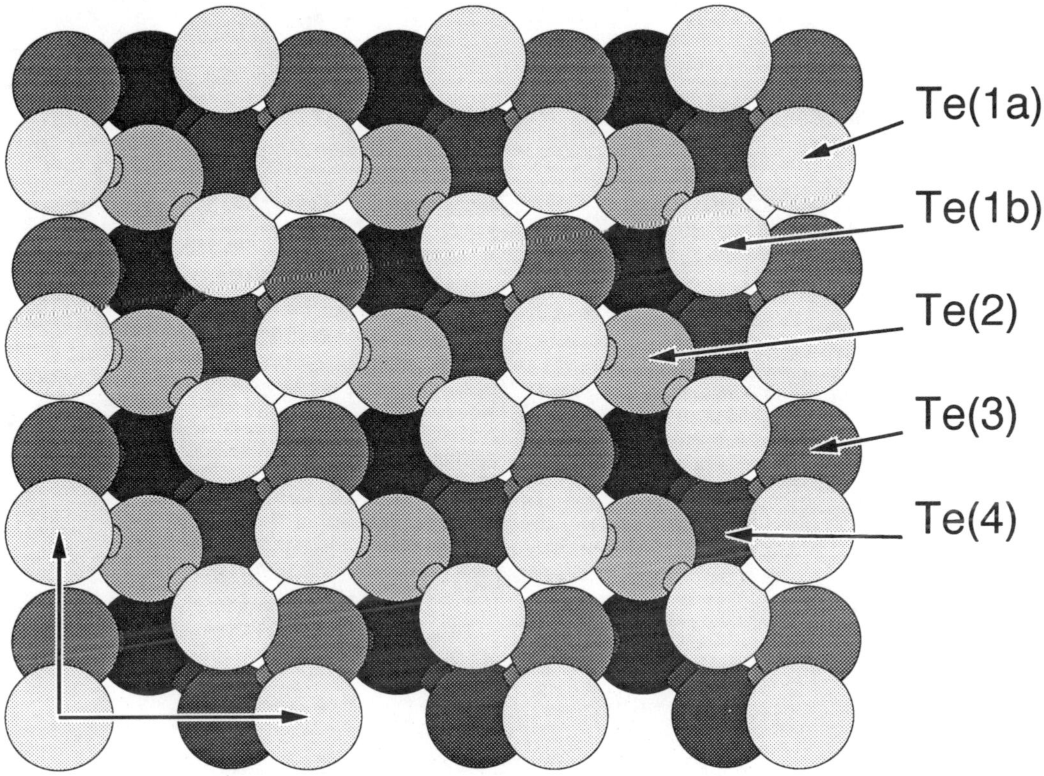

Fig. 158a : Te(10-10)-(1x1) (top view)

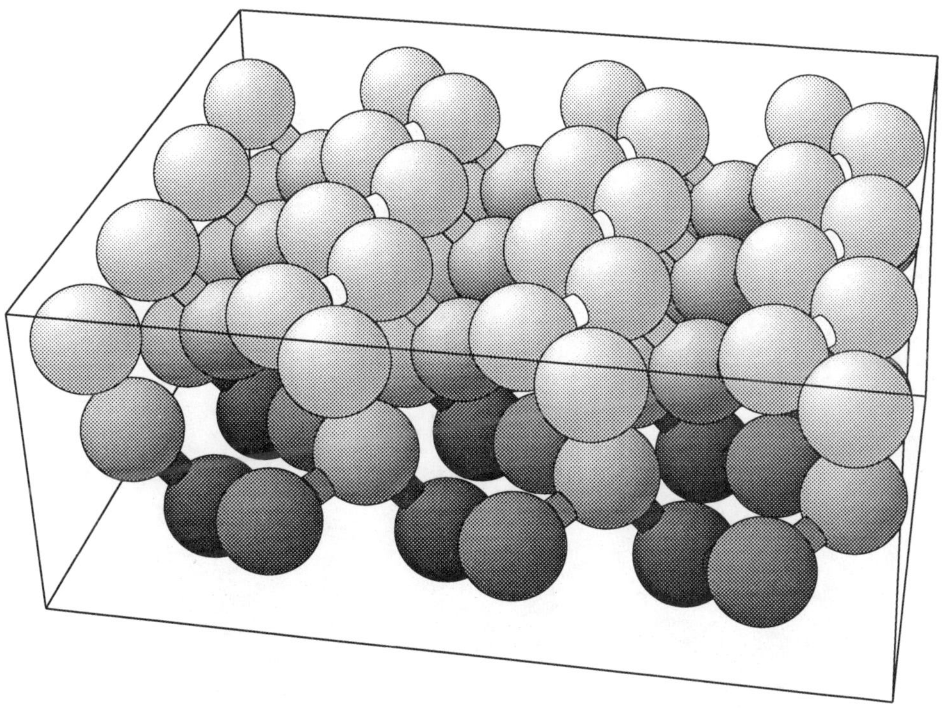

Fig. 158b : Te(10-10)-(1x1) (perspective)

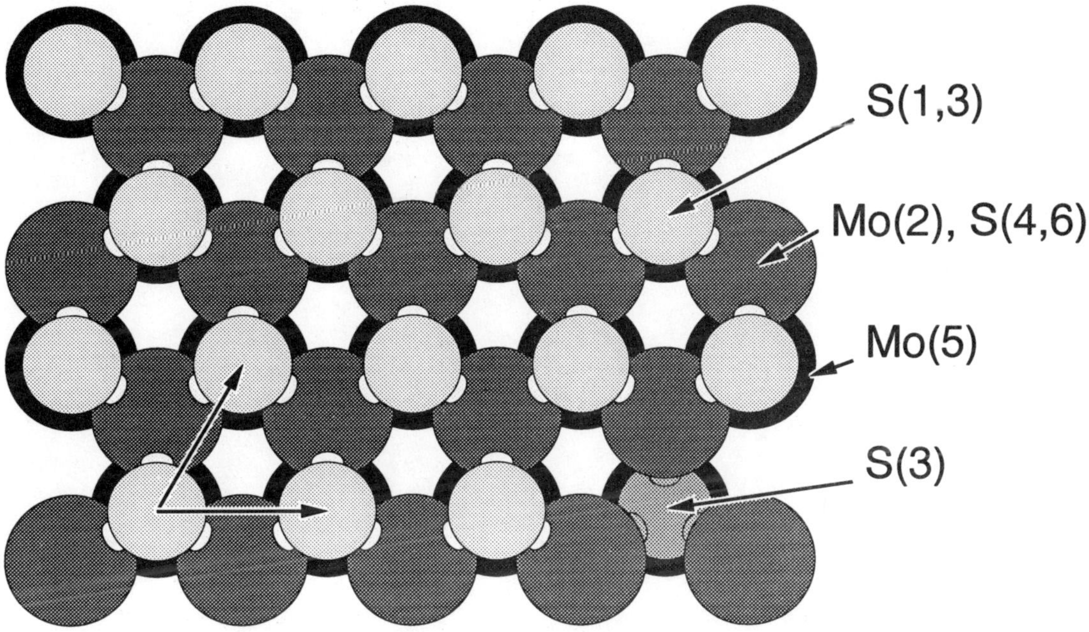

Fig. 159a : MoS2(0001)-(1x1) (top view)

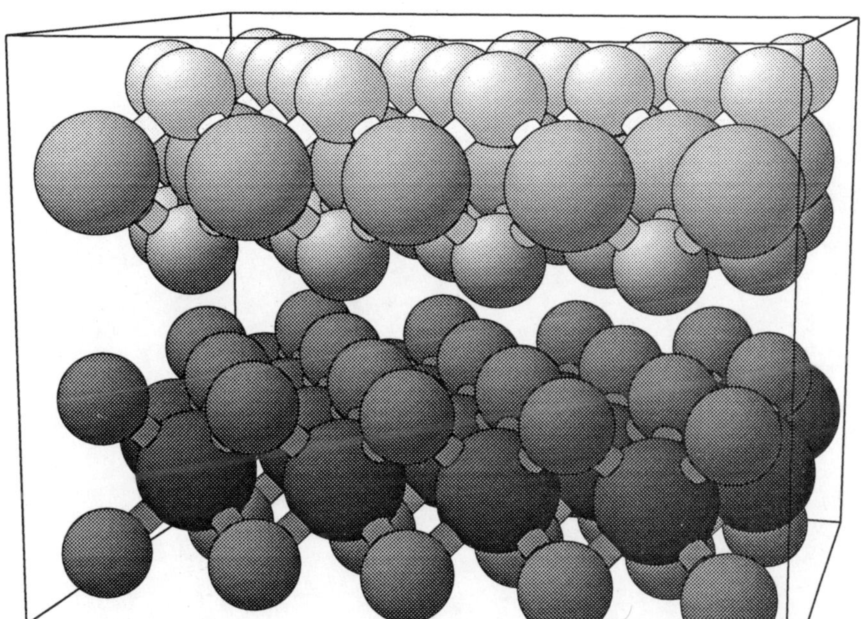

Fig. 159b : MoS2(0001)-(1x1) (perspective)

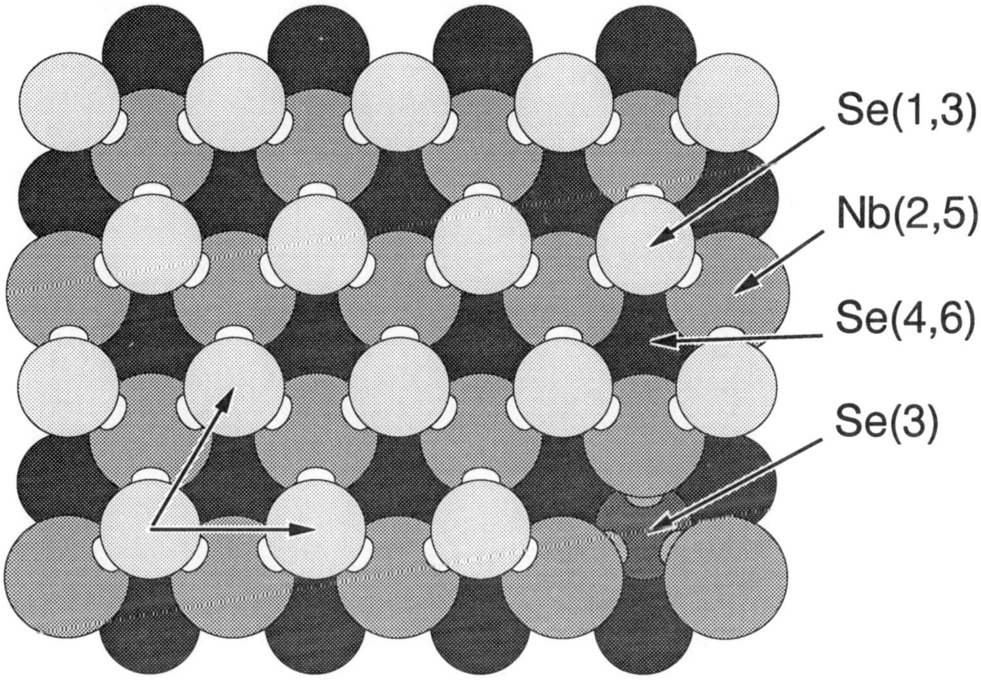

Fig. 160a : NbSe2(0001)-(1x1) (top view)

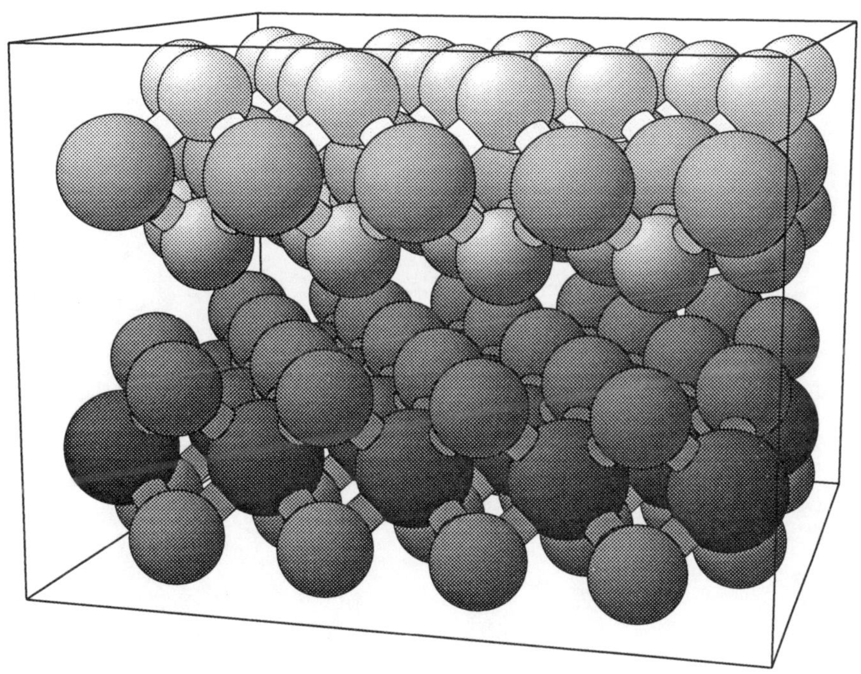

Fig. 160b : NbSe2(0001)-(1x1) (perspective)

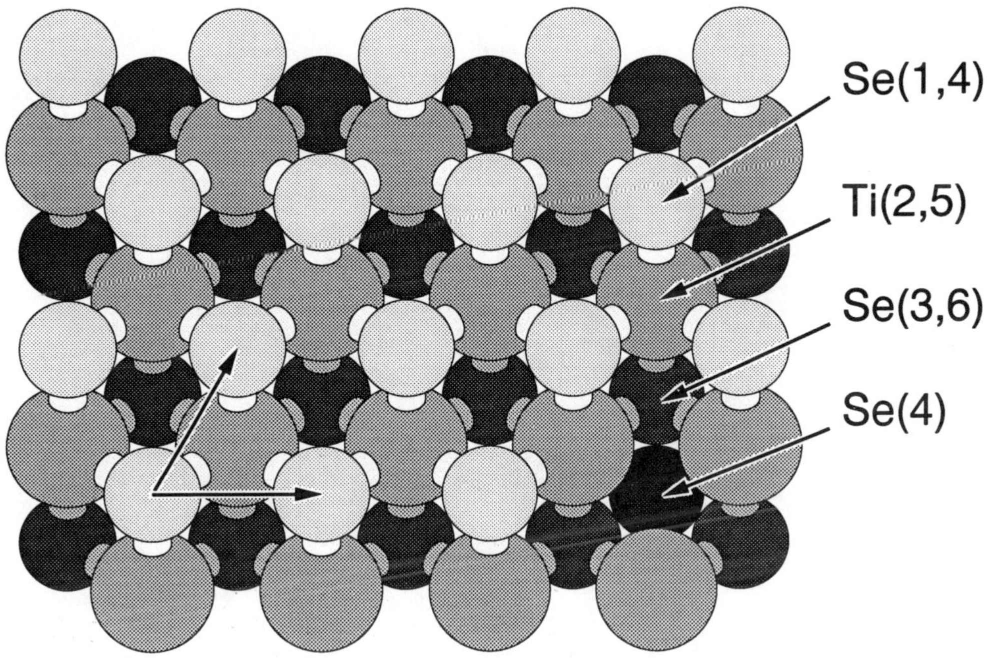

Fig. 161a : TiSe2(0001)-(1x1) (top view)

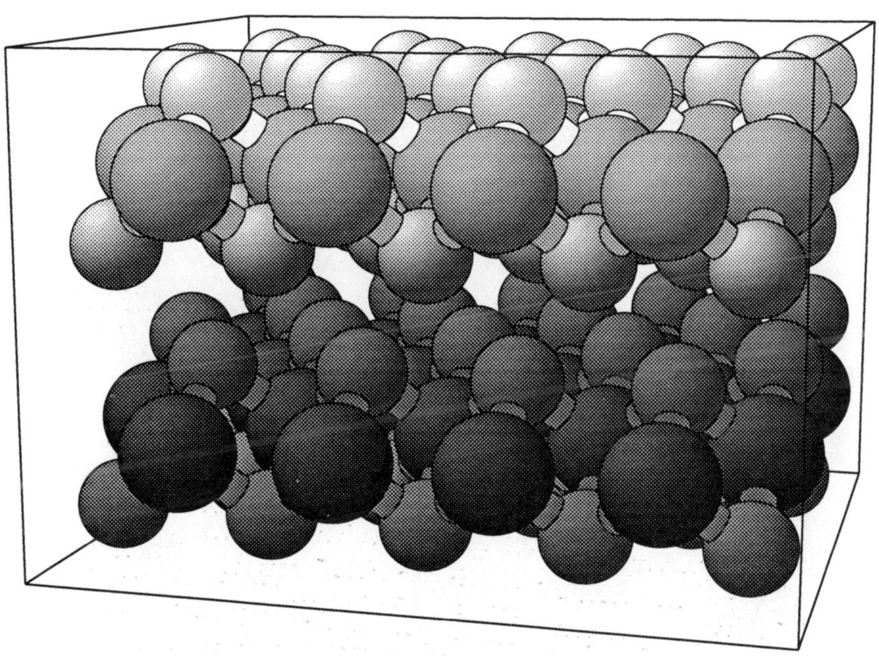

Fig. 161b : TiSe2(0001)-(1x1) (perspective)

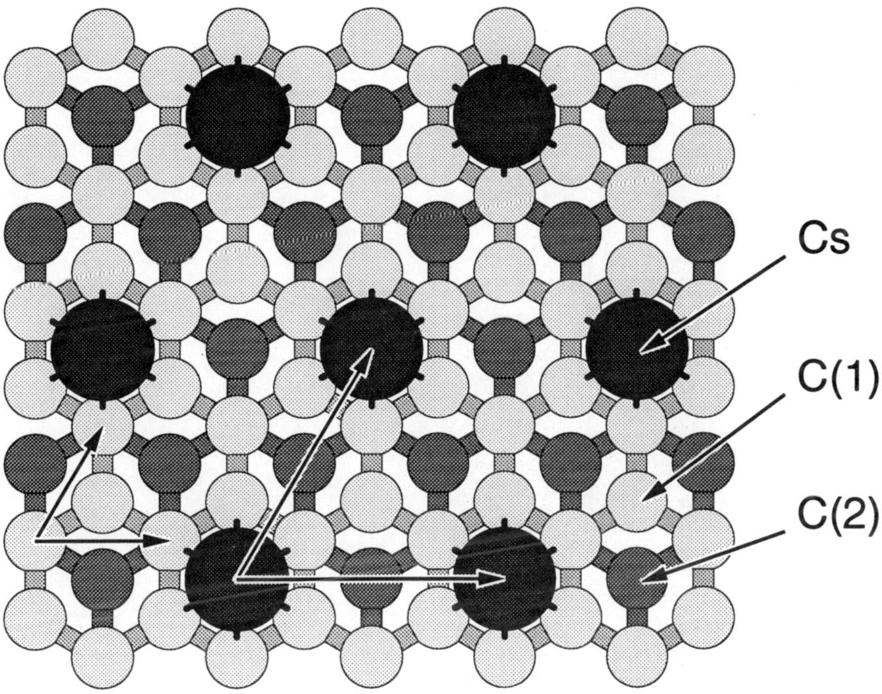

Fig. 162a : C(0001)-(2x2)-Cs (top view)

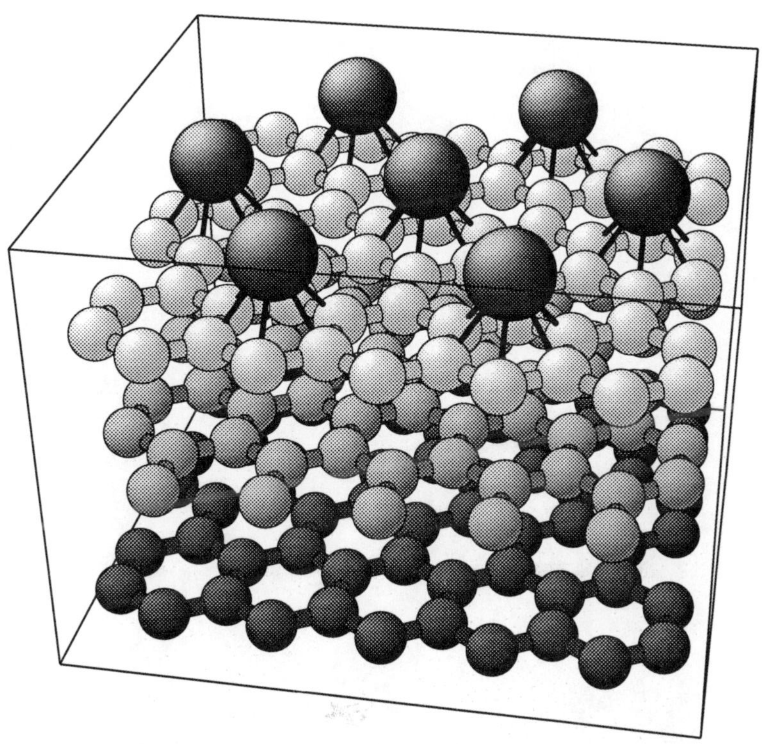

Fig. 162b : C(0001)-(2x2)-Cs (perspective)

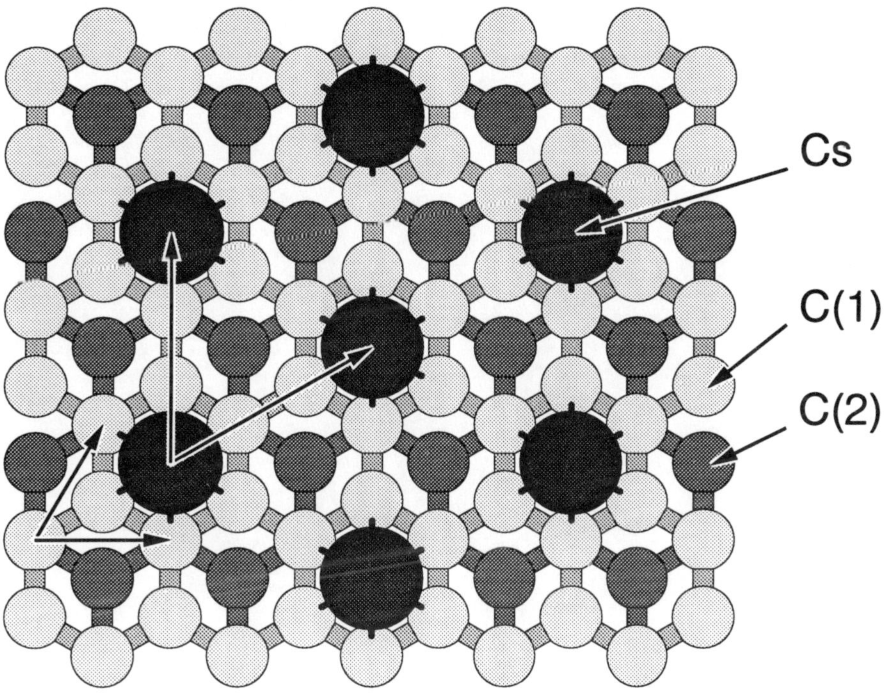

Fig. 163a : C(0001)-(√3x√3)R30°-Cs (top view)

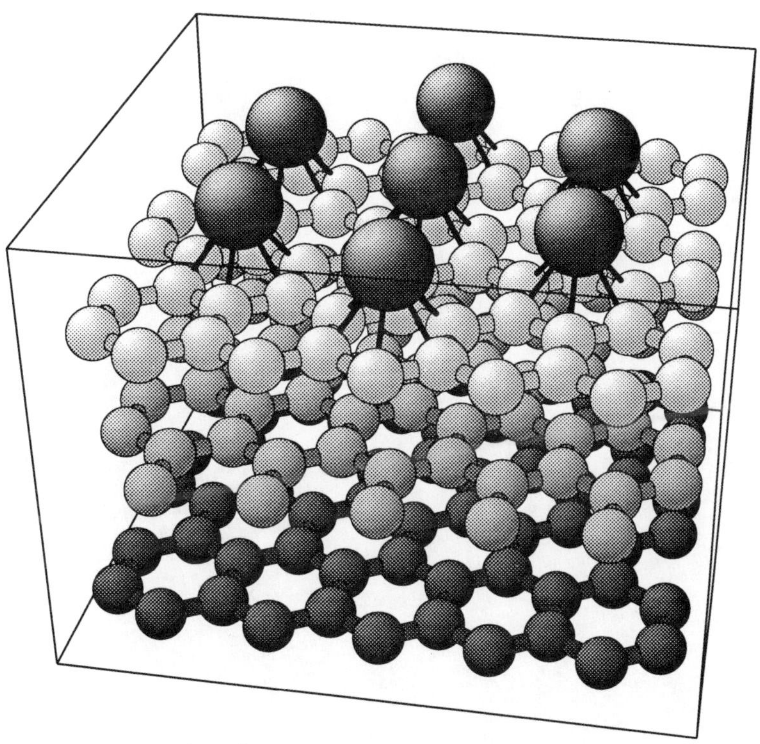

Fig. 163b : C(0001)-(√3x√3)R30°-Cs (perspective)

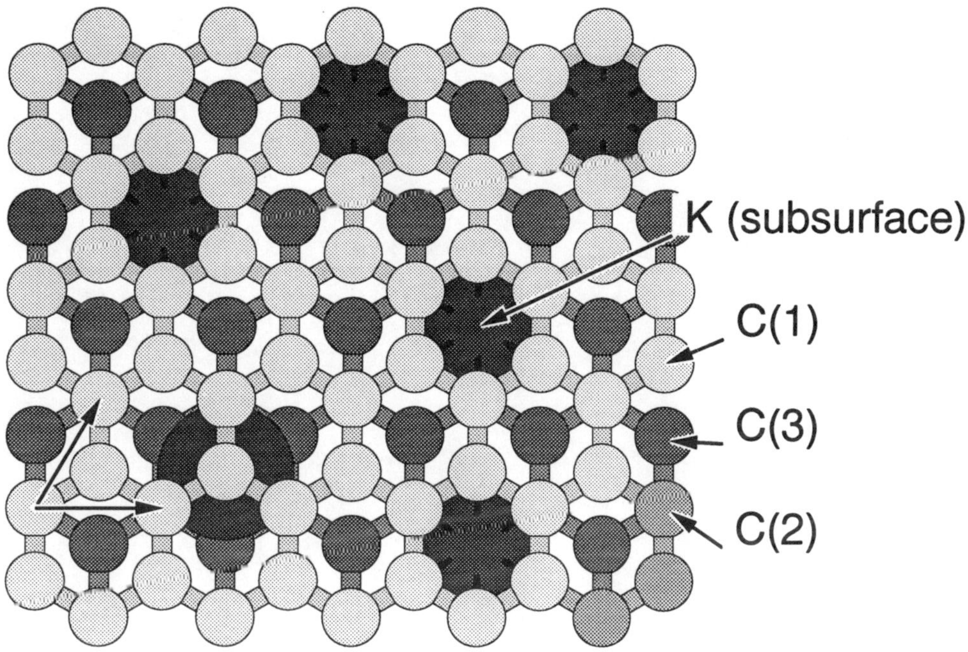

Fig. 164a : C(0001)-1K disordered underlayer (top view)

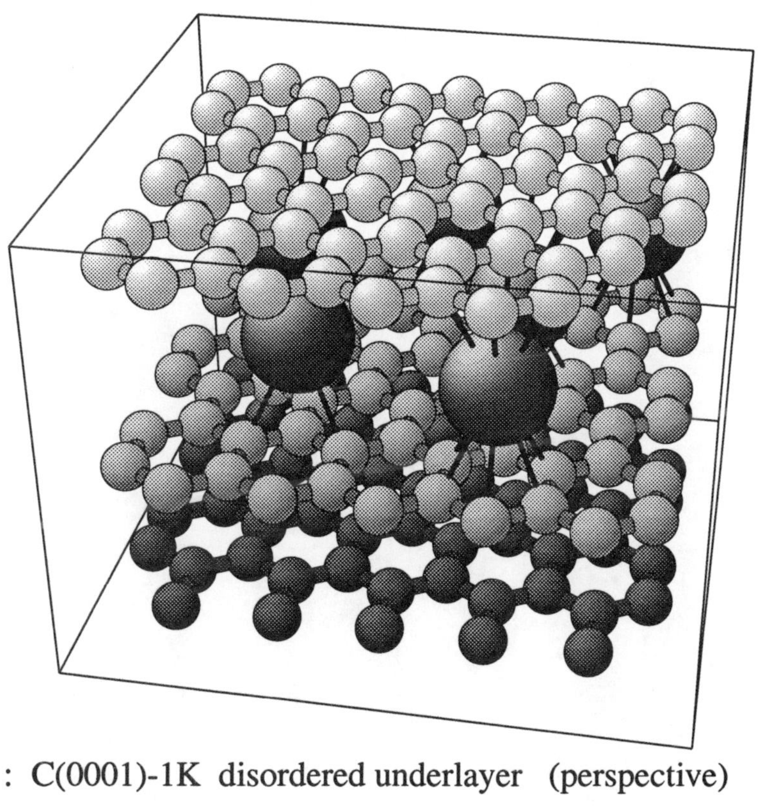

Fig. 164b : C(0001)-1K disordered underlayer (perspective)

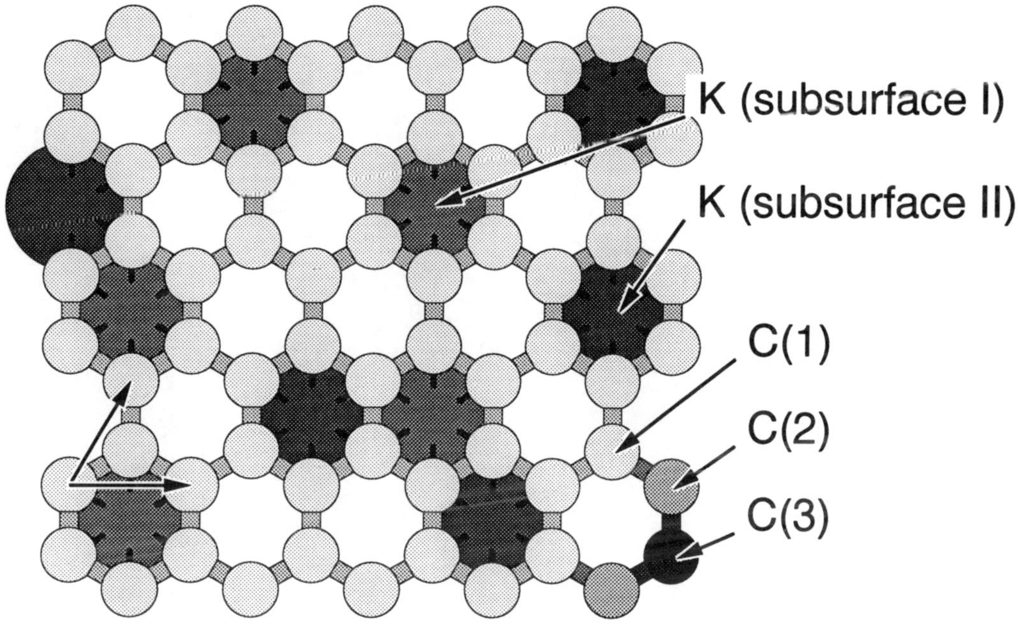

Fig. 165a : C(0001)-2K disordered underlayers (top view)

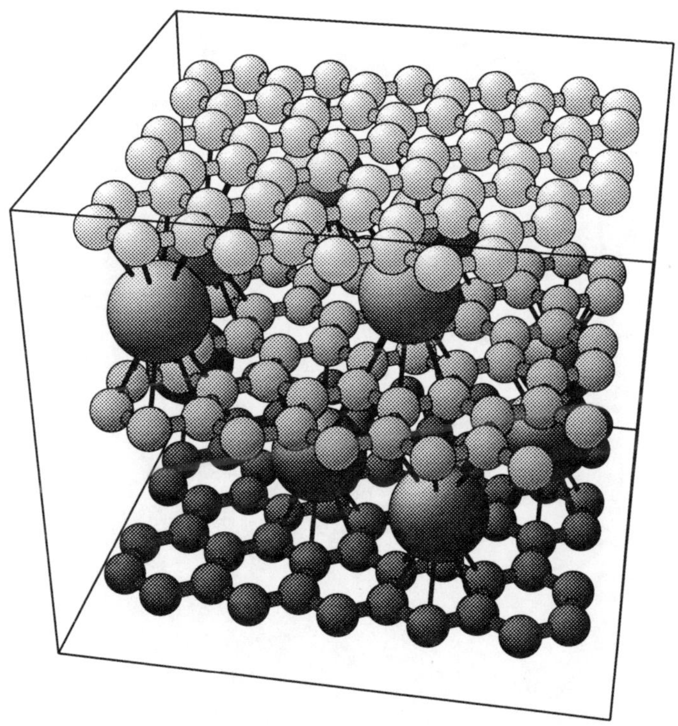

Fig. 165b : C(0001)-2K disordered underlayers (perspective)

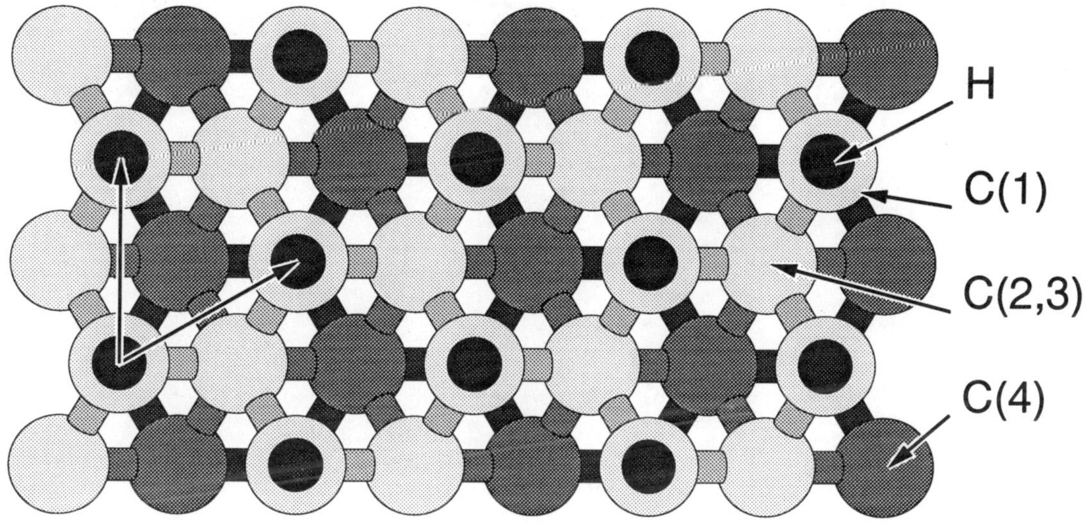

Fig. 166a : C(111)-(1x1)-H diamond (top view)

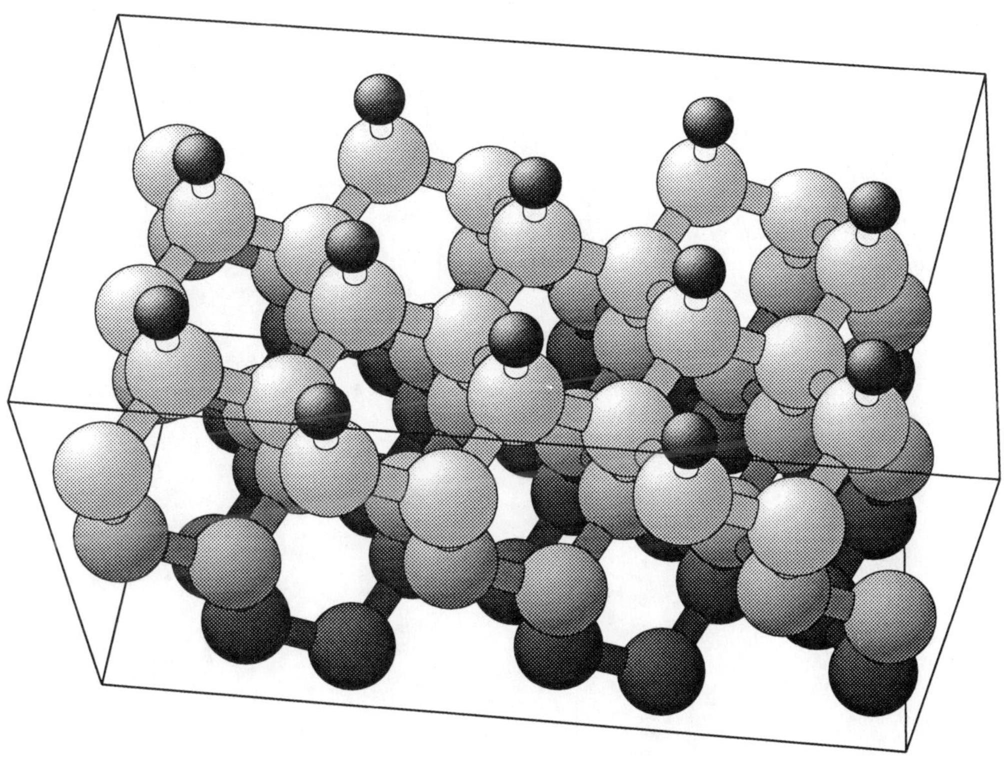

Fig. 166b : C(111)-(1x1)-H diamond (perspective)

3. Acknowledgements

The authors are grateful to the National Institute of Standards and Technology, particularly to the Standard Reference Data Program, for undertaking the support of the electronic Surface Structure Database and of this atlas, which is its printed version. We are also much indebted to our consultants, Prof. J.B. Pendry and Prof. G.A. Somorjai, for helpful advice and much needed organizational support.

Much of the hard work of collecting data and putting them into the required format was accomplished by the individuals shown as "major contributors".

A considerable amount of compiling was also accomplished by the following persons, whose contributions were absolutely essential to the success of this task:

At Imperial College (London):

James MacLaren, Philip Rous, Dilano Saldin and Dimitri Vvedensky;

At Lawrence Berkeley Laboratory:

Simon Bare, Brian Bent, Gregory Blackman, Istvan Böszörmenyi, Mark Bussell, José Carrazza, Peter Ditlevsen, Morgan Edwards, Sabrina Fu, David Godbey, Brian Hadden, Ian Harrison, Michael Hilton, David Jentz, Chi-Tzu Kao, David Kelly, Colette Knight, Mark Levin, Kenneth Lewis, Bruno Marchon, Mathew Mate, Peter McAnally, Brian Naasz, Pedro Nascente, Roger Nix, Frank Ogletree, Hiroko Ohtani, Pedro Pereira, Jim Powers, Philip Rous, Thomas Rucker, Erik Sowa, Daniel Strongin, Gil Vandentop, Gerard Vurens, Kevin Williams and Mu-Liang Xu.

4. Bibliography and Related Software

4.1. References

J.M. MacLaren, J.B. Pendry, P.J. Rous, D.K. Saldin, G.A. Somorjai, M.A. Van Hove and D.D. Vvedensky, "Surface Crystallographic Information Service, A Handbook of Surface Structures, Software and User's Manual" (D. Reidel, Dordrecht) 1987.

J.F. Nicholas, "An Atlas of Models of Crystal Surfaces" (Gordon and Breach, New York) 1965.

H. Ohtani, C.-T. Kao, M.A. Van Hove and G.A. Somorjai, "A Tabulation and Classification of Clean Solid Surfaces and of Adsorbed Atomic and Molecular Monolayers as Determined from Low-Energy Electron Diffraction", Progr. Surf. Sci. **23**, 155 (1986).

P.R. Watson, M.A. Van Hove and K. Hermann, SSD electronic database: "NIST Surface Structure Database – Ver. 1", NIST Standard Reference Data Program, Gaithersburg, Maryland (1993).

P.R. Watson, "Critical Compilation of Surface Structures Determined By LEED Crystallography" J. Phys. Chem. Ref. Data **16**, 953 (1987).

P.R. Watson, "Critical Compilation of Surface Structures Determined By Ion Scattering Compared with Those from LEED Crystallography" J. Phys. Chem. Ref. Data **19**, 85 (1990).

P.R. Watson, "Critical Compilation of Surface Structures Determined By SEXAFS Compared with Those from LEED and Ion Scattering" J. Phys. Chem. Ref. Data **21**, 123 (1992).

E.A. Wood, "The 80 Diperiodic Groups in Three Dimensions", Bell System Techn. Journ. **43** (Part 2), 541 (1964).

4.2. Software

SCIS (Surface Crystallographic Information Service)
marketed by Reidel Publishing (Dordrecht 1987, The Netherlands, or Kluwer Academic Publishers Group, USA).

SARCH/LATUSE/PLOT3D (Surface ARCHitect / LATtice USEr / PLOT 3 Dimensional systems)
to obtain copies contact either M.A. Van Hove or K. Hermann (see Appendix G).

BALSAC (Build and Analyze Lattices, Surfaces, and Clusters)
to obtain copies contact K. Hermann (see Appendix G).

American Institute of Physics

500 Sunnyside Boulevard
Woodbury, NY 11797-2999

Roland W. Schmitt	Chair, Governing Board
Marc H. Brodsky	Executive Director and CEO
Roderick M. Grant	Secretary
Arthur T. Bent	Treasurer and CFO
John S. Rigden	Director of Physics Programs
Darlene Carlin-Walters	Vice President, Publishing

The American Institute of Physics was founded in 1931; it is chartered as a membership corporation with leading societies in the fields of physics and astronomy as members. The Institute combines into one operating agency those functions which can best be done by the societies jointly. Its purpose is the advancement and diffusion of the knowledge of physics and its application to human welfare. About 100 000 physicists belong to its ten member societies: The American Physical Society, Optical Society of America, Acoustical Society of America, Society of Rheology, American Association of Physics Teachers, American Crystallographic Association, American Astronomical Society, American Association of Physicists in Medicine, American Vacuum Society, American Geophysical Union. In addition, there are 20 Affiliated Societies, 84 Corporate Associates, and 6500 members of the Society of Physics Students.

For itself and its member societies, AIP publishes 20 archival journals, 17 translated Soviet journals, a general interest journal, conference proceedings, handbooks, and bulletin-programs, as well as an integrated package of current awareness journals, magnetic tapes, and microfilm. AIP also serves the public by making available to the press and other channels of public information reliable communications on physics and astronomy and their progress; carries on extensive career planning and placement activities; encourages and assists in the documentation and study of the history and philosophy of recent physics; cooperates with local, national, and international organizations devoted to physics and the related sciences; and fosters the relations of the science of physics to other sciences and to the arts and industries.

American Chemical Society

1155 Sixteenth Street, NW
Washington, DC 20036-9976

Paul H. L. Walter	Chairman of the Board
Ned D. Heindel	President
John K. Crum	Executive Director
Brian A. Bernstein	Treasurer
D.H. Michael Bowen	Deputy Executive Director and Secretary
Robert J. Massie	Director, Chemical Abstracts Service
Robert H. Marks	Director, Publications Division*
Halley A. Merrell	Director, Membership Division
Sylvia A. Ware	Director, Education Division

The American Chemical Society was founded in 1876 as a scientific and educational society, and in 1937 was granted a national charter by an act of Congress. Its objectives include the advancement of chemistry, the promotion of research in science and industry, improvement of qualifications and usefulness of chemists, increased diffusion of chemical knowledge, and the promotion of scientific interest and inquiry.

ACS publishes 26 magazines, journals, and other scientific periodicals in addition to its participation in publication of the *Journal of Physical and Chemical Reference Data*. It also publishes books and other nonperiodical reports and operates the Chemical Abstracts Service, the key to the world's chemical literature. It pursues its objectives through its services to chemists and chemical engineers and to education, industry, and government.

*Inquiries regarding ACS participation in JPCRD should be directed to this division.

National Institute of Standards and Technology

U.S. Department of Commerce
Gaithersburg, MD 20899-0001

U.S. Department of Commerce	Ronald H. Brown, Secretary
Technology Administration	Mary L. Good, Under Secretary for Technology
National Institute of Standards and Technology	Arati Prabhakar, Director

The National Institute of Standards and Technology (NIST), formerly the National Bureau of Standards, was established by an act of Congress 3 March 1901. NIST works to strengthen U.S. industry's international competitiveness, advance science and engineering, and improve public health, safety, and the environment. With an annual budget of more than $350 million and about 3100 staff members, NIST conducts science and engineering research in commercially important fields such as advanced materials, information systems, biotechnology, optoelectronics, computer-integrated manufacturing, and sensor technology.

NIST's laboratory research is designed to support development of critical emerging technologies and the new measurement methods and standards necessary to make them commercially viable. The ability of U.S.-based industries to exploit these new technologies determines in large part the health of the U.S. economy.

Several new NIST programs are designed to help spur innovation at U.S. businesses through seed money for development of generic technologies; grants to states for support of technology transfer programs; and financial and technical assistance to help small and mid-sized companies adopt more efficient manufacturing methods.

Throughout its history, NIST has helped industry to develop high-quality products and processes by supplying standard reference materials and data; by continuously improving technologies for precision measurements of mass, time, length, voltage, radiation, and other units; and through active participation in voluntary standards-setting bodies.